Elemental Germans

Elemental Germans

Klaus Fuchs, Rudolf Peierls and the Making of British Nuclear Culture 1939–59

Christoph Laucht
Lecturer in 20th Century British History, University of Leeds

First published 2012 by
PALGRAVE MACMILLAN

Palgrave Macmillan in the UK is an imprint of Macmillan Publishers Limited, registered in England, company number 785998, of Houndmills, Basingstoke, Hampshire RG21 6XS.

Palgrave Macmillan in the US is a division of St Martin's Press LLC, 175 Fifth Avenue, New York, NY 10010.

Palgrave Macmillan is the global academic imprint of the above companies and has companies and representatives throughout the world.

Palgrave® and Macmillan® are registered trademarks in the United States, the United Kingdom, Europe and other countries.

ISBN 978-0-230-35487-6

This book is printed on paper suitable for recycling and made from fully managed and sustained forest sources. Logging, pulping and manufacturing processes are expected to conform to the environmental regulations of the country of origin.

A catalogue record for this book is available from the British Library.

A catalog record for this book is available from the Library of Congress.

10   9   8   7   6   5   4   3   2   1
21   20   19   18   17   16   15   14   13   12

In memory of Dr med Reinhard Laucht
(1943–2010)

Table of Contents

List of Figures   viii

Preface   ix

List of Abbreviations   xii

Introduction   1

1   Difficult Beginnings: Social Integration between
    Survival and Internment   12

2   Almost Accidental Beginnings: Professional Integration
    between Marginalization and British–American Nuclear
    Cooperation   31

3   American Interlude: The Manhattan Project, the
    Atom Bomb and the Emergence of a New Approach
    to Nuclear Research   56

4   A Nation Betrayed? The Klaus Fuchs Atomic Espionage
    Case Reconsidered   82

5   Subject to Suspicion: Rudolf Peierls and the Klaus
    Fuchs Espionage Case   110

6   The Responsible Scientist: Rudolf Peierls and the
    Formation of the Atomic Scientists' Association   125

7   The 'Unpolitical' Scientist: Rudolf Peierls, the Concept
    of 'Objective' Science and the End of the Atomic
    Scientists' Association   151

Conclusions and Afterthoughts   172

Notes and References   178

Bibliography   237

Index   257

List of Figures

1.1 Genia and Rudolf Peierls, date unknown. Photograph by
    Francis Simon, *Source*: American Institute of Physics (AIP)
    Emilio Segre Visual Archives, Francis Simon Collection,
    Peierls Rudolf G 1   14

2.1 Rudolf Peierls (centre) at the University of Leipzig, 1931,
    Source: AIP Emilio Segre Visual Archives, Rudolf Peierls
    Collection, Peierls Rudolf D 5   35

3.1 The Los Alamos Trading Post, October 1945, *Source*:
    Churchill Archives Centre, Cambridge, United Kingdom,
    The Papers and Correspondence of Egon Bretscher, BRER
    A.62.a/0418. Courtesy of the Bretscher Family. Reprinted
    with kind permission   59

4.1 Klaus Fuchs, 1960, DEFA-Photo, German Democratic
    Republic, *Source*: AIP Emilio Segre Visual Archives, Physics
    Today Collection, Fuchs Klaus B 1   104

6.1 Rudolf Peierls, Sir James Chadwick and Geoffrey
    Ingram Taylor, Cambridge, date unknown, *Source*: AIP
    Emilio Segre Visual Archives, Gift of Dr Andrew Brown,
    Peierls Rudolf C 10   133

7.1 Peierls on a visit to the West German capital of
    Bonn in 1979, *Source*: AIP Emilio Segre Visual Archives,
    Rudolf Peierls Collection, Weizsäcker Carl D 3   155

Preface

The famed physicist and so-called father of the atom bomb, J. Robert Oppenheimer, once wrote to a friend that he was extremely fond of physics and the arid wilderness of the American Southwest, but doubted whether he would ever be able to bring them together. In 1943, he was able to combine them when he became the scientific director of the Manhattan Project's central Los Alamos laboratory in the state of New Mexico. While my mathematical skills are far from enabling me to even remotely share Oppenheimer's fondness of physics, I have developed a considerable interest in the history of nuclear physics over the years, and share his love for the American Southwest. For it was during a visit to Los Alamos in September 2001 that I started my investigation of the histories of nuclear culture and the Land of Enchantment, as the state of New Mexico is also known. And, to a large extent, this book is the product of these interests.

A large project like this book requires the help of many people from various phases of my academic life. At the University of Kiel, Dr Jens-Peter Becker introduced me to American cultural history. During a one-year exchange programme at the University of Arizona in 1998–9, Professor Michael Schaller subsequently sharpened my awareness of Cold War – and in particular nuclear – history. After my return to Germany, Professor Brigitte Fleischmann supervised my first investigation of American atomic culture and established contact with the University of New Mexico (UNM) where I continued with my examination of the topic as part of an MA course in German Studies. At UNM, I am enormously indebted to a number of people who supported me on my voyage into the world of nuclear culture, in particular Dr Anne-Marie Werner-Smith, Professors Emeriti Warren Smith and Peter Pabisch as well as the late Professors Ferenc Szasz and Timothy Moy who offered very useful advice in the early stages of my research but sadly passed away before this book was finished.

The focus of my research then changed in 2005 when I started my Ph.D. degree in the University of Liverpool's School of Cultures, Languages and Area Studies. I am extremely grateful to my doctoral supervisor Professor Eve Rosenhaft for her help, support and encouragement, and my second supervisor Professor David Seed for valuable advice. My Ph.D. examiners Professor Martin Daunton and Dr Will

Ashworth also offered very helpful feedback. At the University of Manchester's Centre for the History of Science, Technology and Medicine, Drs Jeff Hughes and Simone Turchetti offered some very insightful atomic chats. Professor John Krige of the Georgia Institute of Technology also provided highly valuable comments and suggestions.

The present book then took shape during my time as a lecturer in the University of Liverpool's School of History, where my colleagues provided an intellectually stimulating environment. I would like to thank especially Dr Jonathan Hogg, who shares my passion for anything nuclear, Dr Dmitri van den Bersselaar, who shared both his office and house with me, and Professor Mark Peel for offering helpful advice on countless rides on the 107 'history line' bus. The School of History also generously provided funds to purchase photographs for this book. At the University of Liverpool's Physics Department, I am also extremely grateful to Dr Peter Rowlands, who introduced me to the world of theoretical physics. During my time at the University of Liverpool, I also received a generous small research grant from the British Academy (SG090835) that enabled me to carry out additional archival research for the present study.

The staff at numerous libraries and institutions in Britain, Germany, Russia and the United States were extremely helpful in providing or tracking down materials. I particularly wish to thank the staff at the National Archives in Kew, the Western Manuscript Division at the Bodleian Library in Oxford, the Churchill Archives Centre in Cambridge, the University of Liverpool's Sydney Jones Library – especially Adrian Allan –, Dr Brian Pollard of the University of Bristol's H.H. Physics Laboratory who supplied important documents, the Landesarchiv Schleswig-Holstein, the Stadtarchiv Kiel, the Federal Commissioner for the Records of the State Security Service of the former German Democratic Republic, the University of New Mexico's Center for Southwest Research and the Los Alamos National Laboratory Archives, especially Alan Carr, for their generous supply of materials. I am greatly indebted to the Council for Assisting Refugee Academics (CARA) for granting me permission to use the Society for the Protection of Science and Learning Papers, deposited at the Bodleian Library. Professor Robert Hinde of the British Pugwash Group also kindly provided documents.

The support of family and friends over the years was crucial in finishing this project. In particular I want to thank my family, especially my mother, my brother and his wife, my grandmother and my uncles, aunts and cousins and their families in Kiel, Hamburg, Berlin and Bavaria for their patience, understanding and support. Numerous friends from all over the world have accompanied me on my journey over the past years, in particular Sven Sindt, Axel Taschner, Robert Osmers, Carsten Groene, Professor Tobias Hochscherf and family, Tim Bridstrup, Svenja Klust, Colette Uhlmann, Serge Simonov, Tim Maschuw, Andy Dubach, Dr Jaime Cuadros Valle, Susie and Charlie Knoblauch, Dr Robbie Aitken and family, Dr Carmen Gómez Galisteo, Ana Reimao, Dr Frank Brunssen, Dr Manos Gkeredakis, Katharina Müller, Dr Mark Lawrence and family, and Captain Brian Marren. I am infinitely grateful to Depeche Mode for providing the soundtrack to my life and offering consolation and hope in difficult times. Finally, I would like to dedicate this book to the memory of my father Dr med Reinhard Laucht, who always encouraged me to pursue a career in history and unexpectedly passed away in 2010. We sorely miss him.

Christoph Laucht
Leeds, October 2011

List of Abbreviations

AAC	Academic Assistance Council
AB	Records of the Atomic Weapons Establishment and Predecessors
AEC	Atomic Energy Commission (US)
AERE	Atomic Energy Research Establishment, Harwell, Oxfordshire
AIP	American Institute of Physics
ALAS	Association of Los Alamos Scientists (US)
AM	Air Ministry
ASA	Atomic Scientists' Association
ASC	Atomic Scientists' Committee
AScW	Association of Scientific Workers
ASJ	*Atomic Scientists' Journal*
ASN	*Atomic Scientists' News*
AWE	Atomic Weapons Establishment, Aldermaston, Berkshire (from 1987)
AWRE	Atomic Weapons Research Establishment, Aldermaston, Berkshire (1950–87)
BA	British Association for the Advancement of Science
BAS	*Bulletin of the Atomic Scientists* (US)
Ber. Wiss-enschafts-gesch	*Berichte zur Wissenschaftsgeschichte*
BMFRS	*Biographical Memoirs of Fellows of the Royal Society*
BRER	The Papers and Correspondence of Egon Bretscher
BStU	Die Bundesbeauftragte für die Unterlagen des Staats-sicherheitsdienstes der ehemaligen Deutschen Demokratischen Republik (Germany)
CAB	Records of the Cabinet Office
CARA	Council for Assisting Refugee Academics

CHAD	Papers of Sir James Chadwick, 1914–74
CIA	Central Intelligence Agency (US)
CND	Campaign for Nuclear Disarmament
CSWR	Center for Southwest Research (US)
DSIR	Department of Scientific and Industrial Research
ECAS	Emergency Committee of Atomic Scientists
ETH	Eidgenössische Technische Hochschule (Switzerland)
F-Division	Fermi Division (Los Alamos)
FAS	Federation of American Scientists
FBI	Federal Bureau of Investigation (US)
FDR	Franklin Delano Roosevelt
FO	Foreign Office
FREEZE	Nuclear Weapons Freeze Campaign
FRG	Federal Republic of Germany
G-Division	Gadget Division (Los Alamos)
GDR	German Democratic Republic
Gestapo	Geheime Staatspolizei (Germany)
HSPS	*Historical Studies in the Physical and Biological Sciences*
HUAC	House on Un-American Activities Committee
ICI	Imperial Chemical Industries
KGB	Soviet Secret Police
KPD	Kommunistische Partei Deutschlands (Communist Party of Germany)
KJVD	Kommunistischer Jugendverband Deutschlands (Communist Youth Association of Germany)
KV	Security Service: Personal (PF Series) Files
LAHM	Los Alamos Historical Museum Archive
LANL	Los Alamos National Laboratory
MAP	Ministry of Aircraft Production
MED	Manhattan Engineer District; the Manhattan Project
Met Lab	Metallurgical Laboratory, University of Chicago (US)
MfS	Ministerium für Staatssicherheit (Ministry for State Security, GDR)

MI5	Security Service (Military Intelligence, Section 5)
MRC	Medical Research Council
NATO	North Atlantic Treaty Organization
NCAI	National Committee on Atomic Information (US)
NHS	National Health Service
NSDAP	Nationalsozialistische Deutsche Arbeiter Partei (National Socialist German Workers' Party)
NDRC	National Defense Research Committee (US)
OSRD	Office of Scientific Research and Development (US)
PAPS	*Proceedings of the American Philosophical Society*
PREM	Records of the Prime Minister's Office
RARDE	Royal Armament Research and Development Establishment, Fort Halsted, Kent
RAI	Radiotelevisione Italiana (Italy)
RGASPI	Russian State Archive of Socio-Political History
RSG	Revolutionäre Studentengruppe (Revolutionary Group of Students, Germany)
RTBT	Papers of Professor Sir Joseph Rotblat
SED	Sozialistische Einheitspartei Deutschlands (Socialist Unity Party, Germany)
SJL	Special Collections and Archives, Sydney Jones Library, University of Liverpool, Liverpool
SPD	Sozialdemokratische Partei Deutschlands (Social Democratic Party of Germany)
SPSL	Society for the Protection of Science and Learning
T-Division	Theoretical Division (Los Alamos)
TA	Tube Alloys
TNA	The National Archives, Kew, Richmond, Surrey
UKAEA	United Kingdom Atomic Energy Authority
UN	United Nations

Introduction

Adolf Hitler's appointment as German chancellor on 30 January 1933 not only marked the end of the Weimar Republic and the beginning of the 12-year period of Nationalist Socialist rule over Germany and large portions of Continental Europe, but had far-reaching consequences for the evolution of British nuclear culture. During the following months and years, the Hitler regime imposed numerous restrictive measures in order to consolidate its power.[1] Among these was the notorious Law for the Restoration of the Career Civil Service (*Gesetz zur Wiederherstellung des Berufsbeamtentums*) that aimed at the 'Germanification' of academia.[2] The introduction of so-called *Deutsche Physik* (German physics) whose conduct was exclusively reserved for 'Aryans' had a devastating effect on the German physics community and ended a golden age of internationally acclaimed physics in Germany. The Nazi policies resulted in an exodus of scientists from Germany and other European countries that was unprecedented in history. In Germany alone, the racist legislation and persecution affected some 875,000 people and an estimated 500,000 people fled Nazi-controlled parts of Central Europe as a consequence of the National Socialist seizure of power.[3] Between 1933 and 1939, about 90,000 people emigrated to the United Kingdom, of whom some 2,200 were scholars who had left German universities and polytechnics by 1938.[4] Klaus Fuchs and Rudolf Peierls were among those physicists who emigrated to or – in the case of Peierls, who was on a Rockefeller Fellowship at the time of Hitler's coming to power – stayed on in the United Kingdom.

Klaus Fuchs and Rudolf Peierls

This book considers the role of these two key figures in the development of British nuclear culture. As outsiders coming to the United Kingdom,

1

the experiences of Klaus Fuchs and Rudolf Peierls offer points of access to key features of British nuclear culture, in particular its scientific foundations and the social, cultural and political consequences of the atomic scientist's work. Fuchs's and Peierls's ethnicity, their socialization and schooling in Germany along with their exposure to German culture before coming to the United Kingdom were instrumental in shaping nuclear culture in their host country.

Born in Germany in 1911 and 1907 respectively, Fuchs's and Peierls's German descent prohibited them from working in areas such as radar and proximity fuse that were initially believed to be of greater significance to the Allied war effort than nuclear research. Consequently, they were pushed almost accidentally into the direction of atomic arms research. Their schooling in Germany, with its strong preference for theoretical physics, had equipped them with skills which were urgently needed in the United Kingdom during the war, and eventually facilitated their integration into the British physics community. And, what is more, their German education formed the basis of their important roles in establishing a new approach to nuclear science during the Second World War that built on a close collaboration of theoretical and experimental atomic scientists. Rudolf Peierls in particular became a key player in the early British nuclear weapons project, Tube Alloys (TA). In the course of the Second World War, Fuchs and Peierls spent some time in the United States where they worked on the joint Anglo-American-Canadian Manhattan Project, at first in New York City and later at the central Los Alamos Laboratory in New Mexico.

Alongside the two scientists' unique skills, their experiences with National Socialism either personally or through family members and loved ones led to a strong motivation to engage in atomic arms research in both of them. Peierls, for example, co-authored with the Viennese émigré Otto Frisch the seminal 'Frisch-Peierls Memorandum' that was crucial for starting both the British and Allied nuclear weapons projects. In Fuchs's case, however, these experiences had a particularly strong effect: they radicalized him politically so that he would eventually reveal the secrets of both the British and Allied nuclear weapons projects to the Soviet Union. After Fuchs's confession, the impact of his earlier radicalization in Germany on democracy and political cultures could be felt in Britain and beyond. In a similar fashion, Rudolf Peierls's socialization in the academic milieu of inter-war Germany informed his understanding of the relationship between science and politics and the role he envisaged scientists taking in public education and advising political decisionmakers. This would later be crucial for his involvement with the chief organization of the British atomic scientists' movement – the Atomic Scientists' Association (ASA).

At the same time, Fuchs's and Peierls's German origin made them the target of defamatory attacks and suspicion. While this distrust by their British hosts had played a significant part in bringing them into nuclear research, it also led to Fuchs's internment as an 'enemy alien' early in the war. Peierls, who was of Jewish origin, later stated that he never faced anti-Semitism. In the aftermath of Fuchs's confession, however, the British Security Service (Military Intelligence, Section 5; MI5) kept Peierls under surveillance because he was German-born, had recruited Fuchs for TA work and had been Fuchs's mentor.

Fuchs and Peierls as access points to British nuclear culture were chosen first, and most basically, because their lives were reciprocally entangled: not only did Peierls recruit Fuchs for work on TA and the Manhattan Project, and served as his mentor, but the revelation of Fuchs's atomic espionage for the Soviet Union made Peierls a prime subject of suspicion by journalists and British and American homeland security agencies. At the same time, contrasts in their careers bring out and help illuminate different aspects of their German backgrounds and thus their experiences in their host country. Peierls formed part of the Jewish emigration, whereas Fuchs belonged to the political emigration. And, what is more, Klaus Fuchs was still fairly junior in his career (he completed his higher education in Britain). Peierls, by contrast, was quite well established at the time of his emigration. Finally, the choice of a different set of scientists – say Otto Frisch and Max Born – would have failed to provide such a thorough examination of key aspects of British nuclear culture as in the present study, in particular the early British nuclear weapons project, Anglo-American nuclear cooperation, the emerging research culture of Big Science, perceptions of the effectiveness of British homeland security in the Cold War and the social responsibility of atomic scientists.

Rudolf Peierls and Klaus Fuchs, as the underlying case studies, are woven into the analysis of British nuclear culture through the application of an 'eco-biographical approach', as David Cassidy has coined it.[5] This places crucial episodes from their lives within their broader social, cultural and political contexts in order to examine their roles in the development of key elements of British nuclear culture. Since Fuchs was tried for his espionage in March 1950 and spent the time after his trial until his release and subsequent move to the German Democratic Republic (GDR) in 1959 in prison, he became a sideshow after 1950.

Alongside its exploration of key features in the making of British nuclear culture, *Elemental Germans* contributes to understanding the role of German-speaking émigré atomic scientists in the United Kingdom which has not been appropriately foregrounded before. Although this book is not an orthodox biography of Peierls and Fuchs, it aims to add to their biographical study. Despite the fact that Peierls has recently caught considerable attention and Sabine Lee has edited a substantial selection of his correspondence,[6] no biography of Rudolf Peierls has been produced to date, apart from his autobiography, a collection of some of his articles and two short biographical sketches.[7]

By contrast, Fuchs's case is much more complicated than Peierls's. While several biographies of Fuchs have been produced over the years, they are by and large highly biased.[8] Early studies of the Fuchs case by Alan Moorehead, Oliver Pilat and Rebecca West were written shortly after his confession and under its direct impact.[9] Later biographies by Montgomery Hyde, Norman Moss and Robert Chadwell Williams were produced with timely distance to the actual events of the Fuchs case, but they were still written under the dictum of the Cold War and without access to crucial primary sources, especially the MI5 files on the spy case.[10] Like these British and American works, short biographical sketches of Klaus Fuchs that appeared in the GDR have to be read with a critical eye.[11] The same holds true for Ronald Friedmann's and Eberhard Panitz's recent Fuchs biographies.[12]

Although Fuchs and Peierls were German-born, this book does not exclusively restrict its focus to them, but makes references to émigré atomic scientists from various parts of the Germanophone world – Germany, Switzerland as well as the Austro-Hungarian Empire, including Austria, Czechoslovakia and Hungary – who resided in Britain and the United States to arrive at more general conclusions about German-speaking émigré scientists as a cohort.[13] Like Fuchs and Peierls, most of these nuclear scientists had received considerable parts of their higher education in Germany during the country's golden age of international science in the 1920s and early 1930s.[14]

Closely connected to their geographic origin was their individual status as emigrant. Since the scientists came from various migration backgrounds, this book applies the fairly broad term 'émigré'.[15] Its usage allows the incorporation of the majority who were part of the Jewish emigration and included, besides Rudolf Peierls, Max Born, Hans Bethe, Edward Teller and Victor Weisskopf as much as the political émigré Klaus Fuchs and the Swiss-German émigrés Felix Bloch, Egon Bretscher and Haus Staub who were not directly part of the forced emigration from Continental Europe but who had often emigrated for economic reasons.

British nuclear culture

In a similar fashion, this book makes the case for a broad definition of British nuclear culture whose evolution Fuchs and Peierls influenced considerably. While the study of atomic culture has received particular attention within the context of United States history, it has still remained largely untouched in Britain.[16] In the only study to date which has used the term 'British nuclear culture' and examined its origins before the Second World War, Kirk Willis has defined it as 'the knowledge, imagery, and artifacts of applied nuclear physics'.[17] *Elemental Germans*, by contrast, promotes a wider definition of British nuclear culture as the sum of all experiences with regard to civilian and military uses of atomic energy, including such diverse layers as science and technology (both theoretical and applied), society, culture, politics, identity, gender, race, ethnicity and class.

Given the width of the field, this book takes into account Clifford Geertz's cautionary note that 'Cultural analysis is intrinsically incomplete' and does not try to tackle the impossible task of providing an examination of every facet of British nuclear culture.[18] Instead, it intends to inspire and serve as a basis for a debate among scholars leading towards achieving a fuller and more comprehensive examination of the subject. On a basic level, *Elemental Germans* thus limits its focus to two key features in the production of British nuclear culture, which are relevant to Fuchs's and Peierls's lives and work: the practice of nuclear science and the social, political and cultural implications of the atomic scientists' work. In this, it goes beyond existing studies that have either focused on the technical and scientific history of the making of the atom bomb or the cultural, social and political fallout from its development.[19]

These two components of British nuclear culture bear a reciprocal relationship to each other. Peierls and Fuchs exemplify this reciprocity particularly well. Since they helped to design the first atomic bombs, they had a great share in confronting the world with a new source of energy. At the same time, these products of their wartime work had a strong impact on their creators and the culture and society that developed them. The book's structure reflects this reciprocity. The first three chapters address the scientific foundations of British nuclear culture and concern what Andrew Pickering has called 'scientific culture'.[20] They look at Fuchs's and Peierls's way into nuclear research, their motivations to engage with it and their actual work on nuclear weapons. Chapters 4–7 then examine the implications 'scientific culture' had for political cultures and perceptions of national security as well as scientists' social responsibility after the war. These four chapters reveal that Fuchs's and Peierls's wartime research, and in Fuchs's case also his espionage, led to the production of what Jutta Weldes and others have defined as 'cultures of insecurity'.[21]

These 'cultures of insecurity' have to be seen against the background of an emerging 'risk society', as Ulrich Beck terms it, and are crucial for understanding the wider contexts in which Fuchs, Peierls and British nuclear culture operated during the Cold War.[22] After Hiroshima, the atomic bomb assumed the role of what David Nye has called 'the technological sublime', prompting an ambivalent response from the world public.[23] In 1949, the first test of a Soviet atomic bomb made the spectre of global nuclear war become more real. Britain's geo-strategic location at the centre of a potential all-out nuclear war between the superpowers created a strong sense of urgency and awareness of the atomic threat among its population. The anti-nuclear movement with its famous march from the London city centre to the Atomic Weapons Research Establishment (AWRE; renamed the Atomic Weapons Establishment [AWE] in 1987) at Aldermaston in Berkshire on Easter 1958 is especially indicative of these concerns. But nuclear accidents, too, appeared to be a real possibility. After all, it was at Windscale, Cumbria, that the world's first major reactor accident occurred in October 1957 when a fire blazed in Pile No. 1 and released considerable amounts of radioactivity. The incident preceded those at Three Mile Island near Harrisburg, Pennsylvania, in the United States, in 1979, the most serious one so far in the Soviet reactor at Chernobyl in April 1986 and the triple meltdown at the Fukushima plant in Japan in March 2011.[24] Fuchs and Peierls coped differently with the responsibilities that they felt were emerging from their wartime work. While Fuchs intended to secure peace by helping to break the United States' monopoly on nuclear weaponry, Peierls and the ASA confronted these post-war insecurities through the promotion of a system of international control of nuclear power, science advising to political decisionmakers and public nuclear education campaigns.

In their endeavours, Klaus Fuchs and Rudolf Peierls acted within the context of British culture and 'Britishness', and this framework is key to any attempt to analyse their roles in the evolution of British nuclear culture.[25] The adjective 'nuclear', however, added a peculiar, ambivalent dimension to it, as it did in other countries, especially the United States. Before Peierls and Fuchs actively engaged in work on nuclear weaponry, speculation about possible uses of atomic energy had formed part of the popular imagination and evoked ambivalent associations between hope for the atom's peaceful applications and fears about nuclear devastation.[26] The tensions between the conflicting meanings of nuclear utopia and dystopia were deeply rooted in British culture. They formed part of a tradition of how large portions of British society have confronted and eventually approved of technological progress and change regardless of considerable doubts and scepticism since the late nineteenth century.[27] British popular culture was one of the first to envisage risks involved in nuclear technologies. H. G. Wells, for example, referred to atomic warfare in his novel *The World Set Free* (1914). In director Maurice Elvey's filmic adaptation of Bernhard Kellermann's novel *The Tunnel* (1935) an accident results in the fatal contamination of workers with radioactivity when an atomic drill that is used to dig a transatlantic tunnel malfunctions.[28]

Although British nuclear culture was partly built on tradition and people had imagined the benefits and dangers of atomic energy long before it became a reality, it was not a planned development but the result of a long process of scientific investigation and discovery. Fuchs's and Peierls's almost accidental push into the direction of atomic arms research thus not only reveals their unintentional entanglement in the making of British nuclear culture, but also points to its unpremeditated emergence in general.

Fuchs's and Peierls's involvement in 'scientific culture' during the war also helped to define the distinctly national characteristics of British nuclear culture. What differentiated nuclear culture in Britain from that in other countries was the peculiar definition of 'modern' that it took on, particularly by contrast with the United States.[29] After the Second World War, British nuclear culture faced an inherent tension between – on the one hand – traditional and conservative values and symbols linked to the past, perhaps most apparent in the constitutional monarchy and the Empire, which had by then already dissolved in vast areas; and – on the other hand – current events, above all, the Cold War which relegated the country to the second division of world powers.[30]

When confronted with a rapid deterioration of Britain's role in the world shortly after the Second World War, Prime Minister Clement Attlee vigorously started to pursue an ambitious project to develop an independent British nuclear deterrent to redress the global power balance in Britain's favour in a way that was perhaps reminiscent of the days of the British Empire. This led the United Kingdom to become both the world's third nuclear and thermonuclear power after the United States and the Soviet Union in 1952 and 1957 respectively.[31] In this calculation, nuclear power became an integral part of 'technopolitics', as Gabrielle Hecht has called it in the French context: 'the strategic practice of designing or using technology to constitute, embody, or enact political goals'.[32] The fact that Britain's determination to develop its own independent deterrent represented a key factor in achieving the post-war cross-party consensus between the Labour and Conservative Parties underlines how deeply this British form of 'technopolitics' pervaded the thinking of British political decisionmakers.[33]

At the same time, British nuclear culture comprised modern elements which appeared to resemble progressive social projects, above all, the reform of the welfare state through the creation of the National Health Service (NHS) in 1946.[34] The launch of one of the world's first programmes to use atomic power to generate electricity under the supervision of the newly established United Kingdom Atomic Energy Authority (UKAEA) in 1955 was perhaps the most significant one among these modern elements.[35] With his work at the Atomic Energy Research Establishment (AERE) at Harwell in Oxfordshire until 1950, Fuchs helped to establish the foundations of Britain's civilian nuclear energy programme. Likewise, Peierls and the ASA attempted to prepare Britons' nuclear mindframe through their education campaign that focused in particular on peaceful prospects of atomic energy. Yet Queen Elizabeth II's attendance at the opening ceremony for the first of the two reactors at Calder Hall in Cumbria in October 1956 demonstrated that these modern elements were embedded in the traditional political culture of constitutional monarchy. While the reactors at Calder Hall which adjoined the existing piles at Windscale were the second in the world after the Soviet reactor at Obninsk to generate electricity for the national grid, they simultaneously produced plutonium for the country's aspiring atomic weapons project.[36]

That Britons accepted these modern elements lay rooted in the fact that they generally were more at ease with change and modernity because British culture was pluralistic and thus differed considerable from Fuchs's and Peierls's native Germany where anything novel was tested against an almost monolithic definition of *Kultur*.[37] Therefore, British nuclear culture is not homogeneous but a multifaceted entity that can be self-reflective and even self-critical, including contradictory elements or what Margot Henriksen has termed 'cultures of consensus and dissent'.[38] Different preferences in research cultures, Peierls's clash with the emerging British and American national security states over the freedom and internationalism of science, or Fuchs's interpretation of global security as proliferating nuclear secrets to end the United

States' monopoly on nuclear weaponry, are only a few examples of this multifaceted character.

British nuclear culture, as Fuchs and Peierls helped to forge it, extended well beyond the United Kingdom and into the Empire. Again, it built on older representations: since the 1890s, Bernhard Rieger notes, 'The imperial *leitmotif* [had] figured as an important, long-standing theme in British discussions about technology'.[39] Early on, the New Zealander Ernest Rutherford provided key basic research with his discovery of alpha and beta rays in uranium for the creation of the atom bomb. Peierls and Fuchs experienced this imperial connection in the form of a protective xenophobia, as Chapter 2 shows, shortly after their arrival in Britain when they faced considerable difficulties integrating into their new host country's physics community. During the war, they also collaborated with scientists from the British Dominions of Australia, Canada and New Zealand.[40] Between 1952 and 1958, the participation of Australia, New Zealand and the Republic of Kiribati (formerly a part of the Crown colony of Gilbert and Ellice Islands) in British atomic and thermonuclear testing then took this link to a new level that can perhaps best be described as atomic imperialism.[41]

The United States as a nuclear reference culture

Klaus Fuchs's and Rudolf Peierls's role in forming British nuclear culture did not stop there; rather it extended into another key area – Anglo-American nuclear cooperation. But in the case of these 13 former British colonies, the situation was different from Commonwealth nations because the United States exerted a strong political, cultural and techno-logical influence on Britain. Beginning in 1943, the British government collaborated with the United States on the Manhattan Project. Here, Klaus Fuchs and in particular Rudolf Peierls played decisive roles, as Chapters 2 and especially 3 demonstrate. While the first cracks in Anglo-American atomic relations appeared during wartime, British-American nuclear cooperation suffered its first severe crisis in August 1946 when the US Congress passed the McMahon Act. Under the new legislation, it was illegal to share nuclear data with foreign governments and this left the United Kingdom virtually cut off from any US atomic information. It was not until August 1958 that nuclear cooperation between the two countries was restored.[42]

While the relationship between the United States and Britain in atomic affairs had still been more or less reciprocal during the Second World War, when Fuchs and Peierls formed an active part of it, the post-war period saw an increasing reliance and dependence on the United States. Through their wartime work in the Manhattan Project, Fuchs and Peierls had helped to establish the United States' lead in post-war 'scientific culture' in Western Europe, including Britain.[43] In addition, limited resources affected British 'scientific culture' after the war so that it operated on a much smaller scale than its American counterpart. The United States had its atomic cities at Los Alamos, Oak Ridge and Hanford, whereas Britain possessed its 'atom village', as a newspaper article labelled the AERE in 1946.[44] The fact that Britain was still a colonial power by the mid-1950s and engaged simultaneously in the Cold War put a dual strain on the already economically stretched country, limiting its nuclear efforts.[45]

American influence, however, went far beyond science. And Whitehall's decision to create a nuclear-capable Britain has not only to be seen against the background of the country's dwindling role in world affairs but the changing state of atomic relations between the two countries. Marcus Oliphant, a leading scientist in both TA and the Manhattan Project, summarized well the British sentiment in May 1948, arguing that 'America's attitude towards atomic energy and towards war and peace, would be modified in a healthy way if Great Britain also, as a result of our own initiative, possessed atomic weapons'.[46]

Even after Britain had become an atomic power in 1952, the Suez Crisis four years later revealed dramatically the limitations of British influence in the world vis-à-vis the superpowers in general and Anglo-American nuclear relations in particular.[47] It thus appears ironic that in spite of the fact that 'Britain had been the midwife of this bomb', as Margaret Gowing observed, the new weapon epitomized both the decline of Britain as a world power and the emergence of the United States as one of the two superpowers.[48] As a consequence, British nuclear culture cannot be understood separately from larger trends in international and transnational history, especially in the United States but, to a lesser degree, also in the Soviet Union.[49] And the United States clearly assumed the role of *the* nuclear reference culture for British nuclear culture in the period under investigation. Anglo-American nuclear cooperation and comparisons with developments in the United States such as the social and professional situation of German-speaking émigré scientists, perceptions of the Fuchs case or the atomic scientists' movement are therefore an underlying theme throughout this book.

Elemental Germans

Elemental Germans consists of seven main chapters that are organized both chronologically and thematically. The first three chapters cover the

Second World War from its outbreak in 1939 when Rudolf Peierls first started to give serious thought to military applications of atomic energy to about one year after hostilities ended and the last British scientists, including Klaus Fuchs, returned home from the United States where they had worked on the Manhattan Project in 1946. These chapters deal with the process of – and events surrounding – the making of the first atomic bombs, and focus in particular on 'scientific culture'. Chapters 1 and 2 examine Peierls's and especially Fuchs's difficulties in integrating socially and professionally into their new host country and society. While the first chapter is crucial for understanding the motivations of Fuchs, Peierls and other German-speaking émigré scientists for engaging in atomic weapons research, the subsequent chapter illuminates the almost accidental nature of their involvement in work on atomic arms. The third chapter then follows them to the United States and examines their contributions to the acceleration of Big Science in the Manhattan Project that left a strong imprint on post-war nuclear science.

Chapters 4–7 follow Fuchs and Peierls back to Britain where they ceased to engage actively in atomic weapons research. Fuchs held a senior administrative appointment at the AERE Harwell, while Rudolf Peierls resumed his professorship at the University of Birmingham. These chapters concern the impact Fuchs's and Peierls's work on nuclear weapons had on them and British culture and society in the post-war period. Chapter 4 examines the impact of the Klaus Fuchs atomic espionage case on perceptions regarding the efficiency of British national security agencies in defending the democratic state at home and abroad. Fuchs was tried shortly after his confession and sentenced to 14 years' imprisonment, nine of which he served. After his release from prison in 1959, he emigrated to the GDR. As the subsequent chapter shows, the Fuchs case also had serious repercussions for his former mentor Rudolf Peierls and other German-speaking émigré nuclear scientists in Britain and the United States, where the news of his confession coincided with the notorious anti-communist witch-hunts of Senator Joseph McCarthy. Chapters 6 and 7 then deal with Peierls's crucial role in both establishing and disbanding the ASA. Chapter 6 examines his part in shaping the organization's policies and ideology regarding science and politics, as well as its role as a body of experts offering public education and science advice to political decisionmakers, and its advocacy of international control of nuclear energy. Chapter 7 analyses the reasons for the ASA's decline and ends with its disbandment in 1959 when it gave way to an anti-nuclear mass movement, above all, the Campaign for Nuclear Disarmament (CND). It deals too with how British atomic scientists increasingly operated within transnational networks, especially the Pugwash movement.

1
Difficult Beginnings: Social Integration between Survival and Internment

After their arrival in their new host country, Rudolf Peierls and Klaus Fuchs experienced an ambivalent atmosphere between political crises and forced emigration, on the one hand, and major advances in the physical sciences, in particular nuclear physics and solid state physics, on the other. Hans Bethe later described this peculiar ambience in a lecture under the somewhat ambiguous title 'The Happy Thirties'.[1] In retrospect, Herbert Fröhlich saw 'the particular events of his own life', especially his experience as an émigré in Britain, 'as a fairly amusing adventure film'.[2]

Most émigrés fled to countries which shared borders with Germany such as France, the Netherlands, Belgium or Czechoslovakia in the immediate aftermath of Hitler's rise to power. Peierls and Fuchs were among the few who came to Britain quite early on.[3] While the scientists' emigration, like the mass exodus of film people or medical personnel, formed, by and large, part of the Jewish emigration, it differed from that of other professions – such as writers, publishers and politicians – whose vocational groups included a higher proportion of political émigrés.[4] Here, the case of Rudolf Peierls was much more typical than that of the political émigré Klaus Fuchs.

Fuchs's and Peierls's German origin proved to be a key factor in their attempts to integrate socially, culturally and professionally into their host society. What it meant to be a German-speaking émigré in Britain at the time is important in understanding how they shaped British nuclear culture. Because of their direct and indirect experience of National Socialism, their German background also represented a crucial component in their motivations to engage in atomic weapons research with far-reaching consequences for British nuclear culture. At the same time, their German origin made them objects of suspicion and even led to Fuchs's internment.

Initial experiences

Klaus Fuchs, who was still a student at the time, landed at Folkestone on 24 September 1933, coming from Germany via France.[5] Rudolf Peierls, by contrast, had already come to the United Kingdom before Hitler's takeover. When his contract as assistant to Wolfgang Pauli in Zürich ran out after three years in 1932, Pauli strongly urged Peierls to apply for a Rockefeller fellowship, which he obtained.[6] Like Hans Bethe before him, Rudolf Peierls chose to divide the one-year Rockefeller fellowship between Rome, where he worked with the renowned Italian theoretical physicist Enrico Fermi, and Cambridge, where he worked at the famous Cavendish Laboratory. As the Rockefeller fellowship was about to end, and since Adolf Hitler had come into power in Germany, it became obvious to Peierls that he could not return to his native Germany. Lawrence Bragg (later Sir Lawrence) from Manchester University then arranged a two-year grant so that Peierls could stay in the United Kingdom.[7]

Since the changed political situation did not allow Peierls to return home, he had to turn down an offer to join Otto Stern's laboratory at Hamburg University in early March 1933.[8] He later declared that he had made the decision not to return to Germany even before Adolf Hitler became chancellor. Peierls 'saw the red light', he claimed after the war, when the Government of Kurt von Schleicher was replaced by that of Franz von Papen.[9] While Peierls had acted farsightedly, other German-speaking émigré atomic scientists did not yet perceive the immediate danger posed by Hitler. 'Well, chancellors come and chancellors go, and he will be no worse than the rest of them', Otto Frisch, who was working as Stern's assistant in Hamburg recalled his judgement of Hitler at the time.[10] By June 1933, however, Frisch had lost this job (*wissenschaftlicher Hilfsarbeiter*) under the new anti-Jewish legislation.[11] And it became clear that Peierls had made the right decision.

Émigrés from National Socialism often went through traumatic experiences, losing family members or relatives. What made it particularly hard for the émigrés was that it was often difficult for them to learn about news from Germany. Rudolf Peierls received most of his information from occasional telephone calls to his parents and from newspapers.[12] The rise of the National Socialist regime strongly affected Peierls's family. While his brother Alfred managed to emigrate to the United Kingdom and his father, with his second wife as well as his sister Annie and her husband, left Germany for the United States, Peierls lost relatives who did not escape National Socialist persecution. To make matters worse, Rudolf Peierls and his wife Genia were also separated from their children, Ronnie and Gaby, in the summer of 1940 when the two children were evacuated to Toronto, Canada, as a precaution against a dreaded German invasion of the British Isles (Figure 1.1).[13] Although Rudolf Peierls referred to himself as 'a former German' after the war, he called for moderation in the treatment of Germans and advised against an interruption of scientific exchange between German and British scientists despite the involvement of many German scientists with the National Socialist regime.[14]

The case of Klaus Fuchs's family also reveals a great deal of tragedy. In 1933, Fuchs's father Emil was among the first professors in Germany to lose his job and was even arrested for his active engagement in the fight against Nazism. Just two years earlier, Emil Fuchs, who was a Quaker, had been the first Social Democrat to be appointed professor at the Pedagogical Academy in the northern German city of Kiel. At the University of Kiel, Emil Fuchs had openly advocated liberal ideas when he served as the president of the Republican Club, a university group that brought him into contact with liberals such as Otto Baumgarten or Walther Schücking, who would later be targeted by fascist students for their progressive views. Like their father, Klaus Fuchs's siblings Elisabeth and Gerhard faced serious reprisals after Hitler's coming to power. Since they were activists in the Kommunistische Partei Deutschlands

Figure 1.1   Genia and Rudolf Peierls, date unknown

(Communist Party of Germany; KPD), they both spent some time in prison. While Elisabeth Fuchs committed suicide on 7 August 1939, Gerhard Fuchs managed to emigrate via Prague to Switzerland.[15] So disturbed was Emil Fuchs that he tried to work through his traumatic experience in the Third Reich with the publication of his booklet *Christ in Catastrophe* in 1949.[16]

The loss of home, however, with all its psychological consequences, represented only one facet of a complex set of problems which Klaus Fuchs and Rudolf Peierls encountered during and after their migration. And it was, above all, the reception in their new home country that proved almost equally challenging for the two theoretical physicists.[17] Upon their arrival in the United Kingdom, Peierls and Fuchs confronted an ambiguous atmosphere: while, on the one hand, they were out of the reach of the National Socialists, and aid societies and fellow émigrés offered support to relieve their situation; on the other hand, they found themselves struggling to find employment in order to have a steady income, faced difficulties in integrating into their new host country's society and academic world as well as severe reprisals as so-called enemy aliens. In many cases, these ambivalences resulted in what Thomas Elsaesser has called a 'two-fold estrangement', a separation of the émigrés both from their homelands and from the attitudes held by many of their hosts towards them and their native lands.[18]

Fellow German-born émigré physicist Max Born, under whom Klaus Fuchs worked at the University of Edinburgh, expressed such a 'two-fold estrangement' in his autobiography. Despite the atrocities committed by the National Socialist regime, Born recorded that 'there remained an extinguishable homesickness for the German language and landscape'. While 'Scotland had invited and accepted us, given us nothing but kindness, opened our minds to the ways of democracy and political fairness, and widened our horizon by making us members of the great British community of nations, the Commonwealth', he felt that 'we were not Scots and would never be'. In conclusion, Max Born noted on this almost Faustian symbolism of a German and a Scottish soul resting in his breast: 'Germany meant for us a struggle between hatred and love, Scotland between love and strangeness'.[19] Here, Born referred to two key factors that affected many émigrés. The language barrier constituted indeed a major obstacle for many German-speaking émigrés.[20] Apart from the language hurdle, Born's insuperable homesickness (*unüberwindliches Heimweh*)[21] for the German landscape represented another chief aspect of many émigrés' uprooting and emotional disorientation between the societies of home and host country.[22]

Closely linked to the estrangement from both home and host country was the culture shock many émigré scientists underwent upon their arrival in Britain.[23] It often occurred on the level of everyday life. Rudolf Peierls was, for example, astonished by the lack of flavour in British foods.[24] James M. Ritchie has even argued that the United Kingdom was not a first-choice immigration country because many émigrés 'preferred the continental café culture of Prague to the unknown hazards of English beer and English cooking'.[25] With regard to these cultural differences, Marion Berghahn has also referred to German-Jewish refugees as 'Continental Britons'.[26]

The German-speaking émigré atomic scientists' first experiences of their new host country and its culture differed significantly. In the wake of Hitler's anti-Jewish laws, Otto Frisch had left Germany for England in October 1933, 'the land which Goethe had so much admired that I expected it to be inhabited almost entirely by supermen'.[27] Immediately after his landing at the London docks, the harsh realities shattered Frisch's Goethe-influenced view of England. It was in particular the 'sort of general messiness and untidiness of everything' that shocked him.[28] Kurt Mendelssohn, by contrast, felt a sense of safety and security after his arrival in the United Kingdom. Decades later, he still reminisced about the first night he had spent in his new host country after his flight from Germany: 'I had slept deep, soundly and long – for the first time in many weeks. ... [T]his was England, sanity, peace and security, and I was deeply grateful for being alive and free'.[29]

With the beginning of the Second World War, this feeling of safety and security soon vanished and the United Kingdom came under attack from the German air force. For the first time in their lives, many of the émigré atomic scientists encountered and shared with their British hosts the direct experience of war. Rudolf Peierls witnessed the destruction caused by air raids in Birmingham.[30] Otto Frisch experienced the bombardment of Liverpool by the Luftwaffe.[31] Frisch even suffered from depression at the time and believed he had 'only got a few more months to live' and that he 'would be hit by the bombs when the bombing started'.[32] That a German invasion of the British Isles appeared to be imminent by May 1940 put additional strain on many émigré atomic scientists.[33]

Finding a steady source of income

Apart from these physical threats to life and limb, Fuchs and Peierls – like many German-speaking émigrés – faced other existential problems, in particular finding a steady source of income. What aggravated their situation was that émigrés were officially not allowed, as it said on the Alien Registration Form, to 'enter any employment, either paid or unpaid while in the United Kingdom'.[34] They even needed official approval if they wanted, as in Klaus Fuchs's case, to accept a research scholarship at Edinburgh University.[35]

Even as an established scientist, Rudolf Peierls encountered a difficult situation on the job market, which was chiefly the result of two major factors: first, Britain suffered at the time from an economic depression, which aggravated the émigrés' situation significantly. Though previously there had been only very few openings in university departments, there was now even less funding available for higher education. To make matters worse for the émigrés, posts at British universities usually had a higher teaching load and the teaching system varied considerably from the German one. For many German-speaking émigré atomic scientists, these fundamental differences, coupled with their, in many cases, poor command of the English language, constituted a seemingly unbridgeable gap.[36]

Second, academic protectionism prevailed in many universities and made it harder for Peierls and many other émigrés to find permanent employment.[37] This attitude manifested itself in the form of a kind of academic protectionism towards scientists from non-English-speaking countries, while university departments generally gave priority to non-British scientists from English-speaking countries of the Commonwealth such as the Australians Harrie Massey and Marcus Oliphant, Canadians like Jack Allen or New Zealanders such as Ernest Rutherford.[38] Despite the fact that Peierls's list of references read like a *Who's Who* of theoretical physics, including Werner Heisenberg, Niels Bohr, Max Born, Erwin Schrödinger and Paul Dirac, he was subject to this sentiment.[39] Here, the situation in Britain differed significantly from that in the United States where German-speaking émigré atomic scientists were more widely accepted. Hans Bethe, who had re-emigrated from Britain to the United States, tried to explain this difference by emphasizing the United States' traditional role as a country of immigration.[40]

Peierls faced this form of protective xenophobia in British academia when he applied for an assistant lectureship at Manchester University. Although Lawrence Bragg was delighted about Peierls's application, he could not hire him. It was explained to Peierls that the university had had such bad press after appointing the German-speaking émigré Michael Polanyi as Chair of Physical Chemistry and giving a temporary contract to Hans Bethe that they were unable to hire him in spite of his qualification. Bragg, however, managed to get Peierls a two-year university grant of £250 per year.[41] The size of the stipend was similar to those issued by the Academic Assistance Council – the chief aid organization for displaced scientists. In 1936, the Council was renamed the Society for the Protection of Science and Learning (and given its rebranding, the organization is jointly referred to as AAC/SPSL hereafter).[42] Although the University of Manchester even extended his contract for a third year until October 1936,[43] Peierls accepted the offer to join the Mond Laboratory in Cambridge in 1935 where he stayed until 1937, when he became a permanent member of the faculty of Birmingham University.[44]

Although Peierls never received a grant from the AAC/SPSL, he had completed the general admission form in October 1934 in order to become part of their network.[45] In 1936, his name thus featured on an AAC/SPSL list of displaced scientists who resided in the United Kingdom among the likes of Herbert Freundlich, Walter Heitler, Heinrich Kuhn, Nicholas Kurti (Kürti),[46] the brothers Fritz and Heinz London, Kurt Mendelssohn, Michael Polanyi, Eugene (Eugen) Rabinowitch, Erwin Schrödinger, Franz Simon and Leo Szilard.[47] The AAC/SPSL also kept personal files on a number of German-speaking émigré atomic scientists, including, apart from Peierls and Fuchs: Hans Bethe, Max Born, Otto Frisch, Herbert Fröhlich, Walter Heitler, Nikolai Kemmer, Heinrich Kuhn, Nicholas Kurti, Fritz and Heinz London, Lothar Nordheim, Erwin Schrödinger and Franz Simon.[48]

With regard to academia, the AAC/SPSL was the most important aid organization. It was created at a meeting of émigrés from National Socialism held at the Royal Albert Hall in 1933, chaired by Lord Rutherford and addressed and attended by such famous émigrés as Albert Einstein. Shortly after its formation, its membership mounted to about 2,000.[49] Out of his gratitude, Rudolf Peierls became both a member of and a donor to the AAC/SPSL in 1937 when he joined Birmingham University and had a steady income.[50] The aid organization focused in particular on two primary objectives: first, it supplied émigré academics with information on jobs in the United Kingdom and elsewhere, especially the United States. Second, it allocated so-called temporary maintenance grants.[51]

Given the limited opportunities the British job market had to offer German-speaking émigré atomic scientists, the AAC/SPSL strongly encouraged their re-emigration, in particular to the United States but also to other, more exotic places.[52] In January 1935, Rudolf Peierls received an offer to apply for a job in Quito, Ecuador, which he declined for monetary reasons because, as he argued, 'I suppose the physicist will not have the same opportunities of earning money in other ways as a Cabinet Minister of Ecuador and will therefore have to live on his salary'.[53] The same job was also offered to Heinrich Kuhn, who had previously turned down the opportunity to join the physics department at the University of Rangoon in Burma.[54]

Owing to the lack of opportunities in Britain, many émigré scientists who had initially come to the United Kingdom re-emigrated, like Hans Bethe, Albert Einstein, Fritz London and Edward Teller to the United States, or like Erwin Schrödinger and Walter Heitler to Ireland.[55] The AAC/SPSL's engagement, however, was not restricted to émigrés residing in the United Kingdom. The Council helped, for example, Lothar Nordheim, who later worked at the Manhattan Project installation at Oak Ridge, Tennessee, to find employment in the United States, while he was working in France and the Netherlands.[56] With about 10,000 arrivals in the years between 1939 and 1945, the United States represented the main destination for re-emigrants.[57] In the years between 1933 and 1941, about 100 physicists found refuge there.[58] Like Hollywood, which became the main destination of German-speaking émigré film personnel, the United States eventually became the main haven for most émigré nuclear physicists.[59]

To those émigré scientists who stayed in the United Kingdom, the AAC/SPSL tried to give as much support as possible, in particular awarding temporary maintenance grants. Klaus Fuchs was among the recipients of such an award. His mentor, Max Born, actively tried to make funds available for Fuchs through the AAC/SPSL.[60] In November 1937, Fuchs thus registered with the AAC/SPSL.[61] As a result of Born's engagement, Fuchs was awarded an assistance grant of £42 per year over a fixed period of 12 months.[62] When Fuchs received the news of this scholarship, he praised the ideals of the AAC/SPSL, writing to its General Secretary, Walter Adams, that 'the work of your society means to people in my circumstances more than the material benefit we may draw from it'.[63] The Notgemeinschaft Deutscher Wissenschaftler im Ausland (Emergency Society of German Scientists Abroad) also became involved in the matter of Klaus Fuchs, and approached the AAC/SPSL to negotiate an extension of his supplementary grant – in vain.[64] Other beneficiaries who received financial support from the organization included Walter Heitler and Herbert Fröhlich.[65]

Apart from helping displaced scientists find funding and employment, the AAC/SPSL also offered legal advice when Peierls's parents, for example, wanted to leave Germany.[66] In addition, the organization provided legal support for Klaus Fuchs, Herbert Fröhlich and Egon

Orowan.[67] Like many German-speaking émigré atomic scientists, Klaus Fuchs faced serious difficulty with the immigration service. After the German Consulate in Bristol – where Fuchs was studying under Nevill Mott – informed Fuchs that the German Embassy in London refused to issue him a new passport, but only a temporary one which allowed him to return to Germany in October 1934, Mott immediately contacted the AAC/SPSL to ensure that his student could stay in the United Kingdom.[68] In return, the AAC/SPSL reacted instantly and contacted the authorities.[69] The AAC/SPSL also assisted Klaus Fuchs and Heinz London during their naturalization processes.[70]

Besides the AAC/SPSL, several other aid organizations existed to alleviate the serious conditions that many émigrés faced. These included Jewish groups such as the Central British Fund for German Jewry and the Jewish Refugee Committee, the Lord Baldwin Fund, the Refugee Children's Movement and the Council for German Jewry.[71] Furthermore, a number of non-Jewish relief organizations were set up, above all, the Germany Emergency Committee.[72] British industry also funded several displaced physicists. Immediately after the imposition of the infamous Law for the Restoration of the Career Civil Service by the National Socialist regime in Germany, Imperial Chemical Industries (ICI) designed a temporary support scheme for émigré scientists from Nazi Germany under the auspices of Frederick Lindemann, later Viscount Cherwell. Recipients of such fellowships included Erwin Schrödinger at Oxford as well as the atomic scientists Heinrich Kuhn, Nicholas Kurti, Fritz London, Kurt Mendelssohn and Franz Simon at the Clarendon Laboratory in Oxford.[73] In the years from 1930 to 1938, industrial laboratories received a threefold increase in funding. Although this development did not necessarily imply by itself that more German-speaking émigré physicists found employment in industrial research, an occupation many of them did not regard as an adequate surrogate for a university position in any case, some like Wolfgang Berg, Dennis Gabor, Otto Klemperer or Walter Zehden were hired by industry.[74]

Apart from aid organizations and industry, much support was also given on the personal level through the initiative of individuals, often fellow émigrés. Frederick Lindemann represented perhaps the most famous of these individuals. Born in Germany but raised in England, Lindemann attended high school in Germany and even obtained a Ph.D. in physics from the Physikalisch Technisches Institut in Berlin where he studied under Walther Nernst before returning to the United Kingdom in 1914. Once the Second World War broke out, Lindemann, as, in the words of Kurt Mendelssohn, a 'famous one-man relief organization', helped many German-speaking scientists – including Mendelssohn, Franz Simon, Nicholas Kurti and Heinz London – to move to the United Kingdom.[75] It was through his special knowledge of German science, his many contacts with German colleagues, and his connections with ICI that Lindemann became so crucial in helping and supporting numerous German-speaking émigrés.[76]

Rudolf Peierls also became involved in supporting fellow German-speaking émigré scientists. Together with Max Born, he helped Klaus Fuchs find a temporary job. After his AAC/SPSL grant had run out and before his internment in 1940, Fuchs had held a Carnegie Foundation fellowship worth over £250 per annum at the University of Edinburgh from October 1939.[77] When the Carnegie Foundation introduced new regulations under which they no longer funded aliens and consequently stopped supporting Klaus Fuchs, Max Born became active again and alerted the AAC/SPSL to the new situation which threatened to leave Fuchs without an income once he was released from internment.[78] Prior to Fuchs's dismissal from internment, Max Born tried to find a job for his former student, although he was not sure about the exact date. He approached Rudolf Peierls, who showed an interest in hiring Fuchs as a temporary, part-time lecturer to relieve him of his teaching load, but also expressed doubts about the feasibility of his plan, in particular on account of the uncertainty of Fuchs's release date and a hostile climate he faced in Birmingham.[79]

Although it looked initially as if Fuchs would not make the move to the University of Birmingham, Peierls remained tireless in his effort to find a job for the talented physicist at his university. In the end, he succeeded and offered Fuchs a temporary position.[80] Apart from giving professional aid, German-speaking émigré atomic scientists also socialized in their private lives, as the case of Peierls reveals, for at one time or another, Bethe, Frisch, Fröhlich and Fuchs stayed at the Peierls's family home for longer durations.[81] Rudolf Peierls also assisted Herbert Fröhlich in his attempt to obtain funding from the AAC/SPSL.[82] Moreover, Peierls tried to help Fröhlich return to the United Kingdom from Leningrad (now St Petersburg) in the Soviet Union where he had spent some time working when his visa was suddenly not renewed.[83]

Facing Germanophobia and National Socialist persecution abroad

While personal intervention by individual scientists, aid organizations and industry helped ease the severe situation that Fuchs, Peierls and many émigrés faced on the job market, there were other obstacles and hardships in the new host country that were much harder to overcome: after the beginning of Hitler's invasion of Poland on 1 September 1939, many émigrés faced a general sentiment of Germanophobia.[84] Max Born and Franz Simon, for example, were affected by anti-German attitudes as proposed by Sir Robert Vansittart in his sixpenny pamphlet *Black Record: Germans Past and Present*.[85] In a letter to Rudolf Peierls, Born expressed his and Simon's relief about the publication of Victor Gollancz's book *Shall Our Children Live or Die?* in 1942, which challenged 'Vansittartism'.[86]

To make matters worse, German-speaking émigrés often faced Germanophobia mixed with anti-Semitism.[87] But the émigrés' individual experiences differed considerably. Despite his German-Jewish origin, Rudolf Peierls did not encounter any kind of 'general xenophobia', as he later wrote in his autobiography.[88] 'The man in the street', stated Kurt Mendelssohn affirmatively, 'fully realized that the refugee scientists living among them were even greater enemies of Hitler himself'.[89] Mendelssohn's generalization did not apply to all German-speaking émigrés. On the contrary, numerous Jewish and political émigrés encountered a very hostile climate.

Like many political émigrés of the same political *couleur*, Klaus Fuchs was not wholeheartedly welcome in Britain, and communists faced many difficulties.[90] This became obvious for the first time when he applied for a new German passport needed to renew his visa in 1934. The Chief Constable of the Bristol Central Police Office informed the Aliens Department about information his agency had received from the German Consulate that indicated that 'Fuchs is a notorious Communist'. This was substantiated by enclosed documents.[91] These documents included crucial pieces of correspondence: an internal letter from the Police Chief of the German city of Kiel where Fuchs had been registered before his flight from Germany, as well as another letter to the German Consulate in Bristol in which the Police Chief voiced 'political doubts' against issuing Fuchs a new passport.[92]

Moreover, the Geheime Staatspolizei (Secret State Police; Gestapo) had planted further evidence of Fuchs's alleged communist ties in the accompanying correspondence. It contained another, earlier letter from Klaus Fuchs to the Registration Office in Kiel on the back of which the Gestapo Field Office in Kiel had noted (translated) remarks, providing background information on his communist affiliations.[93] That the Gestapo effectively used the evidence supplied through the German Consulate in Bristol to the British authorities in order to denounce

Fuchs, damage his reputation, and perhaps even have him extradited is substantiated by the fact that the secret police forwarded further copies of the correspondence directly to the Home Office.[94]

In Fuchs's case, the Gestapo was following one of its common practices by providing governments of countries that harboured émigrés from Nazi Germany with documents which undermined the expatriates' reputation. In their attempt to silence and eradicate any opposition and political agitation in exile, they not only created a network of agents in the respective countries but also enlisted members of the Nationalsozialistische Deutsche Arbeiter Partei (National Socialist German Workers' Party; NSDAP) who lived abroad and, as in Fuchs's case, the German diplomatic service. In order to keep a close eye on the émigrés, the Gestapo had ordered the Prussian police as early as May 1933 to collect a list of all persons who had left Germany after 30 January 1933. These local lists of émigrés (*Emigrantenlisten*) were then kept in the Gestapo's central Émigré Archive in Berlin.[95]

Both Fuchs and Peierls were 'emigrants' (*Emigranten*) in the official diction of the Gestapo, which applied the term to Jews and political émigrés (including members of the KPD, the Sozialdemokratische Partei Deutschlands [Social Democratic Party of Germany; SPD] and Zentrum [Centre Party]) who were living abroad for 'political reasons', regardless of whether they had emigrated before or after Hitler's coming to power.[96] Along with that of his brother Gerhard, the British authorities later also found Klaus Fuchs's name on a seized document that the Gestapo Field Office in Kiel had compiled in 1941. It had been prepared for the Wehrmacht in anticipation of the attack on the Soviet Union, and listed specific people to arrest.[97] In what was perhaps one of its most notorious cases, the Gestapo Field Office in Kiel also initiated withdrawing the German citizenship of Herbert Frahm (alias the later West German Chancellor Willy Brandt) after the latter had fled to Norway.[98]

Becoming 'enemy aliens' and confronting internment

To compound the harassment by the Gestapo, Britain imposed strict regulations on the émigrés' official status. At the time, British immigration policy was, above all, influenced by a perceived national self-interest, and, as Louise London has argued, 'escape to Britain was an exception for a lucky few; exclusion was the fate of the majority'.[99] Even as members of the relatively small and privileged group of intellectual emigrants, Fuchs and Peierls were not spared restrictive measures imposed by the British authorities.[100]

In Birmingham, Peierls faced problems because he was an 'enemy alien', but fortunately he was not interned.[101] On 26 March 1940, he became a naturalized British citizen and was subsequently relieved of the fear of internment that seemed to hang like the sword of Damocles over many non-naturalized foreigners in Britain after the outbreak of the Second World War.[102] Frisch, for example, managed to evade internment because he could prove, with the support of his colleague Philip Moon, that he was involved in important war work on the separation of uranium isotopes.[103] Other émigrés were less fortunate. Rudolf Peierls's brother Alfred and the latter's wife were interned on the Isle of Man.[104] Klaus Fuchs faced similar problems when he applied to become a British citizen. His first application for British citizenship in July 1939 was denied and it would take Fuchs until 1942 to become a naturalized British subject.[105] In 1938, however, the Home Office had changed his visa status, lifting any time limits that could affect his stay in the United Kingdom.[106]

When war with Germany broke out, so-called enemy aliens like Klaus Fuchs and Rudolf Peierls were subject to severe restrictions and had to live with the constant fear of being sent off to an internment camp. The limitations in place for 'enemy aliens' were much more humiliating than those effective in the First World War.[107] They needed, for example, permission to own cars, bicycles and large-scale maps, and they were not allowed to move around the country freely. They also had to obey a curfew. These were serious restrictions for people who had to travel considerably between different universities and were used to working late hours in their laboratories.[108]

As early as April 1940, Klaus Fuchs thus had to file an application to be granted permission to reside in Edinburgh, which had by then become a 'protected area'.[109] Even after he had moved to Birmingham, he still had to get official authorization even if he only intended to pay a short visit to people living in restricted areas.[110] Before his naturalization as a British citizen, Rudolf Peierls had also shared the burden of being an 'enemy alien'. When he wanted to do his part in the war against Nazi Germany by joining the Civil Defence Corps, his application was rejected because of his status as an 'enemy alien'. He was forced to join the Auxiliary Fire Service as the least possible form in which he could contribute to the defence of his new host country.[111]

Once war broke out in September 1939, Whitehall put into effect internment measures. The concept of internment was not entirely new as it had already been applied in the First World War.[112] During that period, some 30,000 'enemy aliens' had been interned. The Second

World War saw the imprisonment of over 25,000, with thousands being deported to far-away Dominions such as Canada or even Australia.[113] The road to internment had been paved by anti-alienism, which enjoyed a long tradition in Britain and was crucial in shaping public opinion.[114] In a Gallup poll which was conducted in July 1939, the overwhelming majority of the respondents (70 per cent) agreed that émigrés should be allowed to enter Britain but, at the same time, 84 per cent of the interviewees who had answered in the affirmative also agreed that restrictions should be imposed on their movements. In another poll taken in May 1940, only 2 per cent of the respondents thought that Whitehall's treatment of aliens residing in the United Kingdom was 'Too strict', while 64 per cent regarded it as 'Too lenient' and 25 per cent as 'About right'. In another survey carried out in the following month, public opinion was divided over the internment of 'enemy aliens': 48 per cent stated that 'only those who may be unfriendly and dangerous' should be interned, and 43 per cent called for the internment of all 'enemy aliens'.[115]

Until today, the internment of 'enemy aliens' remains, by and large, as Tony Kushner and David Cesarani have argued, 'a hidden feature of British history', for it would blemish the myth of 'the Good War' or 'the People's War' as it has also become known in popular memory.[116] François Lafitte concluded retrospectively that 'The only blessing for which we can thank Britain's rounding up of its "enemy aliens" in 1940 is that it unintentionally accomplished the genius of the Amadeus Quartet'.[117]

In line with British government regulations, the 'enemy aliens' Klaus Fuchs and Rudolf Peierls had to appear before a tribunal, which was to determine their status of loyalty to their new host country. The émigrés were divided into three classes (A, B and C), with those in Class A being the ones the tribunals were most doubtful of and those in Class C being regarded as the most loyal among the 'enemy aliens'.[118] Both Fuchs and Peierls were placed in Class C and thus exempted from internment.[119] In November 1939, Fuchs was summoned to appear before a tribunal in Edinburgh.[120] As for most émigrés placed in Class C, the reason given for his exemption from internment was that he was indeed a 'Refugee from Nazi oppression'.[121] In order to be granted this status, Fuchs – like all other 'enemy aliens' – had to provide two letters of support. He received one from Paul D. Sturge, the General Secretary of the Friends Service Council, and a second from his boss Max Born. Apart from praising Fuchs's scientific merits, the latter described him as a 'man of excellent character, deeply devoted not only to his science, but to all human ideals and humanitarian activities'. Born went on: 'He is passionately opposed to the present German government and hopes for the victory of the Allies'.[122] Klaus Fuchs and Rudolf Peierls were thus two of over 73,800 aliens whose cases were examined during the first months of the war. Less than 1 per cent of those screened by tribunals were interned and about 64,200 were placed in Class C.[123]

Things changed dramatically for Klaus Fuchs in the immediate aftermath of the German invasion of Belgium and the Netherlands, when the government finally adopted a policy of mass internment. The Home Secretary, Sir John Anderson, had previously declined calls for internment of aliens as indefensible. With Winston Churchill, an ardent supporter of the mass internment of 'enemy aliens', taking Neville Chamberlain's place as prime minister, the Home Office reluctantly implemented this policy.[124] Winston Churchill's infamous order 'Collar the lot!' summarized – perhaps most graphically – this new course of action.[125] The press also played an important part in creating a climate hostile to the émigrés. Highly conservative papers like the *Pictorial*, *Dispatch* and *Sunday Express* launched a major campaign against émigrés, warning that they served Hitler as Fifth Columnists. As Tony Kushner has argued, this xenophobic climate represented a shift to the 'world of "Clubland Heroes", the inter-war thrillers and detective novels of John Buchan, Sapper, Dornford Yates and Agatha Christie'.[126]

Following the newly implemented policy of mass internment, Klaus Fuchs was, at first, detained at Donaldson's Hospital Internment Camp in Edinburgh on 12 May 1940, before he was transferred to an internment camp on the Isle of Man and finally deported to Canada.[127] Apart from Fuchs, Walter Kellermann, a German-born émigré and collaborator of Max Born at the University of Edinburgh, was interned. Born remarked on their appearances, 'it is rather curious that the Rabbi's son Kellermann, looks 100% Aryan whereas the Pastor's son Fuchs, is of the super-intelligent type frequently found amongst educated Jews'.[128] In his autobiography, Max Born later described how he found the two men missing from their workplaces one morning and gradually learnt what had happened to them.[129] Born immediately intervened on behalf of Fuchs and Kellermann and contacted Esther Simpson, the AAC/SPSL secretary, assuring her that both scientists 'would be prepared to do work of national importance in any place ascribed to them'. He underlined, 'I can strongly affirm that Dr. Fuchs and Dr. E. W. Kellermann are not only prepared, but extremely keen to do such work, as their fate entirely depends on the victory of this country'.[130] In return, Simpson reassured Born 'that we shall do our best on behalf of the scholars and scientists registered with us'.[131]

Other German-speaking émigré physicists who were interned included Walter Kohn and Hans Kronberger, as well as eight members of the physics department at Bristol University: Walter Heitler and his brother Hans, Herbert Fröhlich, Kurt Hoselitz, Philipp Gross and Heinz London, and two of their students Robert Arno Sack and G. Eichholz.[132] With a total of eight internees, the University of Bristol's physics department was particularly hard hit by what Arthur M. Tyndall appropriately described as a 'bombshell'.[133] Tyndall, who was the head of the department at the time, took a proactive role on behalf of his émigré colleagues early on because he had realized their unique qualifications and significance for TA work.[134] As Tyndall graphically put it in a letter to George Thomson (later Sir George), 'a spanner is thrown in the works if all these friendly aliens are excluded' from TA work.[135]

Not all senior scientists in the British nuclear weapons project shared Tyndall's view of German-speaking, in particular German-born, scientists working in sensitive areas such as TA. Thomson was 'a little troubled about the number of people of German origin who are getting to know of the work' and 'hope[d] in particular Peierls has not been indiscreet'. He harboured a deep feeling of suspicion towards German-speaking, émigré scientists and therefore suggested that 'it is very important that as few of the German refugees as possible should be concerned in this work, as in the present state of things I do not, to speak frankly, feel too confident of any of them'. As a possible solution to the problem, Thomson suggested sensitive work should only be given to naturalized British subjects as they have 'given pledges which [they] cannot revoke'.[136]

While the example of the University of Bristol's physics department illustrates the occasional magnitude of the internment measures, Hans Kronberger's case represents a peculiar personal tragedy. Born in the Austrian city of Linz in 1920, the future collaborator on the post-war British atomic weapons programme was still a student during the war. Unlike his parents and his only sister who were murdered in the Holocaust, he had managed to emigrate to the United Kingdom in 1938 when Germany invaded Austria. He was then interned from 1940 until 1942, first on the Isle of Man, and later deported to Australia.[137]

Another émigré, Oskar Bünemann, was interned in Canada, and returned to Liverpool in January 1941.[138] The case of Bünemann represented another injustice of the British practice of interning 'enemy aliens'. Born of German parents in Milan, Italy, Oskar Bünemann attended high school in Hamburg where he also began to study mathematics at the university. After spending three months during the summer of 1933 in a labour camp, he engaged in clandestine political action against the National Socialist regime. Once exposed as an opponent of the 'Third Reich', he served 18 months in prison. Upon his release in October 1935, Bünemann emigrated to the United Kingdom where he resumed his studies under Douglas R. Hartree at the University of Manchester.[139] He was later engaged in atomic weapons research during the war and joined the Theoretical Division of the AERE Harwell after the war.[140] Even famous scientists such as the Austrian-born Hermann Bondi were not spared the internment experience. Bondi was also deported, but released after six months to conduct crucial work on the improvement of radar.[141]

On 3 July 1940, only a day after about 700 internees had perished on board the Newfoundland-bound *Arandora Star* which was torpedoed and sunk by a German submarine, Fuchs boarded the *Ettrick* at Liverpool and embarked on the two-week-long journey across the Atlantic.[142] The sinking of the *Arandora Star* generated fears and worries among the families and relatives of many internees.[143] Like Fuchs, the later Nobel laureate Max Perutz also travelled on the *Ettrick*. He vividly described the unbearable conditions onboard the ship: 'Suspended like bats from the mess decks' ceilings, row upon row of men swayed to and fro in their hammocks. In heavy seas, their eruptions turned the floors into quagmires emitting a sickening stench'. To make matters worse, 'Cockroaches asserted their prior tenancy of the ship', Perutz recalled, and 'The commanding colonel called us scum of the earth all the same, and once, in a temper, ordered his soldiers to set their bayonets upon us'.[144] But the conditions on board the *Ettrick* were only a kind of foretaste of what awaited the internees in Canada.

Klaus Fuchs, who bore internee number 417, arrived in Quebec, Canada, on 13 July 1940. He was, at first, interned at Camp L until 16 October 1940 when he was transferred to Camp N where he stayed until he was returned to the United Kingdom on 17 December 1940.[145] Despite all hardships, a university was set up in Camp L, and Klaus Fuchs even lectured physics classes. At Camp N, sanitary conditions deteriorated dramatically. The internees, who numbered about 720, were housed in train shacks and had only six toilets and five taps at their disposal.[146] Herbert Fröhlich recalled the sudden occurrences of panic and hysteria on the part of the guards in internment camps. These could sometimes take bizarre forms as, for example, in the case of Walter Heitler whom they accused of giving secret signals to German pilots when his wife hung some of his light clothes on a line to dry after she had washed them.[147] In spite of the harsh conditions, Klaus Fuchs later noted on his internment experience that 'I felt no bitterness by the internment because I could understand that it was necessary and that at that time England could not spare good people to look after the internees, but it did deprive me of the chance of learning more about the real character of the British people'.[148] Robert Sack expressed well the ambiguous feelings he had about his internment in a letter to Arthur Tyndall. 'It is a strange feeling, though, to be guarded by soldiers within an enclosure of barbed wire', he wrote, 'having a Polish mother, a brother serving actively against Hitler – God knows what has happened to my people in France – and having offered my services so wholeheartedly and repeatedly'.[149] Sack's lines sum up well the paradoxical situation that many German-speaking émigré scientists had to endure in the face of internment.

An inmate who spent about four months in the same camps with Fuchs in Canada from August 1940 until Klaus Fuchs's return to the United Kingdom stated that Fuchs had become involved in administrative matters there. The anonymous man portrayed Fuchs's character as 'very far from easy, but not a hermit in the sort of camp life'. The man described the internees at Camps L and N as 'predominantly Jewish', and the atmosphere there, in particular in the latter camp, as 'predominantly anti-Nazi'.[150] During the time of his internment in Canada, Fuchs, for example, offered moral support to Jewish fellow inmates whom he called retrospectively the 'most displaced' people after Hitler's coming to power. Not only did he try to protect his fellow internees from being accidentally sent to a camp of Nazi POWs, but also from being exchanged for Canadian POWs held by the Third Reich, a fear that caused much anxiety among the inmates of the camp at the time. Moreover, he appealed to the authorities not to appoint a son of the former German crown prince as camp leader.[151]

From July 1940 onwards, public opinion had become more benevolent towards the internees because of the tragedy of the *Arandora Star* and after complaints from internees about unbearable conditions on the ships and in the camps had reached the outside world. As a consequence of criticism from all ranks of British society, Whitehall was forced to reconsider and finally to abandon its internment policy.[152] Klaus Fuchs, like many of his fellow internees, was thus officially released from internment on 19 October 1940 and granted permission to return to the United Kingdom and continue his research at Edinburgh University. It was, however, not until 11 January 1941 that he landed in Liverpool and was, by order of the Secretary of State, exempted from internment.[153]

After his return to the United Kingdom, Klaus Fuchs stayed in Edinburgh for a short period until he moved to Birmingham to start his new job under Rudolf Peierls in May 1941. The latter was as impressed with the young German émigré's work as Max Born, who praised Fuchs in 1939 as his 'best and most efficient collaborator' as well as being 'at present the best theoretical physicist of the younger generation in Scotland'.[154] It was through Rudolf Peierls that Fuchs eventually became an integral part of the British effort to develop the atom bomb.

Conclusion

Rudolf Peierls's and especially Klaus Fuchs's paths towards integration into their host country's society were rocky and unpaved. After Peierls and in particular Fuchs had overcome the many difficulties related to their German origin, especially their status as 'enemy aliens', they eventually integrated comparatively well into their host society. Like many of their fellow German-speaking émigrés from Nazism, Fuchs and Peierls often faced existential problems. They had, for instance, a hard time finding employment or, in the case of Fuchs, a body to fund his studies. Furthermore, they lived under the constant threat of being deported to an internment camp. While Rudolf Peierls was spared the internment, Klaus Fuchs spent a considerable amount of time in an internment camp in Canada under unbearable conditions. To make matters worse, they were subjected to bombardments by the German air force. But these experiences had a strong impact on their motivations to engage in nuclear weapons research and ultimately shape British nuclear culture.

2
Almost Accidental Beginnings: Professional Integration between Marginalization and British–American Nuclear Cooperation

If Fuchs's and Peierls's integration into their host country's society had proved at times very difficult, they faced similar obstacles with regard to their professional integration. After their arrival in the United Kingdom, German-speaking émigré atomic scientists did not become involved in nuclear weapons research immediately. Like the many existential problems they often had to master during the early days of their lives in their new host country, many of them also had to overcome serious obstacles in order to integrate professionally into the British nuclear physics community. First and foremost, émigré atomic scientists faced fundamental differences in national preferences in research and teaching styles between Continental Europe and the United Kingdom.

Their German-speaking backgrounds and their origins and schooling in German or Continental European centres of theoretical physics prevented them from engaging in sensitive war work and pushed them – almost by chance – into atomic weapons research which was seen as being of very little value at the time. As before, their experience with the National Socialist regime – either directly or indirectly through family members and loved ones – translated once again into a stronger motivation to pursue atomic weapons research than among their British-born colleagues. This would form a key part in their roles in forging British nuclear culture.

Initial problems

Despite its positive long-term effect, Klaus Fuchs's and Rudolf Peierls's training in German centres of theoretical physics initially proved to be disadvantageous in the British university system. While German universities had a strong emphasis on theory, British schools and physics departments, in general, leaned more towards empirical research.[1] To understand these differences in stylistic preferences, which were rooted in both the historical development and geographical distribution of the main centres of theoretical physics, it is necessary to point to Germany's role in the development of the new physics; above all, quantum physics. During the era of the Weimar Republic, three major centres of modern physics developed at Göttingen, Berlin and Munich.[2] Cambridge, Copenhagen, Leiden and Zürich were the four remaining world-leading physics institutes outside Germany.[3] Rudolf Peierls, like most German-speaking émigré atomic scientists at the time, attended more than one of these well-known centres: he studied at the universities of Berlin and Munich, as well as at the ETH (Eidgenössische Technische Hochschule [the Federal Institute of Technology]) in Zürich, and spent half of his Rockefeller fellowship at the Cavendish Laboratory in Cambridge.[4]

In Britain, the situation was quite different. Here, the Cavendish Laboratory in Cambridge was an isolated centre of theoretical physics. Aside from the Cavendish, the University of Bristol's H. H. Wills Laboratory, with its focus on theoretical solid-state physics, represented the only other exception from the primarily experimentalist scheme found at British physics institutes at the time. The award of a major research grant to John E. Lennard-Jones at Bristol University by the governmental Department of Scientific and Industrial Research (DSIR) in June 1930 enabled the formation of a theoretical research school there. This was the first endowment to acknowledge theoretical physics on such a large scale in Britain. At one time or another, Lennard-Jones's successor, Nevill Mott, employed several German-speaking émigré atomic scientists, including, apart from Klaus Fuchs: Hans Bethe, Walter Heitler, Herbert Fröhlich, Lorenz Frank, Kurt Hoselitz, Heinz London, Philipp Gross and Robert Arno Sack at the H. H. Wills Laboratory.[5]

Although the traditional 'British' leaning towards empiricism hampered, in general, the integration of émigré atomic scientists, Rudolf Peierls and Klaus Fuchs faced fewer problems than most members of their cohort. As an established physicist at the time of his departure from Germany, Peierls integrated relatively quickly into the British physics community. Still, he encountered initial problems and, during his early days in Britain, he felt his stay in the United Kingdom had a temporary quality. As a consequence, the Peierls called their daughter Gaby which was a name 'pronounceable in any language of any country where we might eventually live'.[6]

After some fixed-term appointments, Peierls was eventually appointed to a tenured position at the University of Birmingham. The fact that, before the war, Max Born and Franz Simon had been the only other émigrés to be granted permanent positions at institutions like Birkbeck College and the Universities of Birmingham, Bristol, Edinburgh and Oxford (where strong clusters of émigré physicists were found) underlined both Peierls's high calibre and the exceptionality of his case. In a different way, Klaus Fuchs's academic career also varied significantly from that of most émigré physicists. As a junior scientist, Fuchs, like other émigré students such as Hans Kronberger, had not completed his higher education at the time of his arrival in the United Kingdom. As graduates of institutions of higher education in the United Kingdom, Fuchs and Kronberger held British degrees, which helped them pursue impressive careers in their new host country where they became involved in the atomic energy project after the war.[7]

The group of German-speaking émigré atomic scientists that came to Britain was, of course, not exclusively composed of theoreticians but also of experimentalists such as Egon Bretscher and Otto Frisch. It was certainly easier for experimentalists to integrate into the British physics community with its strong orientation towards experimentation. And of all the German-speaking émigré theoretical physicists who attended Cambridge at one time or another, including Rudolf Peierls, Hans Bethe, Max Born, Richard Courant, Albert Einstein, Paul Ewald, Leopold Infeld and Victor Weisskopf, none eventually settled there. But it was the experimental physicist Otto Frisch who was appointed professor at Cambridge after the war.[8] In spite of their better initial position in their host country's job market, experimentalists still faced considerable problems finding jobs.

Theoreticians like Peierls and Fuchs and experimentalists like Frisch shared another major obstacle that lay rooted in the structure and sociology of the British physics community, particularly given its relatively small size.[9] Their assimilation depended heavily on establishing networks with universities, laboratories, funding bodies and so forth, and this further complicated the situation for many émigré nuclear scientists, regardless of their qualification.[10] And, indeed, there existed a reciprocal relationship between the individual physicists and their particular scientific environments.[11]

Erwin Schrödinger's failure to adapt to the research culture at Oxford University serves perhaps as the most prominent negative example of the complex processes revolving around the integration of émigré atomic scientists.[12] The Viennese émigré philosopher Karl Popper, who met him in Oxford, described Schrödinger as 'very unhappy in Oxford'. Previously, Schrödinger had worked among such illustrious theoretical physicists as Albert Einstein, Max von Laue, Walther Nernst and Max Planck in Berlin. Although Schrödinger 'had been very hospitably received' at Oxford University, Popper observed that he missed 'the passionate interest in theoretical physics, among students and teachers alike'.[13] Oxford, however, was not the only example of Schrödinger failing to integrate into his scientific environment: in 1946, he unsuccessfully applied to become chair of theoretical physics at Liverpool University, which was ultimately given to Herbert Fröhlich instead.[14]

Although many German-speaking émigré atomic scientists initially encountered difficulties in integrating into the British science community, their presence provided an innovative impulse: since their coming to the United Kingdom coincided with an increase in physics centres, they helped spread the advancement of new, interdisciplinary subfields within the physical sciences.[15] Klaus Fuchs's and, in particular, Rudolf Peierls's participation in the British nuclear weapons programme is a good indicator of their successful integration process.

Motivations for engaging in work on the atom bomb

That Fuchs and Peierls assumed crucial roles in the British atomic energy project was by and large the consequence of two factors: their legal status as 'enemy aliens' and their insider knowledge of the German nuclear physics community, and the scientific potential available to the National Socialist regime translated into an alarm and urgency greater than was felt by their British-born colleagues. When the Second World War broke out, Peierls's and Fuchs's status as 'enemy aliens' not only brought with it the peril of internment but it also affected significantly the direction of their research. Since this status did not permit them to participate in secret war research on radar or the proximity fuse, they were almost accidentally pushed in the direction of nuclear arms research, which was not regarded, at the time, as viable for the outcome of the war, and consequently not classified as secret war work.[16]

Apart from this legal factor, their personal experiences with National Socialism and their insider knowledge of the German nuclear physics community served as strong motivations for them to actively engage in atomic arms research. Both Peierls and Fuchs, for example, had studied under Werner Heisenberg, who now played a key role in Adolf Hitler's atom bomb project (Figure 2.1).[17] Peierls and Fuchs used their insider knowledge and drafted reports on current atomic-related German-language publications as well as on the activities of physicists inside the Third Reich, covering relevant research, university appointments, and so on.[18] Peierls carried on with this scientific intelligence work during the

Figure 2.1 Rudolf Peierls (centre) at the University of Leipzig, 1931, with Werner Heisenberg (to his right) and three future Manhattan Project scientists (from left to right): George Placzek (seated left from Peierls), Felix Bloch (seated behind Heisenberg) and Victor Weisskopf (seated in the background behind Heisenberg's left shoulder)

war. He provided James Chadwick (later Sir James Chadwick), for instance, with a list of physicists inside Nazi Germany who might conduct nuclear weapons research, including the names of Werner Heisenberg, Karl Wirtz, Manfred von Ardenne and Paul Harteck.[19] After the war, Rudolf Peierls cited the fear of an atomic bomb in Hitler's hands as his chief motivation for engaging in work on nuclear weaponry, stressing: 'There was no question that that must be prevented.'[20] In 1985, Rudolf Peierls admitted that he had at the time the feeling that 'there was some mild work going on' in Germany, 'no crash program ... but we didn't dare rely on it'.[21]

Although Klaus Fuchs had a double motivation to engage in the British nuclear weapons programme because of his commitment to the Soviets, which will be discussed in Chapter 4 he also shared Peierls's great anxiety about an atom bomb in Hitler's hands. Fuchs stressed this point in a video interview conducted by the East German state security, Staatssicherheit, in the 1980s. The interviewer confronted him with a passage from Max Born's autobiography. Born was among a group of high-profile, German-speaking émigré scientists – including Peter Debye, Albert Einstein, Lise Meitner, Erwin Schrödinger, Fritz London,

Gerhard Herzberg, Hertha Sponer, and Otto Stern – who chose not to get involved in atomic weapons research for various reasons.[22] In his memoirs, Fuchs's former mentor at Edinburgh University described several discussions that he had with Fuchs after the latter had received an invitation to join Rudolf Peierls's team in Birmingham to work on nuclear-weapons-related matters. Born claimed that, in spite of the veil of secrecy, it was obvious to him what Fuchs was going to work on.[23]

When he was confronted with this passage in the 1980s, Klaus Fuchs defended his participation in work on nuclear arms during the war. While he showed respect for Max Born and his wife Hedwig's pacifist views, Fuchs argued that the fear of losing the race for the new weapon to Germany had particularly driven those émigré atomic scientists who had experienced National Socialism directly such as Hans Bethe, Niels Bohr and Victor Weisskopf.[24] Later in the interview, Fuchs objected to a phrase used by the interviewer with reference to Born's statement, suggesting the scientists' work at Los Alamos was driven by a 'collective lack of conscience'.[25]

German-speaking émigré atomic scientists in the United States such as Hans Bethe, Lothar Nordheim, George Placzek, Leo Szilard, Edward Teller and Eugene (Eugen) Wigner shared these concerns over a German nuclear weapon.[26] In the United States, these fears galvanized in the so-called Einstein Letter of 2 August 1939. The three Hungarian-born scientists Szilard, Teller and Wigner were alarmed by the possibility of a German nuclear weapon in the not too distant future and wanted to inform President Roosevelt (FDR) of the impending danger.[27] When finally the opportunity arose to reach FDR through Alexander Sachs, a national economist, the three scientists convinced Albert Einstein that he alone should sign a letter to the president (drafted by Szilard) warning of the perils of a Nazi atomic bomb. It reached Roosevelt after considerable delay, and consequently had no immediate impact on the decision to initiate a serious atomic arms programme in the United States.[28]

The 'Frisch-Peierls Memorandum'

While the so-called Einstein Letter is commonly cited as the starting point of a serious (American) nuclear weapons development programme, in fact Rudolf Peierls played a key role in getting a serious British and American atomic weapons research project under way, especially through his collaboration with Otto Frisch in their seminal 'Frisch-Peierls Memorandum'. Margaret Gowing assessed these British contributions as having high value, arguing that 'there is little doubt that, without the British work, World War II would have ended – for better or worse – before an atomic bomb was dropped'.[29]

Like German-speaking émigré scientists in the United States, Rudolf Peierls and Otto Frisch were deeply concerned about a nuclear weapon in Hitler's hands. Despite the beginning of the Second World War, a strong internationalism still dominated science, and the important developments in atomic physics made by scientists in Nazi Germany and elsewhere were available for everyone to read. Surprisingly, the outbreak of the Second World War in September 1939 did not initially speed up atomic research, but slowed it down.[30] This was largely the result of the British government's focus on radar, with most scientists involved in radar-related or other research which was seen as crucial for the war effort.[31]

As a consequence of his fear of a German atomic bomb and the restrictions imposed on 'enemy aliens', Rudolf Peierls started to give serious thought to the feasibility of a nuclear device by trying to calculate the critical mass of uranium in the summer of 1939.[32] Shortly after Otto Frisch had arrived at the University of Birmingham that same summer, the two scientists started their collaboration on what later became known as the 'Frisch-Peierls Memorandum'.[33] Like Peierls, Frisch was deeply disturbed by the prospect of a nuclear-capable National Socialist regime in Germany.[34]

This pivotal document can only be understood in the context of the scientific discoveries that preceded it, in particular that of nuclear fission and the concept of a chain reaction, which is a significant feature of nuclear fission. About six years earlier, Leo Szilard had already discovered the idea of a chain reaction in theory. A chain reaction is based on the emission of neutrons. Starting with a single uranium nucleus, Szilard assumed that fission could not only produce large quantities of energy but could further result in a self-sustaining 'chain reaction' of ever-increasing amounts of energy. If kept in check, this reaction could serve as an endless power source, while uncontrolled it could cause a very destructive explosion.[35] Consequently, he applied for a patent explaining his findings, which he allocated to the British Admiralty in order to keep his discovery from becoming public knowledge.[36]

Although the patent had been filed in 1934 with the British Patent Office, its transformation into a viable option did not take place until 1938 when nuclear fission was discovered. Otto Frisch and his aunt Lise Meitner played a decisive role in confirming the possibility of nuclear fission. Shortly before Christmas 1938, Otto Hahn sent his long-time collaborator Lise Meitner, who had by then emigrated to Sweden to escape Nazism, a letter with his own and his colleague Fritz Strassmann's latest findings to ask for her interpretation. At Christmas, when Otto Frisch was still at Niels Bohr's Institute in Copenhagen, he visited his aunt in Kungälv near Gothenburg, Sweden.[37] Subsequently, Frisch and Meitner composed two papers dealing with Hahn's and Strassmann's discovery that were published in *Nature* in February/March 1939.[38] The term 'fission', which describes the break-up of a uranium nucleus with the release of energy, was used here for the first time in connection with atoms. The American biologist William Arnold had suggested it to Frisch.[39] Soon after the news broke, several international research installations, including Columbia University, New York; the Carnegie Institution; Johns Hopkins University; the University of California, Berkeley; as well as Frédéric Joliot-Curie's Radium Institute in Paris – confirmed Frisch's and Meitner's theoretical findings.[40] There was, however, only very little discussion of potential military applications of nuclear fission at Niels Bohr's Institute in Copenhagen where Frisch returned to, and stayed, until he moved to Birmingham in July 1939.[41]

As early as 30 April 1939, the *Sunday Express* had run the headline 'Scientists Make an Amazing Discovery' with the highly dramatic sub-title 'Stumble on a Power "Too Great to Trust Humanity with": A Whole Country Might Be Wiped Out in One Second'. The article went into astounding detail, informing readers that the experiments conducted in laboratories in Liverpool, Birmingham, Cambridge and elsewhere in the world 'concern a new way of producing energy inconceivable quantities by splitting the atom of a rare metal, called uranium'.[42]

Rudolf Peierls, who worked at the time at the University of Birmingham, was part of the anonymous group of scientists mentioned in the *Sunday Express*. After calculating that the critical mass of uranium 235 needed for a nuclear device was much smaller than previously anticipated, Rudolf Peierls turned to Otto Frisch for confirmation of his findings. Then, the two physicists began working together on the problem.[43] Because of fears of espionage, they did not even entrust the manuscript to a secretary; Peierls typed it himself.[44]

The so-called Frisch-Peierls Memorandum, which the two émigrés drafted in February 1940, consisted of two sections, titled 'Memorandum on the Properties of a Radioactive Super-bomb' and 'On the Construction of a "Super-bomb"; based on a Nuclear Chain Reaction in Uranium'. The first part detailed the effects of a nuclear weapon and raised the moral implications of such a weapon of mass destruction, while the second part was primarily concerned with technical niceties.[45] Frisch and Peierls complemented one another on calculating the approximate amount of fissile material needed for the manufacture of an atomic bomb.[46] Perhaps the greatest achievement of their seminal memorandum lay in the fact that it assumed that the critical mass of these 'super-bombs' measured 'about one pound'.[47] The 'Einstein Letter', by contrast, envisaged nuclear weapons being 'carried by boat' as they 'might very well prove to be too heavy for transportation by air'.[48] According to Frisch and Peierls's findings, the manufacture of nuclear weapons thus appeared to be feasible. The highly successful cooperation between the theoretician Peierls and the experimentalist Frisch on their memorandum represented, in a nutshell, a new approach to nuclear physics at the time and was a precursor of the path to the creation of the atom bomb that was also be taken at Los Alamos, as Chapter 3 will show.

Although the 'Frisch-Peierls Memorandum' marks a chief document in the history of science in the twentieth century, it was only with the publication of Margaret Gowing's seminal study *Britain and Atomic Energy 1939–1945* in 1964 that the public learnt in more detail about this important document. 'The question "Who really invented the atom bomb?" is officially answered today with the full backing of documents just off the secret list', reported Chapman Pincher in the *Daily Express* and going on to credit Frisch and Peierls.[49] Not only did the 'Frisch-Peierls Memorandum' anticipate, for the first time, the feasibility of a working nuclear weapon and describe its effects, but it also indicated that its authors had indeed been correct in following their intuition, and had asked the appropriate questions. It appears curious that neither scientists in the United States nor in Nazi Germany had asked the same questions, let alone answered them adequately. Not only did Peierls and Frisch state that an atomic bomb was feasible but they detailed the yield, destruction, and the effects of radiation and nuclear fallout. The two émigrés were apparently the first to address the problem of radioactive fallout and to raise the ethical question of using an atomic weapon of mass destruction.[50]

With regard to the fallout issue, Frisch and Peierls argued in their memorandum that 'Owing to the spreading of radioactive substances with the wind, the bomb could probably not be used without killing large numbers of civilians, and this may make it unsuitable as a weapon by this country'.[51] The 'Frisch-Peierls Memorandum' was produced at a time when many felt that the bombing of civilians as in the case of Guernica in 1937 was disgraceful. But the Blitz and the Allied strategic bombing of German cities, which desensitized the public towards large numbers of civilian casualties, were yet to come.[52] Since there appeared to be no defence against the new weapon and Frisch and Peierls condemned its use in war, they suggested in their memorandum that 'The most effective reply would be a counter-threat with a similar bomb'.[53] The 'Frisch-Peierls Memorandum' thus represents, Lorna Arnold has argued, the origin of deterrence theory.[54]

There remains, however, the question of the credibility of their highly abstract concept of deterrence. What the French existentialist Albert Camus postulated in his critique of capital punishment, 'Reflections on the Guillotine', in 1957, arguing that 'Heads are cut off not only to punish but to intimidate, by a frightening example, any who might be tempted to imitate the guilty', applies to the idea of nuclear deterrence as well.[55] Since nuclear weapons extend well beyond theory, the idea of a – not yet existing, let alone tested – atom bomb as a deterrent raises the further question of whether this weapon's existence as such (without any demonstration) would in practice serve as an effective means of deterrence. It was apparently in part this assumption that led the Truman Administration to drop the atomic bomb unannounced on the Japanese city of Hiroshima.[56]

When Frisch and Peierls authored their memorandum, the Trinity Test – the world's first nuclear explosion – and the atomic bombings of Hiroshima and Nagasaki still lay over five years in the future. The following episode, which occurred in March 1941, after Otto Frisch had transferred to the University of Liverpool to work under James Chadwick, illuminates how abstract the nuclear threat was perceived at the time: after a German parachute mine had exploded in the courtyard behind the Victoria Building, destroying the Engineering Building, Chadwick asked a colleague to measure the area for increased radiation levels with a Geiger counter because he feared the device could have been a German nuclear weapon.[57]

Similar incidents that underline the highly abstract understanding of the new weapon and its possible effects occurred shortly before the Trinity Test of 16 July 1945. In spite of Hans Bethe's calculations, which theoretically refuted the possibility that the implosion bomb would ignite the earth's atmosphere as many of the Los Alamos scientists feared at the time, General Leslie Groves called the New Mexico state governor John Dempsey before the test to inform him that there was a chance he might have to declare martial law in central New Mexico. One of the most bizarre arrangements preceding the test was Groves's order to *New York Times* journalist William Laurence, who was the only member of the press allowed to witness the explosion, to write three press releases for three different scenarios: first, a story about an explosion without any damage or casualties; second, an article about an explosion causing severe damage; and third, the obituaries of all persons present at the test, including Laurence.[58] While these accounts reveal a good amount of uncertainty about the atom bomb's destructive power, it was only after the Trinity Test, Hiroshima and Nagasaki and in particular with the superpowers carrying out extensive testing programmes that constantly demonstrated the destructive force of nuclear and, in particular, thermonuclear weapons, that the concept of nuclear deterrence became very credible.

The 'Frisch-Peierls Memorandum' was, with the help of Marcus Oliphant, passed on to Sir Henry Tizard, the chief British military scientific administrator at the time. As one of the first and foremost results of the document, the Chamberlain Government organized a committee of scientists under the chairmanship of George Thomson, initially under the umbrella of the Air Ministry (AM) and later under the Ministry of Aircraft Production (MAP) in April 1940 to further investigate the feasibility of atomic weaponry.[59] James Chadwick soon became the chief engine in this committee, coordinating the research conducted at the different universities and exerting more influence than George Thomson.[60]

The Maud Committee

The 'Thomson Committee' became more famous under the code name Maud Committee. The designation Maud Committee – or M.A.U.D. as it was also spelled to look more official – originated from a telegram that Niels Bohr had sent to Otto Frisch shortly after the German invasion of Denmark, ending with the mysterious line 'TELL COCKCROFT AND MAUD RAY KENT'. After Frisch had passed it on to Thomson, the odd telegram was among the first issues on the committee's agenda. While several attempts to decipher Bohr's message with regard to vital atomic-weapons-related information were fruitless, the committee finally adopted the cover name Maud Committee. It was only years later that the telegram's sender solved the mystery and it became clear that the telegram had simply been addressed to John Cockcroft and Bohr's former housekeeper Maud Ray, who lived in Kent.[61] The fact that Bohr's telegram started wild speculation about the state of the German nuclear research programme is yet another example of how deep-rooted the fear of the Third Reich, and in particular a nuclear-capable Nazi Germany, was at the time.

Meetings of the Maud Committee regularly took place at the Royal Society, and its earliest members included, apart from its chairman George Thomson: James Chadwick, John Cockcroft, Marcus Oliphant, Patrick Blackett, Charles Ellis, William Haworth and Philip Moon as well as a representative of the AM/MAP.[62] It is ironic that, at first, its initiators, Rudolf Peierls and Otto Frisch, were not allowed to participate in the work of the Maud Committee owing to their status as 'enemy aliens' or, in Peierls's case, as a recently naturalized British subject. Only after Peierls had sent a letter to the committee chairman did Thomson change the rules and fully consulted the authors of the 'Frisch-Peierls Memorandum'. To get Peierls and Frisch more involved at the administrative level, Thomson also formed a Technical Sub-Committee on which they served. Other foreign-born atomic scientists faced similar problems prior to the change of government employment policies regarding aliens and ex-aliens. That Hans von Halban, Lew Kowarski, Joseph Rotblat, Franz Simon or Egon Bretscher made considerable contributions to the work of the Maud Committee was only made possible through government contracts with universities because the government itself was still not allowed to hire 'enemy aliens'.[63] But others like Herbert Fröhlich, Heinz London and Walter Heitler failed to become involved in the work of the Maud Committee despite Arthur Tyndall's intervention on their behalf.[64]

As a result of these government regulations, the early Maud work was exclusively carried out at university laboratories across the United Kingdom. It was only in autumn 1940 that the British government started spending money on Maud research. The four main laboratories where atomic-weapons-related work was conducted were located at the Universities of Liverpool, Birmingham, Oxford and Cambridge with occasional consultation of Herbert Fröhlich, Walter Heitler and Heinz London at Bristol University. Like these three German-speaking émigrés, Paul Dirac was infrequently conferred with over theoretical problems but he was never an official member of any Maud team. Because British-born scientists exclusively worked on sensitive projects like radar, a particularly high number of German-speaking émigré nuclear scientists engaged in – or rather were almost accidentally pushed into the direction of – atomic-weapons research.[65]

At the University of Liverpool, James Chadwick, who had taken first steps to investigate the feasibility of an atomic bomb, especially its size and critical mass, as early as the summer of 1939, was in charge of a Maud team. Following Frisch's and Peierls's seminal memorandum of February 1940, and using the newly acquired cyclotron, Chadwick and his collaborators Joseph Rotblat and Otto Frisch, who had transferred there from Birmingham, began a comprehensive test programme in order to prove the theoretical assumptions outlined in the 'Frisch-Peierls Memorandum'. Some of their work was even extended to include Norman Feather as well as Egon Bretscher, Herbert Freundlich and Nikolai Kemmer at the Cavendish Laboratory in Cambridge.

Among other things, Otto Frisch worked especially on isotope separation in uranium through thermal diffusion. This process represented an early attempt to enrich naturally occurring uranium 238 to become uranium 235, the crucial ingredient for an atomic fission bomb. Frisch and Peierls had proposed this method in their memorandum. Every time a uranium nucleus splits, it emits two key fission products and several secondary neutrons that can trigger a self-sustaining chain reaction by causing more uranium nuclei to divide and release yet more neutrons. As previously noted, Leo Szilard had come up with the idea of a chain reaction in 1933, and Otto Hahn and Fritz Strassmann discovered fission five years later with Frisch and his aunt Lise Meitner confirming their discovery. Shortly afterwards, Niels Bohr proposed that fission was much more likely to take place in the lighter uranium isotope 235 with its slow neutrons than in uranium 238. Since the naturally occurring uranium 235 only contains 0.7 per cent of uranium 235, Frisch and his colleagues attempted to come up with ways of enriching uranium 238 so that it turned into 235. In his early work on isotope separation, Frisch tried to make use of the two isotopes' distinct characteristics: while a uranium 238 molecule disperses towards a cold surface, a uranium 235 molecule diffuses towards a hot surface. These features enabled scientists to separate the two. Frisch, however, abandoned his work on thermal diffusion in late 1940.[66]

Another important area of work at Liverpool concerned measuring cross-sections in uranium fission. Cross-sections describe the probability of interaction between the particles of atoms in an imaginary area. Since Frisch and Peierls had so far only theoretically proposed the efficiency of uranium 235 for a fission bomb but no practical proof existed, these measurements of cross-sections represented a pivotal prerequisite to start a full-scale uranium bomb project.[67]

At the Cavendish Laboratory in Cambridge, German-speaking émigré scientists were involved in accomplishing two major achievements. First, Egon Bretscher and Norman Feather reckoned that, apart from uranium, another element, which is now commonly known as plutonium, could also be used as a nuclear explosive. The Polish émigré Joseph Rotblat reached the same conclusion simultaneously yet independently at Liverpool. But their idea had no impact on the British atomic weapons programme at the time.[68] Later plutonium was used in a fusion bomb. Bretscher subsequently complained that Norman Feather tried to claim all the fame for discovering plutonium for himself.[69]

Apart from Bretscher's and Feather's discovery, Hans von Halban and the Russian-born émigré Lew Kowarski conceived the second big achievement at Cambridge. The two men had worked at Fréderic Joliot-Curie's laboratory in Paris until the German invasion of France when they came to the United Kingdom, bringing with them the world's main supply of heavy water. Although their findings would not have any direct impact on the British nuclear weapons project, they concluded that heavy water could be used as a moderator in a kind of 'uranium machine' – today known as a nuclear reactor – to generate energy.[70] In early 1940, Otto Frisch independently suggested the same idea.[71]

At Birmingham, another team headed by Rudolf Peierls dealt with theoretical problems. Klaus Fuchs joined this group and made decisive contributions to its work. Like Chadwick's team, the Birmingham group under Peierls worked on what Ronald Clark called the 'esoteric aspects' of atomic weaponry such as the specific features of a nuclear device that were needed to investigate the practicability of an atom bomb.[72] Peierls's team worked on the theoretical side, interpreting the basic nuclear data from experiments run at Liverpool, Cambridge and Oxford. In order to calculate the critical mass required to construct a nuclear explosive, Peierls and Fuchs analysed the mechanics of the chain reaction and performed calculations on the yield of a nuclear weapon, for instance.[73] Klaus Fuchs performed important calculations regarding thermal diffusion based on Liverpool data.[74] Peierls also engaged in work on theoretical features of the research on separation processes using gaseous diffusion which was carried out by Franz Simon at Oxford and which Peierls had initiated.[75] By September 1940, Peierls favoured gaseous over thermal diffusion as an effective means to separate uranium isotopes on a sufficient scale. Under this principle, a gaseous uranium compound – usually uranium hexafluoride – is diffused through several sets of semi-permeable membranes. Since the uranium 235 isotopes pass through these barriers faster than uranium 238 isotopes, the degree of enrichment of uranium 235 increases from stage to stage.[76] While Liverpool worked in physics as well as chemistry, a second Birmingham group headed by William Haworth conducted almost all the chemical research.[77]

Besides his work at Birmingham, Rudolf Peierls was also instrumental in engaging Franz Simon in Maud Committee work.[78] Simon's crucial role in the early bomb project epitomized particularly well the paradox many of his fellow German-speaking émigré atomic scientists faced between exclusion from supposedly sensitive war work like radar and being – almost accidentally – entrusted with one of the biggest secrets of the Second World War: not only was his Oxford team primarily composed of non-British-born members but he was also to become the first Commander of the British Empire who had previously earned the Iron Cross in action.[79]

At the University of Oxford's Clarendon Laboratory, Simon headed a group which included his former colleagues Nicholas Kurti and Kurt Mendelssohn as well as, initially, his former student Heinz London, from the Technische Hochschule (Polytechnic Institute) in Breslau where he had been chair of physical chemistry.[80] Born in the 1900s, Nicholas Kurti was, like Leo Szilard, Eugene Wigner, John von Neumann, Egon Orowan and Edward Teller, a native of Budapest and a member of a generation of Hungarian-born but German-speaking atomic scientists who later worked on atomic research and subsequently gained international status.[81] Moreover, in the beginning the team also included Heinz London's older brother Fritz who spent only a short period at the Clarendon on a temporary ICI fellowship.[82] Another major German-speaking member, Heinrich Kuhn from the University of Göttingen, also joined Simon's team at Oxford.[83]

David Schoenberg referred to the Clarendon appropriately as the 'Breslau colony in Oxford'.[84] The joint migration of Simon and his collaborators in low-temperature physics from Breslau to Oxford represents a rare case of almost an entire institution moving to a new host country, and has to be seen in connection with Frederick Lindemann's effort to strengthen Oxford physics.[85] This kind of 'institutional migration', which underpinned par excellence the reconstruction of pre-emigration group networks in the new host country, occurred only seldom. The Warburg Library, which emigrated with most of its staff from Hamburg to London where it was eventually integrated into the University of London and renamed the Warburg Institute, constitutes perhaps the chief example of this kind of 'institutional migration' in Britain.[86]

The work at the Clarendon Laboratory focused on isotope separation through the method of gaseous diffusion as well as work on diffusion membranes.[87] Simon proved that isotope separation by means of gaseous diffusion appeared to be much more promising and effective than the thermal diffusion method proposed in the 'Frisch-Peierls Memorandum'.[88] Again, Peierls's team assisted Simon's group in theoretical matters.[89] In 1940, Rudolf Peierls laid out theoretical basics regarding the efficiency of various separation processes that could be used in isotope separation on an industrial scale in a technical report.[90] He and Fuchs also jointly tackled theoretical problems concerning uranium separation.[91]

Together, Peierls and Simon even put forward a scheme for uranium 235 production through gaseous diffusion on an industrial scale and thus made pivotal assumptions about the nature of the future project.[92] 'Separate isotopes on a practical, on a macroscopic scale, seems a crazy idea', Rudolf Peierls later remarked on his own and his colleagues' initial thoughts about this process. The idea of large-scale isotope separation had at first appeared so abstract a concept that Peierls compared it to science fiction. As a result, no-one had ever attempted to separate isotopes on that scale, and he stated: 'So to do that with large amounts seemed quite crazy, and therefore, one didn't practically think about what would happen if we separated [uranium] 235'.[93]

Apart from institutions, some individuals were also frequently called upon. Heinz London, who had transferred to Bristol University's H. H. Wills Laboratory in 1936, was sometimes consulted on theoretical matters.[94] Walter Heitler and Herbert Fröhlich, who also worked under Nevill Mott in Bristol, made additional contributions to the work of the Maud Committee. Even before the creation of the committee, at about the same time when Frisch and Peierls drafted their memorandum, Fröhlich and Heitler collaborated on an unpublished paper: 'Chain Reactions in Uranium'.[95] Although the two scientists did not reach a sufficient conclusion in their manuscript, the existence of this document reveals their sense to investigate the right issues related to the development of nuclear arms. It took until December 1942 when a team under the Italian-born émigré Enrico Fermi at the University of Chicago achieved the first controlled chain reaction in natural uranium.[96]

Herbert Fröhlich's and Walter Heitler's work for the Maud Committee was related to spontaneous fission in uranium. This phenomenon was seen as a big problem because it could minimize the yield of a nuclear bomb dramatically.[97] Otto Frisch had measured spontaneous fission, which was emitted from uranium 235, the isotope to be used in the weapon. Rudolf Peierls described the sentiment about Frisch's findings at the time, saying that 'in that case the assembly had to be done at high speed. It was contemplated, if necessary, to use a double gun, shooting two projectiles that would meet half-way down the barrel'.[98] Fröhlich's and Heitler's work concentrated on the spontaneous effect in nuclear fission produced by cosmic rays, and led them to produce a document titled 'Fission Produced by Cosmic Ray Neutrons'.[99] Although he had at no point been a full member of the British atomic weapons programme, Herbert Fröhlich was asked to join the AERE Harwell as head of its theoretical division in 1946, and his previous collaborations with Walter Heitler and Nikolai Kemmer on nuclear matters apparently influenced the decision of the Physics Department at the University of Liverpool to appoint him as first professor of theoretical physics.[100]

The fact that the various Maud Committee laboratories had made such good progress was to a large degree the result of Rudolf Peierls's pivotal part in the Maud Committee work. It is safe to say that – perhaps after Chadwick and Thomson – Rudolf Peierls was the most important contributor and administrator in the Maud Committee work, although he did not have full access to all its administrative councils. Not only was he a chief connecting link between different laboratories through his theoretical work, but long before the official creation of the Maud Committee, a steady exchange of information had existed between himself, James Chadwick and Otto Frisch.[101] That the Maud Committee chairman George Thomson entrusted Peierls with the task of compiling a report on the current state of affairs – particularly regarding uranium-related problems – illustrates his central role in early British atomic arms research. By mid-August 1940, Peierls presented him with a ten-page report that was accompanied by nine papers written by him with the help of Otto Frisch and dealing with mathematical clarification of the most burning problems. Among other things, Peierls convincingly demonstrated that, while slow neutrons would be an efficient means to produce atomic energy in a reactor, they could not be effectively used in a bomb as they would achieve only a fairly small explosion. Similarly, he argued that fast neutrons in uranium 238 would not be of any use in an atomic bomb, for the size of the critical mass would be about 28 tons. Instead, TA work now focused on uranium 235, its enrichment and separation from uranium 238.[102] Rudolf Peierls continued writing progress reports from the theoretical physics point of view on the advancement of the British nuclear weapons programme.[103]

Partly thanks to Peierls's tireless efforts, the Maud Committee concluded its work in October 1941 with two final reports, 'Use of Uranium for a Bomb' and 'Use of Uranium as a Source of Power', which were of great significance for the future of atomic energy research in the United Kingdom and in the United States. In the first report, the committee concluded that an atom bomb was feasible and described the research that had been conducted so far, and the work of the Maud Committee had put the United Kingdom at least temporarily far ahead in the race for an atom bomb.[104] The War Cabinet's Scientific Advisory Committee sanctioned the view put forth in the Maud Committee's two final reports and regarded the atom bomb as crucial to the war effort. Prime Minister Winston Churchill, who had been updated by his chief scientific advisor Frederick Lindemann on the latest developments regarding the Maud Committee, shared these views, declaring that 'Although personally I am quite content with the existing explosives, I feel we must not stand in the path of improvement'.[105]

Tube Alloys and the road to Anglo-American nuclear cooperation

The chief result of the Maud Committee reports was the establishment of an independent British atomic weapons project. In October 1941, it came into existence under the official cover name 'Directorate of Tube Alloys' or simply 'Tube Alloys', and was placed under the DSIR. Wallace Akers (later Sir Wallace) of ICI served as TA director. The Technical Committee regularly updated him on the progress of the project. Peierls served on the committee alongside James Chadwick, John Cockcroft, Sir Charles Galton Darwin, Marcus Oliphant, Norman Feather and the German-speaking émigrés Hans von Halban and Franz Simon.[106]

All committee members realized early on that in the long run a separate British programme could only be brought to fruition on a much smaller scale than operations in the United States. Therefore, they decided to limit the focus of TA to especially important areas of inquiry such as verifying basic nuclear data, conducting theoretical studies on the chain reaction, the size, design and yield of an atomic bomb as well as methods of uranium separation through gaseous diffusion, including the design and construction of machines necessary in the process. Another area that saw the concentration of TA resources was the investigation of a moderator, which was needed in a reactor to slow down neutrons in order to be able to achieve a controlled chain reaction. This research involved also experiments with and the manufacture of heavy water.[107]

In principle, the teams engaged in nuclear weapons research stayed the same as before. The groups at the universities of Liverpool and Cambridge, which worked on the experimental determination of nuclear data, were significantly strengthened, and smaller TA research programmes were started at Bristol and Manchester Universities. James Chadwick supervised all this work. At Cambridge, Hans von Halban and Lew Kowarski, in collaboration with Egon Bretscher, were put in charge of the work on slow neutron systems. Von Halban's and Kowarski's team later moved to Montreal, Canada, where newly recruited British, Canadian and American personnel supplemented their group. At Cambridge University, Egon Orowan also contributed significantly to the TA project through his research on the manufacture of uranium metal. Rudolf Peierls continued as head of his Birmingham team. This group, which also included Klaus Fuchs, dealt with theoretical problems such as the chain reaction. Fuchs and Peierls also developed the theory of the operation and performance of a uranium separation plant. Peierls occasionally consulted Paul Dirac at Cambridge on problems relating to isotope separation. At the Clarendon Laboratory in Oxford, Franz Simon's team performed experimental work on the gaseous diffusion method. Nicholas Kurti and Henrich Kuhn assumed leading positions in Oxford. On the theoretical side, the Clarendon group received support from Rudolf Peierls's team, and the second Birmingham group under Haworth solved any chemical problems that arose in the course of Simon's work. Peierls and Simon even collaborated with Metropolitan-Vickers Electrical Co. Ltd. on the development and manufacture of machinery used in the process.[108] Rudolf Peierls and Otto Frisch were also involved in discussing the experimental work conducted in Montreal.[109]

After his naturalization as a British subject in 1942, Heinz London was also able to make fuller contributions to the work on isotope separation. London initially conducted work at Bristol but moved to ICI laboratories at Witton and Winnington in 1943. When his research into ionic migration in liquid electrolytes proved to be impractical for uranium separation, he switched to investigating the promising method of gaseous diffusion, in particular membranes to be used in the process. Later London moved to Birmingham where he even assumed directorship of H. S. Arms's team and worked closely with Simon's Oxford group, examining the porosity of membranes used in the gaseous diffusion process. At Birmingham, he also worked with Hans Kornberger and the Viennese émigré Franz Mandl.[110]

The organization of TA was much looser than the strictly compartmentalized and hierarchically structured Manhattan Engineering District (MED), which will be discussed in the following chapter. While private corporations like Metroplitan-Vickers Electrical Co. Ltd and ICI (after all, the director of TA, Wallace Akers, worked for ICI) joined TA work, these collaborations between the private and public sectors remained on a much smaller scale than the Anglo-American-Canadian operations in the MED. The Manhattan Project operated on a hitherto unprecedented scale that even dwarfed other large-scale projects such as the development of the jet engine. TA also rested much more on individual university departments than the Manhattan Project, which comprised three chief installations: Los Alamos, New Mexico; Oak Ridge, Tennessee; and Hanford, Washington. Three subcommittees, which were to report to the TA Technical Committee coordinated the various TA programmes. These were the Diffusion Project Committee, which included Rudolf Peierls and Franz Simon; the Chemical Panel, including Simon; and the Metal Panel, on which Simon served and which dealt with uranium metal production and general metallurgical matters.[111] Franz Simon thus also assumed a central position in the TA work.

Although Rudolf Peierls only served on one TA subcommittee, he kept on playing a pivotal role in the early British nuclear weapons project. The fact that Peierls composed a history of the early British programme underlines his high standing among TA administrators and scientists. 'Peierls is one of the few men who can write about this time from his own knowledge and not from hearsay', James Chadwick informed the War Cabinet. To emphasize Peierls's pivotal role, Chadwick added that if he had written this account himself, 'It would not differ very much from Peierls".[112]

Meanwhile, owing to the severe shortage of resources in the United Kingdom, it had become obvious even before the Maud Committee had issued its final report that Britain was unable to pursue a viable atomic research programme unilaterally that would produce a working weapon during the war.[113] Senior TA administrators thus deemed it necessary to start large-scale cooperation with the United Sates. In this context, the impact of the 'Frisch-Peierls Memorandum' and the reports of the Maud Committee cannot be emphasized enough because they represented indeed Britain's chief part in getting the Manhattan Project under way.[114]

Bilateral cooperation already existed in other areas. In 1940, over a year before the Maud Committee released its final reports, the so-called Tizard Mission had embarked to the United States in order to share Britain's latest technological developments. The famous black-box that Sir Henry Tizard brought to America contained, for example, information on the design of the Rolls Royce Merlin aircraft engine that would later also power the American Mustang P-50, the proximity fuse, and important radar-related data.[115]

Although John Cockcroft, who was one of Britain's leading atomic physicists at the time, accompanied Tizard to the United States, the delegation did not discuss Anglo-American nuclear cooperation at this stage.[116] This was partly the result of a fundamental misjudgement of the United States' capability by the British government. Since they completely underestimated the growth of the American atomic project, Britain's nuclear programme was severely damaged in the long run and by mid-1942 the British lead in atomic physics was forever lost.[117]

It was around the same time during another visit to the United States that British scientists, including Rudolf Peierls, realized how far their project had fallen behind the American programme. In November 1941, George Pegram and Harold Urey had toured British nuclear research facilities in order to prepare the sharing of information between the British and American projects. The two American professors had shown particular interest in the work of Franz Simon's team on isotope separation through gaseous diffusion at the Clarendon Laboratory, Oxford. Following their visit, a British delegation which included Rudolf Peierls as well as Hans von Halban and Franz Simon, visited the United States between February and April 1942.[118] Owing to its high percentage of foreign-born members, some American colleagues jokingly remarked that the supposedly British group was indeed not 'very typically British', as Rudolf Peierls recalled.[119] During this trip, Peierls met high-calibre American physicists such as Arthur Compton and J. Robert Oppenheimer.[120] The Americans held Peierls in such high esteem that they requested another visit by him and Chadwick in late 1942.[121] Vannevar Bush, however, the director of the Office of Scientific Research and Development (OSRD) that controlled the MED, declined the request to 'invite Peierls' for another visit because Bush was at the time involved in difficult negotiations with the British government over the exchange of nuclear information.[122] As perhaps the most significant result of this mission, the British scientists, to their dismay, had to face the fact that the American project had gathered momentum much faster than had previously been anticipated.[123]

This painful observation led to the realization that the United Kingdom would not be able to sustain a viable atomic weapons project of its own, let alone one on the same scale as that of the United States. As a consequence, leading atomic scientists and political decision-makers recognized that the country had to fully cooperate with the United States in atomic matters if it were not to fall too far behind in nuclear weapons and energy research. In July 1943, the Technical Committee – which included Rudolf Peierls and Franz Simon – therefore agreed that 'full co-operation with the U.S.A. is the only effective method of realizing the T.A. project under war-time conditions'. The delegates further concurred that the United Kingdom should pursue its own nuclear weapons project until a settlement with the United States was reached. At this meeting, Rudolf Peierls revealed a great deal of concern over the viability of TA in the future. In his view, the vagueness of the current state of Anglo-American relations in atomic affairs could affect the TA project significantly, as in the case of the diffusion separation programme, which had already suffered considerably from this uncertainty.[124]

James Chadwick, too, had fully realized the necessity for Anglo-American nuclear collaboration early on. But he also had ulterior motives that went beyond Anglo-American wartime cooperation: for him this partnership provided also a way to gather vital nuclear data that would be useful to an independent British nuclear programme in the long run, after the end of the Second World War.[125] 'The American effort is on such a scale that we could not compete with it even in peace time', Chadwick argued in early February 1944. He thus called for British scientists to 'acquire the fullest possible knowledge and experience of all phases of the project so that we shall be in a position when the time comes to start work in England on the right lines, profiting by American experience'.[126]

At the same time, reports from Rudolf Peierls and Christopher Frank Kearton on the collaboration with American scientists regarding the diffusion plant brought alarming news. As Wallace Akers informed the Chancellor of the Exchequer, their reports 'indicate that the time may come when the Americans will shut us out from a knowledge of the final stages of design and construction, and also from the operation, of their large-scale diffusion and electro-magnetic plants'.[127] And this realization seemed to confirm Chadwick's concerns.

On the American side, similar fears were present which expressed themselves in acute worries over employing foreign-born scientists in the atomic project, especially those of the Free French movement such as Hans von Halban and Lew Kowarski as well as the many émigrés among the British scientists. As US officials argued, legitimate concerns existed that non-naturalized scientists would return to their home countries after the war and take valuable nuclear data with them, which could then be of use to their governments.[128] Or, even to third governments, as in the case of Klaus Fuchs's espionage for the Soviet Union.

These concerns stood in a sharp contrast to official American foreign policy towards Britain at the time: as early as August 1940, under the so-called Destroyers for Bases Deal, the Roosevelt Administration had given Whitehall 50 First World War destroyers. In return, the Churchill Government had leased land for the construction of naval bases on eight of its overseas territories in the Western Atlantic and Caribbean to the United States. And by March 1941, both the United States Senate and Congress had sanctioned FDR's Lend-Lease policy under which the United States could now support the British war effort through loans and credits.[129]

From the beginning, Anglo-American relations were ambivalent as a mutual sense of distrust hampered the atomic cooperation between the two countries. Britain engaged in an all-out propaganda effort to break the American neutrality and win the United States over as its ally long before the Japanese attack on Pearl Harbor on 7 December 1941.[130] But, at the same time, the Churchill Government was initially highly sceptical and hesitant when it came to nuclear cooperation.[131] Before Urey's and Pegram's visit to the United Kingdom in November 1941, for example, the secretary of the DSIR, Edward Appleton, instructed Michael Perrin 'to take a full record of all the discussions with Urey and Pegram'.[132]

It was especially the heightened security levels which the US military had raised that gave the British a feeling of being excluded from vital nuclear information. 'The joint result of these changes has been to leave some doubt in my mind concerning future relations between work to be carried on in America, on the one hand, and in England and Canada on the other', Wallace Akers wrote to James B. Conant in mid-December 1942. Akers went on to grumble that 'although we appreciate the desire to avoid as much as possible leakage of information ... there is a strong feeling among the British group that the division of the work into watertight compartments can be carried to the point at which inefficiency may be considered to outweigh the gain of secrecy'.[133]

When the Roosevelt Administration became aware of its highly advanced position vis-à-vis Whitehall in nuclear energy-related matters during the latter half of 1942, Anglo-American atomic collaboration appeared as a less pressing need from Washington's point of view than it had done earlier. In Britain, by contrast, the situation had changed considerably. As the threat of National Socialist Germany which had revealed itself so dramatically at Dunkirk in May and June 1940 still loomed over the Churchill Government, Whitehall was now desperate to enter atomic cooperation with the United States. Eventually FDR agreed to Anglo-American atomic collaboration.[134] As Winston Churchill later wrote: 'I strongly urged that we should at once pool all our information, work together on equal terms, and share the results, if any, equally between us'.[135]

After a period of severe problems in nuclear relations between the two countries, plus the realization that Britain herself could not sustain a workable atom bomb project, the Quebec Agreement of August 1943 officially regulated the collaboration between Britain and the United States in nuclear matters, at least for the period of the war.[136] In September 1944, FDR and Churchill made amendments to the Quebec

Agreement during a meeting at Hyde Park, New York.[137] With the new agreement in place, the British government provided 'all possible assistance' to the American project working on the electromagnetic method for isotope separation, as Michael Perrin declared in late January 1944.[138] There were in particular, on the British side, a considerable number of critical voices demanding that Washington share all its secrets with Britain. Since the United States made much bigger contributions to the joint project, the British actually did not come out of the deal too badly after all.[139]

Although nuclear cooperation between the two countries had been formalized under the accords of the Quebec Agreement, relations between Britain and the United States remained far from smooth. The acknowledgement of patents, for example, represented a contentious issue in Anglo-American relations throughout the period of the Manhattan Project. While there is no recorded evidence that Klaus Fuchs and Rudolf Peierls experienced difficulties in receiving proper recognition for patents, other German-speaking scientists such as Egon Bretscher, Hans von Halban and Leo Szilard faced problems.[140] But German-speaking émigré scientists did not encounter such considerable problems as, for instance, Enrico Fermi, who fought for 15 years to receive proper recognition of the ownership of an economically highly valuable patent that dealt with a process applied to slow down neutrons in atomic reactions.[141] Despite the many problems in the Anglo-American wartime partnership, the creation of the atomic bombs at Los Alamos represented, as John Baylis has argued, 'one of the most co-operative ventures of the alliance'.[142]

Conclusion

After Rudolf Peierls and Klaus Fuchs had overcome the many difficulties with regard to their German extraction and schooling, they eventually became key players in the early British nuclear weapons programme. Owing to fundamental differences in national teaching and research styles between the United Kingdom and Continental Europe, Peierls and to a lesser extent Fuchs as a student had to overcome obstacles in order to fit into the British physics community. Since their immigration status as 'enemy aliens' or, in Peierls's case as an ex-alien, prevented them from conducting important war work such as the development and refinement of radar, they were – almost accidentally and most ironically – pushed into nuclear weapons research, which eventually became one of the best guarded secrets of the Second World War.

It was in particular Peierls who, in collaboration with Otto Frisch, composed the seminal 'Frisch-Peierls Memorandum'. Not only was this pivotal yet commonly neglected document decisive in starting a serious atomic arms development project in the United Kingdom, but it was also instrumental in starting the joint Anglo-American-Canadian Manhattan Project, as the next chapter will demonstrate. Rudolf Peierls especially and to a lesser degree Klaus Fuchs played crucial roles in establishing close Anglo-American nuclear cooperation which eventually led to the formation of the joint Anglo-American-Canadian Manhattan Project. By August 1943, Washington and London had finalized Anglo-American nuclear cooperation under the Quebec Agreement.

3
American Interlude: The Manhattan Project, the Atom Bomb and the Emergence of a New Approach to Nuclear Research

Under the new Anglo-American cooperation, Rudolf Peierls and Klaus Fuchs joined the Manhattan Project. In late 1943, Fuchs, Peierls and his wife Genia sailed on the Royal Navy troopship *Andes* from Liverpool to the United States.[1] Fuchs and Peierls worked, at first, in New York City before they moved to the MED's central Los Alamos laboratory. And, as eventually in Britain before, their training in German centres of theoretical physics helped Fuchs and in particular Peierls, along with several other German-speaking émigré scientists who had found employment in American universities after leaving Germany, to foster the research culture of Big Science and ultimately forge a key feature of British nuclear culture.

In the United States, the MED had its origins in the National Defense Research Committee (NDRC) that FDR had created by decree in the immediate aftermath of the Japanese attack on Pearl Harbor. Vannevar Bush, who had gained experience as a science administrator through his appointment as president of the Carnegie Institution for Science, headed the NDRC's executive committee, which consisted of seven additional members including Harvard's president James B. Conant. In 1941, Conant took over the directorship of the NDRC from Bush, when the latter became the head of the newly created Office of Scientific Research and Development (OSRD). The new office now assumed the NDRC's business and duties with regard to the atomic weapons project.[2]

The MED then started to take shape in June 1942 when its directorship was transferred to the US Army. In September 1942, Major General Leslie Groves, who had previously supervised the construction of the Pentagon in Washington, DC, became the commanding officer of the entire Manhattan Project. Alongside the MED's central Los Alamos laboratory where Rudolf Peierls and Klaus Fuchs took leading roles in the development of the first atomic bombs, the operation had two other chief sites as Oak Ridge, Tennessee, where the enrichment of uranium took place, and Hanford, Washington, which dealt with the production of plutonium. In addition, the Manhattan Project included several installations in the United States, Canada and the United Kingdom.[3]

First encounters with American culture

Fuchs and Peierls, together with Christopher Frank Kearton and Tony Hilton Royle Skyrme, first moved to New York City where they joined a team of British scientists who engaged in work on uranium separation at both Columbia University and the Kellex Corporation, a private MED contractor. Here, they continued to work in one of their areas of expertise – the design and operation of a large-scale isotope separation plant for uranium 235.[4] Another German-speaking émigré, Nicholas Kurti, worked at Columbia University on instrumentation to test the membranes that were to be used in the gaseous diffusion process.[5]

While Peierls was much more senior than Fuchs at the time and had conducted significant theoretical work on isotope separation, Klaus Fuchs contributed highly valuable calculations to the design of the gaseous diffusion plant for uranium separation to be built at Oak Ridge, Tennessee, during his stay in New York City.[6] The fact that Fuchs authored ten out of the total of 17 papers produced by the British Mission in New York between January and July 1944 underlines his standing as one of the foremost junior scientists involved in the Manhattan Project.[7]

It was in New York that Fuchs and Peierls confronted for the first time some peculiarities of American culture. They were, for example, appalled when they witnessed a case of racial discrimination. Although an African-American applicant was the strongest candidate for a post as computing assistant in Peierls's team, her application was rejected.[8] It must have felt awkward for émigrés who had previously experienced National Socialism to come across such racist tendencies in the country which had traditionally proclaimed itself the haven of freedom and democracy.

On the positive side, Fuchs and Peierls found an abundance of goods in the United States. Coming from war-torn Britain with its rationing of food and clothes, the two physicists, like many of their British colleagues, were deeply impressed by what they came upon in New York City.[9] 'Then going from the rigor of wartime England to the comfort of a commercial airliner from Montreal to Ottawa', Rudolf Peierls later recalled in his impressions of an earlier visit to America, 'and then coming down in the evening on the brilliantly lit city (from my point of view because we were used to blackouts) was an enormous thrill, and of course then to New York, which I'd never seen before'. In addition, he 'was quite impressed with (although theoretically one had known it) the much greater scale of things and size of laboratories, the number of people, the wealth of equipment and so on; together with a sense of purpose and hard work'.[10] Because of the severe shortages in Britain, the Peierls, for example, took the household items which they had acquired during their stay in the United States back home after the war.[11]

Peierls and Fuchs carried this fascination with them to the secret laboratory at Los Alamos in the southwestern state of New Mexico to which they transferred in 1944. While Rudolf and Genia Peierls came to 'the Hill', as Los Alamosans also called their hometown, in early 1944, Klaus Fuchs did not arrive there until August 1944.[12] The northern New Mexican installation was the Manhattan Project's central scientific laboratory. In April 1943, the first scientists arrived at Los Alamos, which had been home to the famous Los Alamos Ranch School until it was taken over by the US Army earlier in the same year.[13] The laboratory's official designation was Site Y, and after the war it became the Los Alamos Scientific Laboratory.[14]

Located on several isolated mesas overlooking the Rio Grande Valley, Los Alamos was the smallest of the three 'atomic cities' in terms of size and population. Yet it hosted the crème de la crème of nuclear scientists.[15] Alongside the Italians Enrico Fermi and Emilio Segrè, three German-speaking future Nobel laureates worked there: Hans Bethe, Felix Bloch and Maria Göppert-Mayer. With its great concentration of high-calibre scientists, 'the Hill' arguably represented something of a mid-twentieth-century version of the Ancient Greek *mouseion* of Alexandria. Ruth Marshak – a wartime resident of Los Alamos and wife of theoretical physicist Robert Marshak – also appropriately remarked on the town: 'Los Alamos stood for the same sort of thing that Hollywood represents to an aspiring starlet'.[16] At the time, the rocket development facility at Peenemünde where leading German scientists and engineers under Wernher von Braun constructed and tested the first V-1 and V-2 'flying bombs' was perhaps the only other research institution that was remotely comparable to Los Alamos.[17]

On 'the Hill', Rudolf Peierls and Klaus Fuchs came across many former colleagues and friends. Decades later, Peierls still recalled 'a strange sensation to meet so many old friends from various phases of our lives in such an outlandish place as Los Alamos'. These 'old friends' included, apart from Hans Bethe: Egon Bretscher, Otto Frisch, John von Neumann, Georg Placzek and Victor Weisskopf, but also the Italians Enrico Fermi and Emilio Segrè as well as the Danes Niels Bohr and his son Aage.[18] Apparently, 'an enormous international reunion of the atomic physics community', as Edward Teller termed it, took place at Los Alamos.[19]

In a graphic way, Los Alamos also embodied the relatively small size of the nuclear physics community at the time. For instance, many theoreticians among the Los Alamos scientists, including Enrico Fermi, Klaus Fuchs, Maria Göppert-Mayer, Edward Teller, Victor Weisskopf and even the scientific director of Los Alamos J. Robert Oppenheimer had been students of Max Born at one time or another.[20] Such 'old friends' often helped each other prior to their arrival on 'the Hill'. Rudolf Peierls, for example, informed George Placzek in detail about professional and private aspects of life at Los Alamos, warning him (despite the relative abundance of goods on 'the Hill') 'Don't of course expect a Fifth Avenue here'; (Figure 3.1)[21] and Hans Bethe's wife Rose briefed Genia Peierls about the conditions she had to expect.[22]

The émigré scientists whom Peierls and Fuchs encountered at Los Alamos usually came to the secret laboratory by two routes. While Bretscher, Frisch and Placzek, like Fuchs and Peierls, joined as members of the British Mission, Bethe, von Neumann and Weisskopf, together

Figure 3.1 The Los Alamos Trading Post, October 1945

with Felix Bloch, Martin Deutsch, Maria Göppert-Mayer, Rolf Landshoff, Hans Staub and Edward Teller came as naturalized American subjects directly from universities in the United States where they had found employment after their departure from Europe.

In comparison with Oak Ridge and Hanford, German-speaking émigré nuclear scientists were of particular significance at Los Alamos. The theoretical physicist Lothar Nordheim was among the few German-speaking émigrés who were engaged in work at other Manhattan Project installations. At Oak Ridge, he oversaw the X-10 project, a pilot reactor for the plant that was later to be built at Hanford. The X-10 graphite reactor made use of neutrons that uranium-235 produced as fission products to transform uranium 238 into plutonium 239. After the war, he served as director of Oak Ridge's Physics Division from 1945 to 1947.[23] But German émigré scientists also worked at smaller MED installations. At the Chalk River site near Montreal, Canada, where Herbert Freundlich and Nikolai Kemmer were based, Hans von Halban collaborated with Lew Kowarski on reactor theory.[24] And Oskar Bünemann worked as a theoretical physicist at the MED's Berkeley laboratory.[25]

The Los Alamos scientists lived under sparse conditions. Still, compared to the severe shortages at home in the United Kingdom, the supply of food and especially laboratory materials appeared, in their eyes, magnificent and thus made it easier for them to acclimatize in the locale of northern New Mexico.[26] Like large parts of the US Southwest, the Los Alamos area offered a rich mixed heritage of Native American and colonial Spanish and Mexican cultures.[27] Peierls and Fuchs also encountered further peculiarities of American culture that differed considerably from Continental and British cultures at the time. Many of their American-born colleagues joined the square dance club, for example, whereas German-speaking émigré scientists enjoyed 'high culture', with Otto Frisch performing weekly concerts for the local radio station KRS.[28] Fuchs, Peierls, Bethe and Egon Bretscher enjoyed hiking and mountain climbing. American-born scientists, by contrast, delved into Native American ruins in the area and took pleasure in horseback riding.[29] Furthermore, the setting in the iconic landscape of the American Southwest evoked associations with the nineteenth-century Frontier days, and the Frontier myth in fact served as a major source of motivation for many of the Los Alamos scientists.[30]

Perhaps the most defining difference in the perception of everyday life on 'the Hill' between the German-speaking émigrés and British-, Canadian- and American-born scientists was in how these groups perceived the strict security regime that was in place. Brigadier

General Leslie Groves and J. Robert Oppenheimer, the scientific director of the Los Alamos laboratory, had deliberately chosen the site in an isolated part of northern New Mexico for a number of reasons. Not only did the secluded locality allow fairly easy control of access to 'the Hill', but it reduced the risk of espionage and kept the ongoing, often noisy experiments out of the public eye. Los Alamos also appeared to be immune to possible German or Japanese attacks.[31] As at the other Manhattan Project installations, a strict security regime impinged upon virtually every aspect of Los Alamosans' lives. This was especially the case with the military policy of compartmentalization. This principle, which was aimed at preventing espionage by separating both the Manhattan Project as a whole and individual installations like the Los Alamos Laboratory into small compartments, generated a good deal of conflict between the scientists and the military leadership.[32]

The secrecy surrounding the Manhattan Project affected future residents of 'the Hill' long before they arrived at Los Alamos. Once recruited for the secret undertaking, they were forbidden to tell anybody where they were going and most of them literally vanished from the corridors of their university departments. Only after the bombing of Hiroshima were Los Alamos scientists allowed to reveal that they worked at the secret MED installation. It was not until December 1945 that Klaus Fuchs informed the AAC/SPSL more precisely about his work at Los Alamos. While he had written to Esther Simpson, the then AAC/SPSL secretary, in September 1941 in fairly general terms about his new job at the University of Birmingham where he conducted 'research work for a Committee of the Ministry of Aircraft Production',[33] Fuchs went into a little more detail in December 1945, writing: 'Since May, 1941, I have been engaged in research for the development of the atomic bomb; first, on the research team of Professor Peierls in Birmingham and later, in the United States, in New York and Los Alamos'. He added: 'You will appreciate that at present I cannot give any more details'.[34] In his reply, Joseph B. Skemp, the AAC/SPSL secretary, assured Fuchs: 'We understand of course that there can be no details given'.[35] In a similar fashion, Otto Frisch explained the reasons behind his disappearance during the war, writing: 'In November 1943, I became a British Subject and was immediately sent to the United States, where I have been working ... at the big research establishment at Los Alamos, New Mexico ... as described in Dr. Smyth's Report on "Atomic power for military purposes"'.[36] Fuchs's and Frisch's vagueness about their wartime work was in line with guidelines issued by the Directorate of Tube Alloys.[37]

While the stories of Richard Feynman's constant challenges to military security on 'the Hill' (that ranged from crawling under the perimeter fence to safecracking and communicating with his wife in forbidden codes) have now become part of Los Alamos folklore, German-speaking émigrés there experienced the security regime quite differently and primarily negatively.[38] In particular, visible manifestations of laboratory security such as barbed wire, fences, patrol dogs and watchtowers elicited mixed emotions in émigrés, especially those like Klaus Fuchs who had experienced fascist or National Socialist persecution or internment in Britain.[39] Hans Staub asked fellow Los Alamosans: 'Are those big tough MPs [military policemen], with their guns, here to keep us in or to keep the rest of the world out?'[40] In addition, the US Army censored almost every aspect of both private and public language, and all telephone calls and all mail were subject respectively to monitoring and to censorship. The fact that the military leadership only allowed English for telephone calls and for conversations in public places like nearby Santa Fe caused problems for many German-speaking émigrés.[41] And again, as in the case of internment in Britain in 1940, many of the German-speaking Los Alamosans suffered from a 'two-fold estrangement', as Thomas Elsaesser has called it, in the prison-like environment of their adopted home town: not only did many of them feel physically and ideologically separated from their homelands, but the views that many of their American hosts held towards them and their fatherlands appeared to put up a further barrier.[42]

Individual contributions to the making of the atom bomb

While the German-speaking backgrounds of Fuchs, Peierls and most of the émigré scientists made them and their families the target of suspicion and discrimination by the MED's security services, their Germanness also proved to be a highly important prerequisite for their scientific achievements at Los Alamos. It was in particular their schooling in German and Continental European centres of theoretical physics that allowed them to fill important niches within the Los Alamos laboratory. In so doing it helped advance the development of the first atomic bombs, and with it the research culture of Big Science.

The many appointments of German-speaking émigré scientists to senior administrative posts in the laboratory's scientific top management serves as a good indicator of their highly important roles on 'the Hill'. The British Mission to Los Alamos clearly demonstrates this phenomenon, for six of its 24 members were group leaders, including the four German-speaking émigrés Peierls, Bretscher, Frisch and Placzek.[43] Among all Los Alamos group leaders, Rudolf Peierls was perhaps the most important. Not only was he in charge of directing a team, but he also assumed the role of the British group leader when James Chadwick received orders to go to Washington, DC.[44] Soon after his arrival at Los Alamos, Peierls had taken over leadership of Edward Teller's T-1 Implosion Hydrodynamics group, which formed part of the Theoretical Division.[45] Peierls's group used early computational and analytical methods in order to examine the compression of materials under varying impacts. Hans Bethe commented on Peierls's presence at Los Alamos: 'What we needed was the combination between the scientific talent of Peierls' group and the computational facilities'.[46] Since Klaus Fuchs and especially Rudolf Peierls had pioneered work on gaseous diffusion for isotope separation, they continued to work in this area as they had done in New York City before.[47]

Peierls's long-time friend and collaborator Otto Frisch was among the other three German-speaking members of the British Mission who held positions as group leaders. After Peierls, he was one of the most important. With the creation of the Gadget (G) Division in August 1944, Frisch became leader of the group working on critical assemblies to determine the critical mass of both uranium 235 and plutonium 239.[48] It was in particular his so-called Dragon Experiment, which came as close as possible to a nuclear explosion, that demonstrated Frisch's importance. He made a critical mass of uranium 235 subcritical by cutting a hole into its centre. The missing piece which fitted exactly into the opening, was then dropped through a barrel. Although the uranium 235 briefly became critical, Frisch averted an explosion because the piece of uranium 235 fell quickly through the hole and the nuclear reaction declined. Since it averted the need to test the uranium-fission bomb, this test was crucial in the building of a working nuclear weapon. Given the limited amount of fissile material available at this point, Frisch's experiment was a priceless achievement.[49]

George Placzek, another close friend whom Peierls met at Los Alamos, is perhaps the most underestimated German-speaking member of the British team and usually forgotten. Like Peierls and Bethe, the Czech-born theoretical physicist had studied for a short period under Werner Heisenberg at the University of Leipzig.[50] Placzek had been professor at Cornell University, Ithaca, New York, before he became involved in the Manhattan Project, at first in Montreal, Canada where he was leader of the Theoretical Physics Division and then later at Los Alamos as part of the British Mission.[51] He was a distinguished expert on neutron diffusion theory, and played a key part in defining the role of neutrons in a chain reaction.[52] From May 1945, he directed a newly formed group within the Theoretical Division which worked on possibilities of creating a combined plutonium-uranium weapon. Shortly after the war, Placzek even replaced Bethe as the head of the reorganized T-Division, which is a good indicator of his high calibre.[53]

Apart from Frisch and Placzek, Peierls had also previously known Egon Bretscher who, like Fuchs, had not quite yet achieved the same status as Frisch and Peierls but had perhaps been the first one to predict the use of plutonium as a source of energy.[54] Following the laboratory's reorganization with the subsequent creation of the F-Division under the leadership of Enrico Fermi, Bretscher headed a group on Super experimentation.[55]

Among the German-speaking members of the British Mission, Klaus Fuchs was the only one not to hold a senior administrative post as group leader. As in New York City earlier, he was Peierls's assistant.[56] Fuchs was valued for his theoretical skills and his report on the scaling for blast waves became a widely accepted key study.[57] Hans Bethe praised Fuchs as 'perhaps the most hard-working member of our entire division' who 'contributed greatly to the success of the Los Alamos project'.[58] In spite of his great talent and work ethic, Fuchs had not yet advanced into the same league as Peierls and Frisch. Fuchs himself later suggested that many of his colleagues had regarded him together with Richard Feynman as the most gifted junior scientists at Los Alamos.[59]

At Site Y, Fuchs, Peierls and the other members of the British Mission collaborated with German-speaking émigrés who had come to Los Alamos from American universities. Hans Bethe, who was perhaps the most important German-speaking émigré atomic scientists on 'the Hill', was among this group. Alongside the Italian-born Fermi, Bethe was the only European-born division leader at Los Alamos. He had arrived via Cornell University, where he had found employment in 1935. On 'the Hill', Bethe directed one of the initial five divisions, the Theoretical or simply T-Division.[60] Both Fuchs and Peierls worked in Bethe's division. Because of his work ethic, Bethe's colleagues nicknamed him 'The Battleship'.[61] J. Robert Oppenheimer held T-Division in high esteem. He valued in particular their calculations relating to problems of efficiency and critical mass that were crucial on the way towards building the first working atom bombs.[62]

Shortly before the completion of the project, Bethe assumed a pivotal role when he refuted with his calculations substantial fears that the high temperatures and pressures inside the Trinity explosion might trigger a reaction which would result in the creation of a new star, igniting the entire planet's atmosphere.[63] On 28 February 1945, Bethe also attended a crucial meeting – including J. Robert Oppenheimer, General Groves, James B. Conant and George Kistiakowsky – which was to settle key questions regarding the schedule and design of the implosion weapon. From 1 March 1945, Bethe was also a member of the so-called Cowpuncher Committee. Set up by J. Robert Oppenheimer to direct the final stage of the implosion project, this potent board included, besides Bethe and Oppenheimer, George Kistiakowsky, William Parsons, Robert Bacher, Samuel Alison, Cyril Smith and Kenneth Bainbridge.[64]

At Los Alamos, Peierls also met his Viennese friend Victor Weisskopf, whom he had come across in various places before.[65] Weisskopf worked as a group leader in Bethe's Theoretical Division. At first, his task was to interpret experiments which had been conducted by some of the experimental groups in order to determine the critical mass. Later, he assessed the yield of nuclear explosions and efficiency. On account of his very successful, intuitive approach to solving problems, his colleagues nicknamed him 'the Los Alamos Oracle'.[66] Hans Bethe acknowledged Weisskopf's talent by appointing him deputy of the Theoretical Division.[67]

Edward Teller and Hans Staub were the two remaining German-speaking group leaders at Los Alamos. Teller, who had previously worked at the University of Chicago's Metallurgical Laboratory, was one of a group of four Hungarian-born scientists who worked on the Manhattan Project.[68] Born in Budapest during the time of the Austro-Hungarian Empire, Edward Teller, Eugene Wigner, Leo Szilard and John von Neumann had all attended the Lutheran *Gymnasium* (high school) in their hometown. When a severe upsurge of anti-Semitism swept through Hungary in the aftermath of the First World War, the four men left their native country for Germany where they received substantial parts of their higher education. Adolf Hitler's accession to power in early 1933, however, forced them to continue the westward move that eventually led them to the United States and into the Manhattan Project.[69]

While Eugene Wigner and Leo Szilard were engaged in work at the University of Chicago, and John von Neumann only visited 'the Hill' occasionally, Edward Teller and his wife Mici were the only Hungarians permanently residing at Los Alamos.[70] Here, he assumed the role of a spokesperson, communicating on behalf of the laboratory with the Manhattan Project installation at Columbia University, New York City.[71] But soon tensions started to grow between the Hungarian-born émigré and the scientific director of the Los Alamos laboratory which continued through the war years and eventually peaked in Teller's testimony against his former boss in the Oppenheimer security hearings of 1954.[72] Teller had first felt aggravated and disappointed when Oppenheimer picked Bethe over him as head of the Theoretical Division.[73]

In addition, Teller had started to show signs of becoming increasingly obsessed with the idea of a thermonuclear weapon – the so-called Super. Owing to this infatuation, he had become more and more reluctant to fulfil his assigned tasks, especially performing calculations for the implosion weapon. When his division leader, Hans Bethe, complained to Oppenheimer about Teller's behaviour, Oppenheimer stripped him of his duties as group leader. Peierls then took over the leadership of Teller's group in T-Division shortly after his arrival on 'the Hill'.[74] Still, Oppenheimer did not order Teller to be excluded from work at the laboratory. Instead, the scientific director of the Los Alamos laboratory separated Teller's group from T-Division and gave the Hungarian-born émigré the freedom to investigate further into the 'Super', provided that he reported directly to Oppenheimer.[75]

The remaining German-speaking group leader at Los Alamos was Hans Staub. While the majority of German-speaking scientists at Site Y – including Peierls, Fuchs, Placzek, Bethe, Weisskopf and Teller – comprised theoreticians and thus worked in the Theoretical Division, Staub, like Otto Frisch and Egon Bretscher, was an experimental physicist. The Swiss-born Staub, who had previously held a post at Stanford University, worked closely with the Italian-born Bruno Rossi in the Experimental Physics Division. At first, he headed a group which was formed in July 1943 and concerned with the improvement of counters. In September 1943, his team was combined with Rossi's group, which had worked on developing enhanced electronic techniques, under the latter's direction as the Detector Group.[76] Staub's cooperation with Rossi was crucial in constructing instrumentation for Robert Serber's so-called RaLa method, named after the element Radio Lanthanum. Serber's group used it to emit rays for diagnosing implosion in numerous tests. With this method, they hoped to examine the feasibility of a plutonium implosion weapon.[77] When the laboratory leadership decided to merge the Research and F-Divisions in the Physics Division in November 1945, Staub and Bretscher became co-leaders of the P-4 team working on Thermonuclear Reaction.[78]

Other German-speaking émigré atomic scientists at wartime Los Alamos who did not hold positions as group leaders included the Hungarian-born theoretician John von Neumann and the German-born Rolf Landshoff and Maria Göppert-Mayer, the Swiss-born Felix Bloch and the Austrian-born Martin Deutsch. While Bloch and Landshoff resided permanently on 'the Hill', Göppert-Mayer and von Neumann visited only occasionally but nevertheless made pivotal contributions to the creation of the first atomic bombs.[79] Together with the Danish-born Niels Bohr who also occasionally visited Los Alamos and the Italian-born Enrico Fermi, von Neumann ranked in the eyes of Hans Bethe among 'the greatest intellects at Los Alamos'.[80] John von Neumann played a critical role in making the implosion principle work.[81] Maria Göppert-Mayer first engaged in Manhattan-Project-related work on gaseous diffusion at Columbia University. She visited Los Alamos on several occasions to work with Edward Teller on – among other things – studies of the opacity of uranium. These were designed to avoid the unintended occurrence of a critical mass that would have led to an accidental atomic explosion.[82]

Felix Bloch, who belonged to the minority of experimentalists among the German-speaking émigré scientists at Los Alamos, came to the secret MED facility from Stanford University where he had previously focused on analysing the fission spectrum by applying proton recoil studies in ionization chambers. At Site Y, Bloch was a member of the implosion group.[83] Owing to tensions with Oppenheimer and his frustration with the organization of the laboratory, Bloch decided to leave Los Alamos like Edward U. Condon before the final wartime mission was completed.[84]

Edward Teller recruited Rolf Landshoff for his group at Los Alamos. During a visit to Chicago, Eugene Wigner told Teller about Landshoff, one of his former students from his Berlin days. After a few attempts, Teller finally succeeded in recruiting Landshoff to his Los Alamos team in November 1944.[85] In the spring of 1946, Landshoff left Teller's group on the Super and became leader of a group working on the Super and radiation hydrodynamics.[86] Like Landshoff, Martin Deutsch came to Site Y directly from an American university. At Los Alamos, Deutsch worked under Emilio Segrè.[87]

In August 1945, the leadership of the Los Alamos laboratory decided to compile a series of edited volumes on the technical history of its achievements. The Los Alamos Technical Series resembled the German *Handbuch der Physik* (Physics Handbook). German-speaking émigrés, three of whom (Peierls, Frisch and Placzek) were members of the British Mission, edited five of the 24 volumes: Peierls (*Theory of Implosion*), Frisch (*Critical Assemblies*), Placzek (*Neutron Diffusion Theory*). Hans Bethe (*Blast Wave*) and Victor Weisskopf (*Efficiency*) edited two further volumes.[88] Their participation in the Los Alamos Technical Series underlined once more the significance of German-speaking émigré nuclear scientists at wartime Los Alamos in general and within the

British Mission in particular. And, what is more, their contributions to the collection of papers represented but one more example of how their German-speaking backgrounds proved crucial in helping them achieve great success at Los Alamos.

Towards a new approach to nuclear science

Apart from their individual contributions, Klaus Fuchs and especially Rudolf Peierls helped shape a new approach to nuclear physics by accelerating the research culture of Big Science. Within the context of the MED, this new methodology was primarily based on three determining factors: first, the Los Alamos scientists had to work under tremendous time pressure and tight scheduling to achieve their goal of beating the Third Reich in the race for the atom bomb; second, and closely connected to the first point, the fear of nuclear weapons in the hands of the National Socialist regime in Germany secured abundant financial support from government sources; and third, the fact that the making of atomic arms was, above all, an engineering task represented a peculiar novelty. As a result, many theoreticians engaged in experimental work, and the dividing lines between experimental and theoretical physics started blurring and became increasingly porous.[89]

In the making of this new approach, which eventually led to the development of the atom bomb, Fuchs and Peierls and combined their more 'traditionally German' theoretical skills with the Anglo-American leaning towards experimentation.[90] At Los Alamos, Fuchs and Peierls not only came into contact with Anglo-American experimentalism, but they were also exposed to the Italian school of Enrico Fermi, represented by its founder and one of his long-time collaborators, Emilio Segrè.[91]

Fuchs, Peierls and their Los Alamos colleagues set out to explore two different nuclear 'explosives': uranium 235 and plutonium 239. At first, plutonium 239 was the preferred choice as it seemed to require less metal and to be faster to produce. 'The greatest gap in the staff's understanding was the nuclear explosion itself – its nature, the methods of initiating it, and its destructive effect', note Richard Hewlett and Oscar Anderson Jr, adding: 'Here Los Alamos had to rely entirely on theory', as 'Experiments had been impossible'.[92] And this would lead to collaboration between theoreticians such as Peierls, Fuchs and Bethe with experimentalists like Egon Bretscher and Frisch on an unprecedented scale. Since the majority of the German-speaking émigré scientists had backgrounds in theoretical physics, they complemented their American, British and

Canadian colleagues tremendously. Therefore, they had a great share in the successful completion of the Manhattan Project.

Peierls and to a lesser degree Fuchs, like the overwhelming majority of German-speaking scientists at Los Alamos, had received considerable parts of their higher education in Germany during the country's golden age of international science in the 1920s and early 1930s.[93] The list of graduates from German universities included other famous members of the Los Alamos laboratory. The Ukrainian-born explosives expert George Kistiakowsky, for example, had received a Ph.D. in chemistry from Berlin University in 1925, and, in 1927, even the scientific director of the laboratory, J. Robert Oppenheimer, had earned a Ph.D. in physics from Göttingen University. [94]

While a strong separation between experimental and theoretical orientation had dominated nuclear science in Germany before 1933, Fuchs and Peierls were now forced into cooperation with experimentalists.[95] During his time in the United Kingdom, Peierls had already demonstrated his ability to collaborate with experimental physicists; for example, at Birmingham University in his work with Marcus Oliphant as well as in his collaboration with Frisch on the seminal 'Frisch-Peierls Memorandum' (see Chapter 2). Other German-speaking émigré theoreticians also engaged in close cooperation with experimentalists: Hans Bethe, Peierls's long-time personal friend and superior in the Theoretical Division, had previously cooperated with Milton Livingston at Cornell University, for instance.[96]

Theoreticians and experimentalists had collaborated before the Second World War. But it was at the Los Alamos laboratory that this combination of different research cultures was taken to a new, unprecedented level, and played an essential part in the development of the first atomic bombs.[97] German-speaking émigré nuclear scientists had a huge share in promoting this highly interdisciplinary approach to problem solving. Since the Manhattan Project operated under tremendous pressure and an extremely tight schedule, conventional analytic modes of investigation failed to produce results within the given time. Hence, theoretical scientists like Peierls and Fuchs were forced into close collaboration with experimentalists on an unparalleled scale. To meet the extremely tight deadlines, experimental scientists often conducted several experiments simultaneously to verify the validity of hypotheses in close cooperation with theoreticians.[98]

Peierls's and Fuchs's rare qualifications in a field that was at the time underrepresented in the United Kingdom and the United States enabled them to work in interdisciplinary areas such as applied mathematics.[99] Bethe's T-Division, which comprised a high number of German-speaking émigrés, was pivotal in forging the new approach to nuclear science because, as Bethe himself said, it 'had to do with practically everything in the laboratory'.[100] As members of the Theoretical Division, Peierls and Fuchs thus became 'bridge-builders' between the preferred research styles of their home and host countries, as Paul K. Hoch has generally called émigré physicists.[101] In their function as middlemen between German and Anglo-American research cultures, Fuchs and Peierls consequently represented a scientific variant of what Margaret Connell Szasz has termed elsewhere the 'cultural broker'.[102] As the site where this 'cultural brokerage' took place, along with the close collaboration between theoretical and experimental atomic scientists and engineers, Los Alamos was a powerful example of Peter Galison's concept of the 'trading zone'; that is, the 'social and intellectual mortar binding together the disunified traditions of experimenting, theorizing, and instrument building'.[103]

The 'bridge-building' between two different research cultures in the Los Alamos 'trading zone' marked par excellence the process which Roger H. Stuewer has generally referred to as a 'multifaceted symbiosis' between émigré nuclear physicists and their American-born colleagues.[104] As a consequence, nuclear science underwent a denationalization process, when Fuchs's and Peierls's German-influenced research styles amalgamated with those practised in Britain, the United States and Italy.[105] That all Los Alamos scientists spoke the transnational language of nuclear science facilitated the communication between the émigrés and their American-, British-, Canadian- and Italian-born colleagues tremendously.[106] Here, émigré atomic scientists clearly had an advantage over primarily language-based professions such as literati. While Fuchs, Peierls and the other Los Alamos émigré scientists substantially contributed to this denationalization of scientific preferences in style, they also helped set in motion a trend towards the Americanization of science. The fact that the Los Alamos 'trading zone' was located inside the United States and that the United States government by and large funded the MED was a precursor of later developments. For the United States would come to dominate science, especially nuclear physics, in Western Europe, including the United Kingdom, for decades to come after the Second World War.[107]

The plutonium implosion bomb

Klaus Fuchs's and Rudolf Peierls's contributions to the development of the plutonium implosion bomb illustrate particularly well their roles in shaping the innovative approach to nuclear science at Los Alamos. In retrospect, Peierls himself valued his work on the implosion principle as his most significant contribution to accomplishing the mission of the Los Alamos laboratory.[108] General Groves later emphasized the historic dimension of the development of the plutonium bomb, describing it as 'a phenomenal achievement; an even greater venture into the unknown than the first voyage of Columbus'.[109]

In the early stages of the Manhattan Project, the Los Alamos laboratory investigated the feasibility of designing both uranium and plutonium fission weapons that used a gun to shoot a piece of either uranium or plutonium into a sub-critical uranium or plutonium core to achieve a critical mass. Work on implosion, by contrast, was regarded as secondary at the time.[110] While Robert Serber and Richard Tolman proposed some ideas on the implosion principle early on, it was especially Seth Neddermeyer who advanced this work as part of a group within E-Division from June 1943, although this approach was deemed unimportant.[111]

This view changed fundamentally in the summer of 1944 when a group working under Emilio Segrè and including Martin Deutsch discovered that pile-produced plutonium emitted five times more neutrons than anticipated.[112] This high neutron flux meant that spontaneous fission would occur and a gun-type plutonium weapon would pre-detonate, or 'fizzle', before reaching critical mass. Like a so-called dirty bomb, such a device would release substantial amounts of radioactive fallout, but fail to trigger a nuclear explosion. 'The greatest problem', Hans Bethe later reminisced about the plutonium bomb, 'was how to assemble the active material and assemble it in a way that it would not prematurely detonate'.[113]

Here, implosion seemed to offer a promising way out of the crisis. As a result, the primary mission basically changed to pursue what had previously been regarded as a secondary option. The new direction in the Los Alamos laboratory's research prompted a massive reorganization of its divisions and groups. Apart from these administrative alterations, the mission change went hand in hand with a further alteration of the established practice: while implosion research up to that point had explored its feasibility in both uranium and plutonium, it focused now exclusively on a plutonium implosion bomb. By contrast, the programme working on a gun weapon now concentrated solely on uranium. Both Fuchs and Peierls also made important contributions to the research that centred on a uranium bomb, in particular through their conceptual work for the uranium separation plant. Other German-speaking émigré scientists, too, had a great share in advancing work on the uranium gun weapon. In particular, Otto Frisch's aforementioned 'tickling the tail of the dragon' experiment proved a priceless achievement.[114]

Since the Los Alamos scientists basically understood the theoretical considerations regarding the gun assembly at the time, the laboratory's primary aim thus changed to exploring the hitherto uncertain implosion principle. Fuchs and especially Peierls played crucial roles in the development of the implosion weapon. While the making of the atom bomb was chiefly an engineering task, it would be wrong to assume that well-trained engineers themselves could have produced a fission bomb, let alone an implosion bomb, without the input of theoreticians such as Fuchs and Peierls as well as Bethe, von Neumann, Teller and Weisskopf.[115]

Given the Manhattan Project's lavish funding by the United States government and working under the tremendous time constraints of the Second World War, Oppenheimer was able to simultaneously approach specific problems from various angles in order to speed up progress. It was this principle that enabled the scientists to achieve a mission change at Los Alamos so quickly. The case of the implosion method revealed particularly well Oppenheimer's critical role in organizing and scheduling the Los Alamos operation because its perfection not only involved a mission change but also the abandonment of previously taken and often well-established paths of scientific investigation.[116]

In spite of the widespread belief in the early days of the Manhattan Project that the implosion project was considered to be only secondary to the gun-assembly principle, Rudolf Peierls had made significant contributions to its advancement quite early on. The directorship of the Los Alamos laboratory had stepped up the implosion programme after a visit by John von Neumann in September 1943. Above all, von Neumann's trip to 'the Hill' gave a boost to the early work on implosion by Seth Neddermeyer's group, which was regarded as peripheral at the time. In discussions with Edward Teller, von Neumann developed the idea of achieving a faster implosion by placing explosive charges around the bomb core. Soon, Teller, Bethe and even Oppenheimer were convinced that an implosion weapon was far more powerful than a gun-type device, and von Neumann's suggestions and ideas led to the extension of the implosion work.[117] As a result, George Kistiakowsky – the leading explosives expert in the United States at the time – was appointed as consultant in October 1943 and eventually made a full member of the laboratory in February 1944. The following month,

Bethe also assigned a theoretical group under the leadership of Teller to further investigate implosion-related problems.[118]

When Teller's group encountered calculation problems, Peierls, who visited Los Alamos in March 1943, was instrumental in enabling them to use punch-card machines to find numerical solutions to the equations defining the implosion. Inasmuch as this formula was identical to that used by Peierls in his numerical experiments to determine blast waves in air, his experience proved decisive in moving the implosion pro-gramme ahead.[119] During another brief visit to Los Alamos in February 1944, Peierls had provided Oppenheimer with insights into the British approach to integrating blast wave equations to the problem of the hydrodynamics of the implosion. And Oppenheimer subsequently wrote to Groves that he was 'planning to attack the implosion problem along these lines with the highest possible urgency'.[120]

For its work on the implosion principle, Bethe's Theoretical Division, which both Fuchs and Peierls joined after their transfer to Los Alamos, used some of the latest computation technology available at the time, especially IBM machines to calculate the implosion.[121] Since Hans Bethe's T-Division based its mission on a close collaboration of its members with experimentalists in all areas of the laboratory, it played a crucial role in establishing this new methodology. When Peierls replaced Teller as leader of the T-1 'Implosion and Hydrodynamics' group shortly after his move to Los Alamos, he became deeply involved in implosion research on a permanent basis.[122]

One of the chief obstacles that Peierls and his colleagues in the Theoretical Division had to master concerned the question of how the plutonium could be compressed so quickly that it would produce a proper nuclear explosion and not 'fizzle'. Here, so-called explosive lenses offered a solution. These lenses consisted of explosives which were shaped in order to focus the explosion into the desired direction. They were placed around the bomb's plutonium core. But their produc-tion proved to be one of the most difficult tasks at Los Alamos. Early on, Peierls and Bethe had thus been among those scientists at Los Alamos who backed George Kistakowsky against sceptical voices in the labo-ratory by insisting that despite the large amount of manpower and material required to pursue the development of explosive lenses, these were indeed a crucial component of a working implosion weapon.[123]

James Tuck's arrival at Los Alamos in May 1944 was crucial in advancing progress in designing these lenses. Tuck – who had previously conducted research in this field – came to 'the Hill' as a member of the British Mission and brought up the idea of a three-dimensional lens.

Shortly afterwards, Rudolf Peierls and Hans Bethe commenced with their quest for an appropriate design for explosive lenses, but remained unsuccessful. It was then John von Neumann who suggested a first feasible design. When the shape of the lens was finalized in July 1944, Peierls started to explore theoretical aspects of explosive lens design.[124] That von Neumann's design worked marked in Bethe's words 'perhaps the most important invention to make implosion go'.[125]

In order to cope with the highly complex calculations, Peierls's T-1 group received increased assistance from the T-6 'Numerical Calculations' group under Stanley Frankel and Eldred Nelson after August 1944. Since Peierls's team had almost fully completed calculations of an ideal spherical implosion at this point, they started to focus more on incongruities between theoretical data and that obtained from actual tests. Peierls and his group focused in particular on two areas, velocity and density, because they had previously calculated them higher than they had in fact occurred in the experiments.[126]

At the motivational level, Peierls also played an important role in the development of the so-called Christy Gadget. As group leader, he strongly supported one of his team members, Robert Christy, to overcome the problem of asymmetry in the implosion principle. This was crucial as any occurrence of an asymmetrical pattern during the implosion would render the plutonium bomb more or less useless. Christy managed to master this problem by using a solid core rather than hollow spheres in the bomb design.[127] As a member of Peierls's team, Fuchs contributed greatly to the solution of a further problem, namely the development of an implosion initiator. Fuchs, who had previously conducted research on the theory of jets, had considerable input in working out an elementary theory of the so-called urchin design together with Hans Bethe, Paul Stein and Robert Christy. By April 1945, Fuchs had formulated a suitable theory for the initiator design in collaboration with Bethe and Charles Critchfield.[128]

Peierls and his theoretical group worked in other areas of implosion research, too. His work in diagnostics serves particularly well to illustrate the close collaboration between members of T-Division and experimentalists as, for example, in the X-ray project. Here, Peierls made a crucial contribution when he proposed the so-called heap-of-disks experiment. In this test, a mound of metal disks was positioned close to high explosives and later scattered by the blast wave generated by an explosion. Through X-ray photography of each disk's relocation, the scientists gained important insights into the qualities of different explosives. This was not Rudolf Peierls's only contribution to implosion diagnostics.

Together with the experimental physicist Otto Frisch, Peierls was among a group of nuclear scientists who proposed the basic ideas of an electric method of diagnosis called the 'pin method' in July 1944. They used it to study the timing and symmetry of the implosion. To study the timing, this technique comprised several metal pins being connected to electric circuits and placed in close proximity to the device that was to be imploded. When the implosion occurred, it hit the metal pins and an oscilloscope made the bursts visible. By placing several metal pins at the same distance from the imploding object, the Los Alamos scientists could apply the same method to examine the implosion symmetry.[129]

With their input, Klaus Fuchs and, in particular, Rudolf Peierls had thus not only helped secure the successful completion of the Manhattan Project's mission, but also participated in shaping a new approach to nuclear science. The visible product of this new methodology was the so-called Fat Man implosion device that was successfully tested on 16 July 1945 in central New Mexico's Jornada del Muerto region near the town of Socorro.[130] The detonation not only marked the successful completion of the Manhattan Project, but confronted Peierls, Fuchs and their colleagues for the first time directly with the results of their work. 'The ball of fire was intensely bright, equivalent to several suns by general estimate', James Chadwick stated in his initial report on the Trinity Test ,'so that it seems reasonably certain that the explosion was equivalent to more than 1000 tons of T.N.T. and possibly several thousand. This I call completely successful'.[131] Otto Frisch noted the brightness 'as if somebody had turned the sun on with a switch'.[132] James Tuck, another member of the British Mission, noticed that 'A great brilliance like sunshine suddenly appeared, and my left eye, which was completely unprotected, was dazzled, and had nebulous brightness as well as a peculiar dazzle sensation similar to that experienced in looking at a strong ultraviolet source'. Tuck attached drawings of the explosion to his report that arguably represent the first 'artistic' treatments of the mushroom cloud.[133] Although Fuchs, Peierls and their Los Alamos colleagues had achieved their mission, not all scientists left 'the Hill' immediately.

Klaus Fuchs stayed on until the summer of 1946. Because Norris Bradbury, who replaced J. Robert Oppenheimer as scientific director of the Los Alamos Laboratory in the autumn of 1945, held Fuchs's skills in very high esteem, he requested the latter's stay until after the first US post-war atomic tests.[134] 'Of course, I do not wish to express any opinion about the absolute importance of the project work now going on here', wrote George Placzek to Chadwick in early February 1946, 'I merely want to state that Fuchs would be a great help for its successful completion'.[135] The British, too, had realized his talent and, in the summer of 1946, demanded his immediate return to the United Kingdom to resume work on their nuclear energy programme.[136] Before his departure from 'the Hill', Fuchs worked in two major areas. He served as advisor to the atomic tests which the United States government planned to carry out in the Bikini Atoll in the South Pacific in the summer of 1946. Bradbury was keen on Fuchs's 'theoretical advice concerning the predicted effect and methods for determining the efficiency of the atomic weapon'. Additionally, Fuchs helped refine the first atomic bombs, especially through his expertise in hydrodynamics.[137] Despite his very significant role during the war, the laboratory's leadership, by contrast, did not deem Peierls's presence important for post-war work at Los Alamos.[138]

The impact and legacy of Los Alamos

Besides their crucial role in establishing a new scientific approach, which relied heavily on a combination of experimental and theoretical methodologies, Fuchs's and Peierls's work on the Manhattan Project also had an impressive legacy in the post-war era. First and foremost, as 'cultural brokers', they had a considerable impact on the development of the research culture of Big Science, as it is widely known today. Plentiful funding by one or more national governments and private sources, large-scale machinery, huge laboratories and the interdisciplinary collaboration of hundreds of international scientists are chief characteristics of this research mode.[139]

With their work on the design and operation of a large-scale isotope separation plant at Columbia University as well as for the Kellex Corporation during their time in New York City, Fuchs and Peierls provided a significant part of the theoretical foundation for the K-25 uranium separation plant at the MED's Oak Ridge facility in Tennessee. Through their significant contributions to the establishment of the K-25 installation, Fuchs and Peierls aided the advancement of the Big Science culture in two areas.[140] First, the isotope separation plant came to symbolize the new Brobdingnagian dimension of Big Science. 'When history looks at the 20th century, she will see science and technology as its theme', speculated Alvin M. Weinberg, the director of the Oak Ridge National Laboratory, about the legacy of Big Science for future generations in 1961. '[S]he will find in the monuments of Big Science – the huge rockets, the high-energy accelerators, the high-flux research reactors – symbols of our time just as surely as she finds in Notre Dame a symbol of the Middle Ages.'[141]

Second, since Union Carbide operated the K-25 facility, it also embodied the close cooperation between the US government and private contractors which US President Dwight D. Eisenhower dubbed 'the military-industrial complex'.[142] Other companies involved in the Manhattan Project included DuPont.[143] Within the British context, David Edgerton has called this phenomenon 'the warfare state'.[144] The fact that Nazi Germany was not even close to building a working atomic device lay – apart from prioritization – partly rooted in the Hitler regime's failure to enlist resourceful private corporations in the German atomic weapons development programme.[145]

As one of the chief consequences of the emerging culture of Big Science, teamwork became the major scientific production mode in gigantic research endeavours like the Manhattan Project. This development had a great impact on scientific authorship in nuclear weapons research, which became increasingly complex. In a 1955 article in the *Bulletin of Atomic Scientists*, which referred to the creation of the hydrogen bomb but equally applied to the making of the atom bomb, Edward Teller pointed out that the development of nuclear arms was indeed 'the work of many excellent people'. Although 'modern technical and scientific development' depended, in Teller's opinion, on '[h]undreds of ideas and thousands of skills', the public was commonly presented with a different story: 'only too often', he observed, success was attributed to 'a brilliant idea' or 'the name of a single individual'.[146]

Ironically, Teller himself has often been called 'the father' of the hydrogen bomb. The anthropologist Hugh Gusterson has convincingly argued that it was especially in nuclear weapons laboratories where work is carried out in a Big-Science-oriented research mode based on team-work and under a veil of secrecy that, as he phrased it, 'the distinctive contributions of individual scientists have been repressed or gathered together under the sign of sacralized individuals standing for groups'.[147] Gusterson's observation helps explain in part why Oppenheimer has commonly been credited with the development of the atomic bomb and Teller with that of thermonuclear weapons at the expense of 'the work of many excellent people' such as Fuchs and Peierls.

Also closely connected with the emergence of Big Science was the massive financial government support for science projects. Here, Fuchs and especially Peierls had a strong impact on the function and scope of government spending in relation to science, in particular as scientists would increasingly act as lobbyists for massive government support. Since European émigré scientists had long been used to state-funded research in the sciences in their homelands, German-speaking scientists had welcomed government funding early on while their American-born colleagues in particular discussed its effect on science.[148] Peter Bacon Hales has thus concisely summed up the essence of the Manhattan Project, calling it 'one manifestation of a complex and evolving ideology blending corporate capitalism, government social management, and military codes of coercion and obedience'.[149]

As one chief result of the massive government spending on science, a great number of scientists found employment outside universities in the United States and the United Kingdom after the war. The Second World War had a tremendous impact on university physics departments in Britain and the United States and the formation of 'the military-industrial complex'.[150] The Los Alamos laboratory, for example, was turned into a permanent nuclear weapons research establishment.[151] While the Los Alamos Scientific Laboratory, as it became officially known in 1945, and the other two chief MED sites at Oak Ridge and Hanford became permanent facilities and the Argonne National Laboratories near Chicago, Illinois, or the Sandia National Laboratories in Albuquerque, New Mexico, evolved directly out of the Manhattan Project, other atomic-arms-and-energy-related installations like the Lawrence Livermore National Laboratory in Livermore, California, were newly founded a couple of years later.[152]

In the United Kingdom, where both Peierls and Fuchs returned after the war, similar laboratories were founded. Apart from the AERE Harwell, Berkshire, these included the Royal Armament Research and Development Establishment (RARDE) at Fort Halstead, Kent, where initially some nuclear weapons research was conducted, and later the AWRE Aldermaston.[153] During his early days at Harwell, Fuchs worked especially in the area of isotope separation. And Oskar Bünemann, who had been engaged in research at the MED's Berkeley laboratory, worked in various projects in Harwell's Theoretical Physics and Nuclear Physics Divisions including 'Piles of Plutonium & Power Production', 'Slow Fission Reactors', 'Cyclotron', 'Fission Products' and 'Gas Cooled Piles for Production of Plutonium and Power'.[154]

Owing to his attempt to understand all the research areas at Los Alamos, Fuchs proved to be of greatest significance for the British espionage effort on their American allies. James Chadwick ordered Fuchs, for example, to visit the Manhattan Project installation at Chalk River near Montreal in Canada to get a better picture of the current state of research going on there.[155] Chadwick even urged members of the British Mission to assemble a sort of 'nuclear almanac' for future research in the United Kingdom.[156] Joseph Rotblat, for example, used some of the knowledge gained during his stay in the United States to make suggestions for both the future British nuclear energy project and work at the University of Liverpool.[157]

Fuchs's engagement in early work on thermonuclear weapons would be of great significance after the war for his work at the AERE Harwell. At Los Alamos, he collaborated with von Neumann on the Super. On 28 May 1946, the pair filed a joint patent application for the radiation implosion principle to be used in the hydrogen bomb, the so-called classical Super, but it surpassed the mathematical tools available at the time to improve it. In August 1946, Teller thus proposed the so-called Alarm Clock design, which was then pursued.[158] The fact that Fuchs and Egon Bretscher, who later also joined the AERE, attended a conference on the Super organized by Teller at Los Alamos in April 1946 further aided the British post-war hydrogen weapons project.[159] Given the fact that British weapons scientists and engineers had to rely at the start of the British H-bomb programme almost exclusively on information from Los Alamos until the summer of 1946, a clearer picture of Fuchs's importance for the British hydrogen bomb project emerges.[160]

Fuchs remained an important source of information for the British H-bomb project, and in 1952, for example, Sir William Penney, the chief scientist behind the British thermonuclear project, visited him in prison.[161] Since the United Kingdom was far behind the United States in its own H-bomb project, much of this information exceeded present British knowledge of thermonuclear arms and the science involved at the time, and thus severely restricted the successful evaluation of Fuchs's information by the British.[162] An MI5 report stressed Fuchs's pivotal role for the British nuclear energy project: 'To his fellow scientists he had become one of the world's leading mathematical physicists'.[163] The British thermonuclear project also benefited greatly from the experimental work that Bretscher's F-3 group had carried out at Los Alamos. Besides Bretscher, the team included two other British scientists: Anthony French and Michael Poole.[164] Consequently, Egon Bretscher was, as well as Fuchs, a major source of knowledge for an independent British (thermo)nuclear arms project after the war.[165]

However, Britain was unable to compete with the United States in scientific matters in the long run. Alongside Fuchs's and Peierls's role in accelerating the establishment of the research culture of Big Science with its massive government-funded research installations, and as a consequence of their work for the MED, the two scientists and the other members of the community of high-level German-speaking émigré scientists who worked on the Manhattan Project proved significant for changing the global positioning of modern physics in favour of the United States.[166] Although Fuchs and Peierls returned to the United Kingdom where they both engaged in physical research after the war, they could certainly not alter or halt – let alone reverse – this general trend, even if they had intended to invert it. After all, British nuclear culture operated on a much smaller scale than atomic culture in the United States with 'atomic villages' rather than 'atomic cities'.

Rudolf Peierls returned to his professorship at the University of Birmingham, and declined an offer to join Cambridge University. 'I think the chief argument that finally tipped the balance', he wrote to Chadwick in February 1946, 'was the rather attractive prospects of experimental physics in Birmingham as compared with the uncertain situation in Cambridge'. In an altruistic way, he also attributed his decision to 'the need, in general, to build up the modern universities to get a fairer share of the good students, not to the detriment of Cambridge, but to re-establish fair proportions'.[167] Apart from his university post, Rudolf Peierls also served as consultant to the AERE until 1957 and again from 1964.[168]

While Fuchs and Peierls returned to Britain without any major complications, the reintegration of German-speaking émigré atomic scientists was not always smooth as Egon Bretscher's case reveals. So frustrated was he with the way senior British TA officials and his former employer, the Cavendish Laboratory in Cambridge, treated him that he briefly considered staying on in the United States and working for an American university.[169] Eventually, Bretscher joined Harwell. In spite of his complaints about, as well as his reluctance to join, Harwell, Egon Bretscher also saw an advantage in returning to the United Kingdom: 'One attraction Britain can offer', he argued, 'lies in the possibility to keep in close contact with the future development of [the] physical and chemical side of T.A.'[170]

Conclusion

With their participation in the MED, Fuchs and Peierls further established themselves in the field of nuclear physics. The two scientists made pivotal contributions to the making of the atomic bomb, at first in New York City and later, in particular, at the Los Alamos laboratory. While Fuchs was still among the junior scientists, Peierls held the rank of group leader in the Theoretical Division and even assumed the leadership of the British Mission to Los Alamos. Both scientists played crucial roles in the advancement of the implosion principle. In their endeavour, as this chapter has shown, Fuchs and Peierls were aided by a number of fellow German-speaking émigré atomic scientists who had come to Los Alamos either as members of the British Mission or from American universities.

Peierls's and Fuchs's involvement in the Manhattan Project had a lasting legacy and helped change the face of nuclear science considerably, in particular by accelerating the emerging research culture of Big Science. The fact that, after the war, the American government recognized Peierls's contributions to the Manhattan Project by awarding him the Presidential Medal of Merit – the highest medal a civilian can receive in the United States – underlined both his exceptional contributions to the making of the first atomic bombs and his distinguished reputation. Besides Peierls, William Penney was the only other member of the British Mission to Los Alamos who received the Presidential Medal of Merit. In Britain, Franz Simon was also presented with it.[171] With their departure from Los Alamos ended both Fuchs's and Peierls's active role in nuclear weapons research. But their wartime work would be of great importance for their roles in shaping post-war British nuclear culture.

4
A Nation Betrayed? The Klaus Fuchs Atomic Espionage Case Reconsidered

After their return from Los Alamos to the United Kingdom, Rudolf Peierls's and Klaus Fuchs's ways parted: while Peierls resumed his professorship at Birmingham University, Fuchs joined the AERE Harwell where he became head of the theoretical physics division. Harwell formed part of the British nuclear energy programme which was established in 1947 with the firm decision to develop and test a British atom bomb and to explore civilian applications of atomic power. While Klaus Fuchs worked at Harwell, which was concerned with civilian applications of atomic power, he also occasionally served as adviser to the British nuclear weapons development project which was at the time headquartered at Fort Halstead in Kent before moving to Aldermaston in Berkshire in 1950.[1]

The fact that Klaus Fuchs had been passing on sensitive nuclear information to the Soviet Union since 1940 had remained unnoticed by the British Security Service. When Fuchs confessed his espionage activities for the Soviets during and after the war to William Skardon in early 1950, he led many Britons and Americans to view the efficiency of British homeland security agencies in the early nuclear age very critically. Although MI5 had at times kept a close eye on Fuchs because of his communist affiliations, they had never gathered any evidence that he was a spy. Instead, he had even become a key player in the MED and made some contributions to Britain's post-war atomic weapons research project. His previous political radicalization in Germany now played a crucial part in influencing an important part of British nuclear culture – the relationship between and impact of science on political cultures and perceptions of homeland security.

The timing of Fuchs's confession coincided with the emergence of the national security state in the early Cold War.[2] But the situation regarding secrecy and security differed considerably in Britain and the United States. Garry Wills argues that in the United States 'the Bomb altered our subsequent history down to its deepest constitutional roots' and 'redefined the government as a National Security State'.[3] Whereas the United States government was still relatively open regarding publicity around nuclear weapons and energy research shortly after the war, Britain saw the emergence of the 'secret state', as Peter Hennessy has termed it.[4] 'Indeed in atomic secrecy the line of rationality was difficult to draw', observed Margaret Gowing on Whitehall's secrecy policies.[5]

Fuchs's radicalization in Germany

Fuchs's German background is central to any attempt at understanding his motivation to engage in atomic espionage for the Soviet Union, for it was in the northern German city of Kiel, as an MI5 report on the case appropriately put it, that 'the seeds were sown' for his turn towards communism.[6] The United States Congress, too, realized the importance of Fuchs's stay there for his political development so that Kiel featured on a 1950 map of Europe titled *The Geographical Focal Points of Espionage*.[7] In his statement during Fuchs's court hearing on 1 March 1950, Fuchs's defence lawyer Curtis Bennett also suggested to Judge Lord Goddard that one 'might be able to understand what was acting in this man's [Fuchs's] brain as a result of what happened in 1932 and 1933'.[8] Given the tremendous impact that the rise of National Socialism had on the lives of Klaus Fuchs and his family, as Chapter 1 has shown, Alan Moorehead rightly argued in *The Traitors: The Double Life of Fuchs, Pontecorvo, and Nunn May* – despite the otherwise fairly strong anti-communist bias of his book – that 'One of the things that must be put down against the Nazis is that they probably did more towards the corruption of Klaus Fuchs' mind than anything the Communists ever achieved.'[9]

Fuchs's time at Kiel University in the northern German province of Schleswig-Holstein, where he studied mathematics and physics from autumn 1931 until spring 1933, coincided with the Weimar Republic's final period of political turmoil, destabilization and violence.[10] The fact that Fuchs himself devoted considerable space in interviews and statements relating to his confession to his time at Kiel University indicates how crucial this period was for his political radicalization and thus for understanding his espionage activities during and after the war.[11] Fuchs had come to the Baltic Sea port of Kiel shortly after his father Emil was appointed professor of religion at the city's Pedagogical Academy in May 1931.[12] Emil Fuchs played an important role in shaping his son's ethical beliefs, especially his conscience. In his confession of atomic espionage activities to William Skardon in early 1950, Klaus Fuchs declared that 'the one thing that most stands out is that my father always did what he believed to be the right thing to do and he always told us that we had to go our own way even if he disagreed'.[13]

While Fuchs's father was thus crucial, it was the highly conservative and authoritarian political environment in Kiel that radicalized him. As early as 1920, Albert Einstein, who had close ties to the maritime city, in particular through his contributions to the development of the gyro compass by Herrmann Anschütz-Kaempfe, experienced this highly conservative and even anti-Semitic atmosphere, when the announcement of a lecture to be given by him on the theory of relativity at the Kiel Autumn Week for the Sciences and Arts sufficed to trigger strong protests.[14] Two years later, in spite of his fondness for the city and for sailing, Einstein declined Anschütz-Kaempfe's suggestion that he buy the house of a famous doctor from Kiel because, he argued, 'The climate in Kiel seems to make the people rather stormy as well'. Einstein added: 'Sometimes one feels among human beings as if one were in a herd of buffalos. In themselves they are not mean, but one must be careful not to be trampled by them'.[15]

The general elections of 31 July 1932 saw a dramatic polarization in the province of Schleswig-Holstein because about 70 per cent of the electorate had voted for oppositional anti-democratic parties. With the best turnout in all of the Weimar Republic, Schleswig-Holstein became the heartland of the NSDAP.[16] Beginning in the latter half of the year 1932, political radicalization and polarization increased in the city of Kiel, resulting in Nazi supporters' verbal and physical violence against political opponents on a daily basis.[17] This general political climate also affected Klaus Fuchs's immediate environment, Kiel University, where numerous professors, like many of their colleagues elsewhere in the Weimar Republic, had never fully internalized the democratic constitution of 1919. This also held true for significant numbers of students, and, as early as 1927, a National Socialist student organization had started to gradually increase its influence among students.[18]

Klaus Fuchs's political development has to be seen against this background, and his membership of political parties serves as a good indicator of his growing radicalization. Having initially joined the SPD before he came to Kiel, he and two of his siblings – Gerhard and Elisabeth, who studied at Kiel, too – finally broke with the SPD over its policy of tacit support for Hindenburg in the presidential elections of 1932. As a result, Klaus, Gerhard and Elisabeth Fuchs joined the

Sozialistische Arbeiterjugend (Socialist Workers' Youth) and the KPD. At Kiel University, Klaus Fuchs also became active in the free socialist student group called the Revolutionäre Studentengruppe (Revolutionary Group of Students; RSG). The RSG members, however, did not restrict their propagandistic efforts to the university campus as they also worked elsewhere close in close conjunction with the KPD and the Kommunistischer Jugendverband Deutschlands (Communist Youth Association of Germany; KJVD).[19] For instance, Fuchs instructed members of the Sozialistische Schülergemeinschaft (Community of Socialist High School Students) in Marxist-Leninist doctrine.[20] Klaus Fuchs and his brother Gerhard enjoyed the reputation as talented public speakers in Kiel's left-wing circles and appeared frequently at KPD meetings. As was standard practice at the time, they also showed up at Nazi gatherings and tried to disturb them.[21] In return, members and sympathizers of the NSDAP repeatedly harassed Fuchs and even made an attempt on his life.[22]

Since Fuchs was a known communist in Kiel, he left the city for Berlin where he enrolled at the Friedrich-Wilhelms-Universität in mathematics and physics after the National Socialist takeover. That he took the train to Berlin very early on the morning after the burning of the Reichstag proved to be the right choice because only a few hours later the Gestapo came to his apartment looking for him.[23] The Nazi students at Kiel University, however, maintained influential political contacts all over Germany so that Fuchs was finally expelled from Berlin University by decree of its rector on 3 October 1933.[24] Fuchs was forced to remain in the underground until he left Germany for France and finally the United Kingdom.[25] As Chapter 1 has shown, several members of his family were less fortunate and were subjected to National Socialist terror.

Therefore, Fuchs's time especially in Kiel and later also Berlin undoubtedly had a very strong impact on the political radicalization that led him eventually to pass nuclear information to the Soviets during and after the Second World War. It remains an impossible task to fully evaluate Fuchs's inner motivations to spy for the Soviet Union, as he presented them in statements to British, American and East German security services. However, the fact that he usually did not accept payments for his services suggests his strong commitment to both the communist cause and the Soviet Union. Apart from his expenses in the early days of his work for the Soviet Union, the only known instance when he received a payment occurred shortly after his return to the United Kingdom in 1946. On this occasion, Fuchs accepted a symbolic payment of £100 as a means of expressing his dedication to the Soviet cause.[26] And Fuchs never lost that commitment. After his arrival in the GDR in 1959, he collaborated with Soviet scientists on fast neutron reactions and was an active member in the German-Soviet Friendship Society as late as the mid-1980s.[27]

Fuchs's espionage, his detection and confession

Klaus Fuchs's dedication to communism led him to pass many of the innermost nuclear secrets to the Soviets and shatter Britons' beliefs in homeland security. Once he joined the TA project, he realized that he had to share all the information which was made available to him with the Soviet Union so that the country would not fall too far behind in the development of nuclear weapons.[28] Fuchs thus established contact with the Soviets through Jürgen Kuczynski who was at the time the leader of a London-based underground KPD cell.[29] MI5 later identified his first contact as Simon Kremer who served as the secretary to the military attaché at the Soviet Embassy at the time.[30] And Kuczynski's sister Ursula also served as Fuchs's courier in Banbury on several occasions.[31] Fuchs worked then under the KGB code names of 'Rest' and 'Charles' respectively.[32]

Shortly before his trial, Fuchs identified Jürgen Kuczynski and Hannah Klopstech to William Skardon as the persons who put him in contact with the Soviets, Kuczynski in 1942 and Klopstech after his return to Britain from the United States after the war.[33] In Kuczynski's case, MI5 started an investigation into his possible involvement in the recruitment of agents for the Soviet Union in the United Kingdom, which concluded in its final report that '[t]he evidence from his file does not enable one to reach any definite conclusion'.[34] Fuchs also identified Harry Gold as his contact 'Raymond' during the interviews conducted by two FBI officers in London.[35] In addition, MI5 investigated the possible involvement of Hans Kahle, a German communist who had fought in the Spanish Civil War, in the 'selection and indoctrination' of Fuchs during his time of internment in Canada.[36]

By 1949, MI5 and the FBI had launched investigations of Fuchs. Within the context of the discovery of the Klaus Fuchs spy case, the so-called Gouzenko Affair was of great importance. In the autumn of 1945, the defection of Soviet cipher expert Igor Gouzenko in Ottawa, Canada, represented an important first step towards uncovering Klaus Fuchs and Alan Nunn May.[37] Fuchs became a chief suspect thanks to the efforts of the FBI, in particular Robert Lamphere and Meredith Gardner. After his appointment as director of the espionage division at the FBI headquarters in Washington, DC, Lamphere assigned the senior cryptographer Gardner to the task of decoding several intercepted Venona messages which had been sent from the Soviet consulate in New York City to the KGB headquarters in Moscow between 1944 and 1945. They soon discovered that someone had passed on top-secret atomic information to the consulate in New York. Early on, the FBI suspected that the KGB spy was among the members of the British Mission.[38]

While Fuchs had authored the intercepted report, all four members of the British team who were based in New York City at the time (apart from Fuchs these were Rudolf Peierls, Tony Skyrme and Christopher Frank Kearton) were initially suspects.[39] Suspicion against Peierls was grounded in the fact that, as a senior scientist in the British Mission, he had access to sensitive data. Moreover, he was a German-born refugee from Nazism married to a Russian-born woman. Once the FBI realized it could not produce any evidence against Peierls, the suspicion vanished.[40]

In late 1949, the evidence gathered by the FBI pointed to Fuchs as the main suspect. The FBI then shared the information with MI5 who started working on a plan to expose him.[41] MI5 decided that Fuchs's telephones at home and work should be tapped, his correspondence monitored, microphones installed in his office, his bank account checked, his movements observed, all his contacts – including Peierls – investigated, and that Wing Commander Henry Arnold, the security officer at Harwell and one of Fuchs's few close friends, should be involved in the investigation.[42] An opportunity appeared when Fuchs approached Arnold to ask for advice on how he should behave because Emil Fuchs (his father) had just accepted a post at the University of Leipzig. Fuchs feared that the British authorities might see him as a security risk owing to his father's move to East Germany.[43] William Skardon, who had also been involved in the interrogation of William Joyce alias Lord Haw Haw, was put in charge of interrogating Klaus Fuchs by mid-December 1949.[44]

During the first of seven interviews with Klaus Fuchs, William Skardon confronted him with the espionage allegations, which Fuchs vehemently denied. Fuchs remained outwardly unimpressed and protested his innocence or, as he later described it himself, 'played the scientist' despite William Skardon's assurance that he was stating a given fact and not probing him about this matter.[45] 'Reviewing all the facts in the light of the interrogation', Skardon wrote in the report after the meeting, 'I feel sure that we have selected the right man, unless by chance someone in the nature of a twin brother was in New York when he was there'.[46] Because of Fuchs's (initial) unwillingness to confess to the charges of atomic espionage on behalf of the Soviet Union, MI5 drew up a contingency plan. It included his surveillance and the wiretapping of Fuchs' phone lines and those of some colleagues and friends at Harwell, as well as his arrest in case he tried to defect.[47]

While Fuchs continued to reject all allegations against him in the second and third interviews, it was in the crucial fourth interrogation on 24 January 1950 that he confessed to William Skardon his espionage activities for the Soviets. Fuchs had requested to see Skardon in his private home at 17 Hillside in Harwell that day, and another meeting had been arranged through Wing Commander Arnold. For the first two hours, Fuchs apparently intended to 'play the scientist' again but was, as William Skardon observed, under 'considerable mental stress'. He gave a lengthy account of his underground activities for the KPD in Germany prior to his flight. Just before they went off for lunch, after Skardon had let Fuchs know that it appeared to him as if he had just provided a 'long story providing a motive for acts' but had not given any hints of the nature of the acts, Fuchs still assured him 'I will never be persuaded by you to talk'.[48]

As Skardon noted in his report, Fuchs seemed to 'be revolving the matter and to be considerably abstracted' over lunch, to such an extent that he urged Skardon to return to his house quickly after the two men had finished their lunch. Upon their arrival at Fuchs's private home, Klaus Fuchs informed Skardon about his decision to cooperate and answer the latter's questions. While he indicated that he had a 'clear conscience at present', Fuchs voiced serious worries about the 'effect of his behaviour upon the friendships [which] he had contracted at Harwell'. Moreover, Fuchs cited his disapproval of Stalinism indirectly as a further reason to confess his activities. As William Skardon recalled, Fuchs declared that while he still believed in the ideals of communism, he rejected it in the form it was currently taking in the Soviet Union, which had transformed it into 'something to fight against'.[49] Rudolf Peierls commented on Fuchs's turn away from communism and his regret for his betrayal of both his host country and his friends that 'From his [Fuchs's] point of view this is perhaps the most tragic: that he now does not even have the satisfaction of suffering for a cause in which he believes'.[50] On 27 January 1950, Fuchs and Skardon met again, this time at the War Office, and Klaus Fuchs dictated to Skardon a statement going into detail about his activities for the Soviet Union.[51]

In his confession, Fuchs explained to Skardon his major reasoning for becoming engaged in nuclear espionage. Besides his 'complete confidence in Russian policy' at the time, Fuchs 'believed that the Western Allies deliberately allowed Russia and Germany to fight each other to the death'. And he regarded his actions as a natural consequence of political and historical developments during the period. In the end, two factors were, according to Fuchs, crucial in forming his decision to confess his spying to the British authorities: first, he felt a growing uneasiness about his betrayal of personal friends as a consequence of his passing of classified information to the Soviets; and, second, he stated, he could no longer tolerate communism as practised in Stalinist Russia. But it was in particular his betrayal of his close friends, Fuchs argued, under which he crumpled in the end. He declared that he had initially intended to control the inner conflict between betraying his friends and spying for the Soviets by establishing 'two separate compartments' in his mind. In one compartment he could 'make friendships, to have personal relations, to help people and to be in all personal ways the kind of man I wanted to be', and in the second one he could be 'completely independent of the surrounding forces of society' and betray his friends. 'Looking back at it now', Fuchs appropriately evaluated his own behaviour, 'the best way of expressing it seems to be to call it a controlled schizophrenia'.[52]

The fact that Fuchs kept a low profile and a rational appearance did not go unnoticed by outsiders. An MI5 surveillance report praised Fuchs's work ethic and described him as 'pre-eminently intellectual, but not a cold-blooded intellectual', concluding: 'His life, in short, is always under the strict control of his intellect'.[53] Gaby Peierls later recalled that Fuchs had come over to her parents' house in Los Alamos frequently on Sundays and that he had been quite popular among the children of Los Alamos.[54] Fuchs's behaviour was not entirely new. In order to resolve what he felt was a clash of interests, he chose a kind of 'mental reservation', as Stephen Toulmin has argued, as intellectuals before him did. But this 'mental reservation', which allowed him to pass on vital nuclear data to the Soviets although he had signed the Official Secrets Act, worked only while the Soviet Union was Britain's ally.[55]

Klaus Fuchs had apparently overestimated his ability to control his feelings for his friends and had to pass, as he later declared, a serious test after the arrest of the Soviet atomic spy Alan Nunn May in 1946.[56] Hanni Bretscher was among the curious Los Alamosans who discussed the espionage case with Carson Mark, Nunn May's former colleague at Montreal. Fuchs was also present at this meeting. Hanni Bretscher later recalled that when Else Placzek – who had also known May personally – replied to a question concerning his character that he was a very nice person 'just like Klaus Fuchs here', Bretscher supposedly detected 'how embarrassed and red Fuchs got' but did not make anything of it.[57] Despite Bretscher's observation, Fuchs himself did not realize, as he later admitted, that Nunn May's arrest was a wake-up call to rethink his behaviour. Instead, he continued to suppress the question of loyalty to his friends.[58]

Fuchs's colleagues, too, noticed that he was apparently under considerable stress. Joseph Rotblat experienced this first hand in an exchange of letters with Fuchs about data the latter was to supply to the AERE Harwell in late 1946/early 1947. What irritated Rotblat in particular was Fuchs's pedantic questioning of the reliability of the experimental results furnished by Rotblat.[59] In another instance, Fuchs accused his Harwell colleague Egon Bretscher in a similar fashion of failing to conduct an experiment at the AERE properly.[60] Hanni Bretscher recalled that she and her husband could not comprehend Fuchs's behaviour at the time, but that 'it all fell into place once he was arrested for treason'. Bretscher suspected that Fuchs might have thought that she and her husband 'guessed he was not trustworthy'. During a dinner party held on the occasion of Emil Fuchs's visit to Harwell, which the AERE security officer Henry Arnold also attended, Hanni Bretscher argued, Klaus Fuchs acted even more strangely, which 'was in fact the first clear indication for the security officer (Henry Arnold) that Fuchs had something to hide'.[61]

When it became known to some of Klaus Fuchs's friends and colleagues that serious allegations of spying for the Soviet Union were brought forward against him, many still did not believe what they heard. Fuchs's colleague and close friend Herbert Skinner assured him that all his Harwell co-workers would support him if he affirmed his innocence. As a result of his friends' support, Fuchs claimed, he was unable to disguise his activities as a Soviet informant any longer.[62] As a consequence, Fuchs not only pleaded guilty on all charges brought forward against him in his trial, but stressed in his final statement that he 'also committed some other crimes which are not crimes in the eyes of the law – crimes against my friends'.[63]

The value and significance of Fuchs's espionage

In order to assess Fuchs's impact on the reputation of MI5, it is necessary to briefly consider the value and significance of his espionage. Like the examination of Fuchs's motives for engaging in and finally abandon his espionage, it remains a difficult task to assess the value of the nuclear data passed on by him to the Soviet Union. Fuchs detailed most of the technical details which he had revealed to MI5 and the FBI in interviews with Michael Perrin, a former senior TA administrator. In the first period of his espionage activities for the Soviet Union from 1942 until December 1943, these included, above all, the results of his own work on theoretical aspects of the uranium 235 isotope separation process through gaseous diffusion at Birmingham University. This work formed part of the so-called MS series. Moreover, he informed his contact in more general terms about progress of the British and to a much lesser extent, as far as he was able to judge it, the US nuclear weapons project. During his time in New York from December 1943 until August 1944, Fuchs started to give the Soviets information which was not only the result of his own work, including data on gaseous diffusion processes and all reports drafted by the British Mission at New York. But it was then during his stay at Los Alamos between August 1944 and the summer of 1946 that he fully grasped the scale of the Manhattan Project for the first time and gave away the most valuable information he ever passed on to the Soviet Union, in particular the principle of the design of the plutonium implosion bomb. In addition, he passed on information on the critical mass of both uranium and plutonium and informed his Soviet handlers about the upcoming Trinity Test.

Fuchs later claimed that, after his return to Harwell in the United Kingdom in the summer of 1946, he restricted the flow of information to the Soviets until he finally ceased his espionage work for them in the spring of 1949. While this period was marked by Fuchs's increasing doubts, as he maintained, about the Stalinist practice of socialism in the Soviet Union, he completed Los Alamos-related information by giving away the so-called Bethe formula, for example, which is used to calculate the efficiency of a nuclear detonation, and he reported on the state of the art of the British atomic arms and energy programmes.[64]

The principal outline of the implosion bomb was perhaps the single most important item Fuchs gave to the Soviets. 'I did what I consider to be the worst I have done', he told William Skardon in his confession, 'namely to give information about the principle of the design of the plutonium bomb'.[65] It is certainly true that the United States government promoted the proliferation of nuclear information, as Fuchs argued in the 1980s, through the publication of Henry DeWolf Smyth's report *Atomic Energy for Military Purposes*.[66] Yet, he failed to mention that the published report omitted the crucial implosion principle and that it was indeed Soviet agents such as himself, Alan Nunn May, Ted Hall and David Greenglass who informed the Soviets about implosion-related matters.[67]

If it has remained a difficult undertaking to assess the value of the information passed on by Fuchs to the Soviet Union, it has remained even more complex – if not altogether impossible – to give approximations of how much time he saved the Soviet Union in its nuclear weapons programme. In an interview by the FBI, Fuchs calculated that his information saved the Soviet Union 'at least one year'.[68] While Fuchs's estimate seems modest, his former KGB case officer, Alexander Feklisov, overemphasized Fuchs's importance as an informer. According to Feklisov, the information Fuchs passed on to him and other handlers allowed the Soviet Union to produce a working nuclear weapon six years earlier than the West expected.[69] Although it is virtually impossible to give an exact amount of time that Fuchs saved the Soviet atomic bomb project, a figure of between one and two years appears plausible.[70] After all, the development of a Soviet plutonium bomb was inevitable and the country's scientists had been working on it since 1943.[71] Yet American and British intelligence services failed to forecast the first Soviet nuclear test accurately.[72] And even Klaus Fuchs himself showed surprise about the fact that the Soviet Union had been able to develop its own atomic bomb so quickly.[73]

Although Fuchs's information was considered to be of 'inestimable value' to the Soviet Union, as an FBI memorandum prepared for J. Edgar Hoover put it, Soviet nuclear scientists still had to do considerable work to produce a working plutonium bomb.[74] To give an appropriate comparison, in spite of the fact that the British Mission had played a pivotal role in designing the first uranium and plutonium bombs at Los Alamos, it took British scientists, after the end of the Anglo-American wartime nuclear alliance in 1946, until 1952 to deliver a working plutonium bomb. And, as they did not know whether the data provided by Fuchs and other spies were correct or not, Soviet scientists were forced to repeat most of the experiments themselves. But the Soviet Union's small supply of uranium ore proved to be the major factor that hampered their plutonium bomb programme until they started to mine it in East Germany.[75]

While the view that Fuchs provided valuable information for the production of the Soviet plutonium bomb has become widely established, recent scholarship, which is based on Russian archives, also suggests that Fuchs played a bigger role than previously assumed in the making of the Soviet hydrogen bomb.[76] These new findings seem to contradict the older view that although Fuchs attended a conference on 'the Super' at Los Alamos in April 1946, he could not have given the Soviets any valuable information on the H-bomb because the scientists on 'the Hill' were on a wrong path at the time. While Oppenheimer played down the significance of the data passed on by Fuchs, Teller was convinced that Fuchs had supplied the Soviets with highly important H-bomb-related information and thus enabled the country to pursue its hydrogen bomb project much more quickly.[77]

Although Fuchs's passing of early work on thermonuclear weapons did not enable atomic scientists in the Soviet Union to create a working H-bomb, his role can perhaps be best described as a kind of catalyst for both the Soviet H- and A-bombs. Still, the German-born Soviet atomic spy had a tremendous share in ensuring that the Soviet Union beat the United Kingdom in the race to become not only the world's second atomic but also a thermonuclear power. While the Soviet Union tested its first hydrogen bomb in 1953, it took the United Kingdom until spring 1957 to detonate its first working thermonuclear device.[78]

The Fuchs case and British perceptions of national security

While Klaus Fuchs had helped the Soviet atomic and thermonuclear projects tremendously, his confession also severely damaged the image of MI5. The fact that the Security Service had failed to detect Fuchs's espionage activities at an earlier stage undermined the confidence in the effectiveness of security agencies in their defence of the democratic order by members of both the British government and the public. As the internal intelligence agency, criticism centred on MI5. In fact, so concerned were MI5 officials about the Security Service's reputation that they circulated a short internal document to staff offering 'some guidance on facts' to prepare their personnel for public discussions, especially of questions regarding MI5's security checks of Fuchs and its knowledge of his communist affiliations. The memorandum, which represents an attempt to establish an official line of argumentation within the Security Service, identified three main questions as the foci on which its authors expected public criticism would centre: 'Why was Fuchs taken on for employment in Atomic Energy?', 'Why was Fuchs' espionage activity not detected?' and 'Why is it that the Americans appear to have known all about Dr. Fuchs' Communist history but not the British?'[79]

The document provided the 'official' answers to all three questions. In response to the first, it stressed Fuchs's significance and expertise in his field that simply made him invaluable for the British and later joint Allied Manhattan Project. While it admitted that the Gestapo, as has been shown in Chapter 1, furnished evidence of Fuchs's alleged communist affiliations, the memorandum rightly questioned the reliability and validity of this source of information, arguing that this was 'an allegation which had been groundlessly made against countless other refugees who later made a notable contribution to the allied war effort'.[80] In other words, Fuchs's skills were simply deemed so important for the British war effort that the authorities were willing to take security risks, especially if they did not appear as such at the time. After all, the fact that the National Socialist regime accused Fuchs of being a communist made it very unlikely that he was spying for Germany – the country with which Britain was engaged in all-out war and which temporarily threatened the very existence of the United Kingdom. In the face of his many contributions to the British nuclear weapons programme and the grave international crisis, MI5 regarded these allegations made by the enemy as being of minor importance. And the benefits of Fuchs's work for the British seemed to outweigh the potential risks of his employment, even after the war when the Security Service shifted its attention to communists as potential threats to national security.[81] It was only after Fuchs's confession and his forced retirement from the AERE Harwell that members of the British nuclear weapons project fully realized the significance of his work, especially since the McMahon Act had cut off British scientists from any American nuclear data and Fuchs had become a major source of information for the British authorities.[82]

The second question – concerning why MI5 had failed to expose Fuchs as a Soviet spy – was an equally thorny one. As the Security Service document convincingly argued, Klaus Fuchs's covert mode of operation 'made the detection of Dr. Fuchs an exceptionally difficult problem in any democratically governed state'.[83] Although MI5's answer appears slightly evasive, it also contains an element of truth. While the British public enjoyed the benefit of hindsight, MI5 had in fact faced many difficulties in its investigations of Fuchs. First and foremost, Fuchs managed to keep a low profile and to disguise his espionage activities well. Consequently, the local police in both Bristol in 1934 (before he engaged in espionage activities) and Edinburgh in 1942 did not have anything against him in their files.[84]

Even the fact that he showed sympathies for communism and the Soviet Union did not necessarily make him a security risk. In his autobiography, Max Born – under whom Fuchs worked at the University of Edinburgh – claimed retrospectively that Nevill Mott had sent Fuchs to Edinburgh because 'He spread communist propaganda among the undergraduates'.[85] Nevill Mott forcefully rejected Born's allegations, and even described Fuchs's leaning towards communism as acceptable at the time because 'anyone who was against the Nazis would have been'.[86] With his statement, Mott captured well an attitude that many Britons shared in the face of the threat that the Third Reich posed to their country during the war. Rudolf Peierls, as one of Fuchs's chief mentors, made a similar point about his political views during the Second World War.[87]

Klaus Fuchs and other scientists such as Nevill Mott openly showed sympathy for the Soviets and were both members of the Bristol branch of the Society for Cultural Relations with the Soviet Union in the 1930s.[88] Still, Fuchs's and Mott's participation in the society did not attract MI5's attention because it was not seen as a potential threat to national security at the time. That MI5 did not expose Fuchs as a Soviet agent earlier lay also in the fact that he acted inconspicuously as a communist. He engaged, for example, in work on a committee set up to help the Republican forces in the Spanish Civil War during his time in Bristol and while in Edinburgh he directed the shipment of propaganda leaflets to Germany on behalf of the KPD.[89] Max Born later recalled that although Fuchs 'never concealed that he was a convinced communist', 'he did not speak about politics very much, except during the time of the Soviet–Finnish War. All of us in the department, including Indians and Chinese, were sympathetic to the Finnish side, while Fuchs was passionately pro-Russian'.[90] At the time of the war between the Soviet Union and Finland, however, Fuchs was not yet a Soviet agent. And, in any case, it was not a criminal offence in a democratic society like the United Kingdom to show sympathy for the Soviet Union and to openly sympathize with communism.

Two connected factors help account for these recriminations against the Security Service: MI5's unpreparedness for the new challenges of the Cold War and the re-emergence of anti-communism in post-war Britain. 'Britain strode into the nuclear age protected by patched clothing designed for a time when dreadnoughts constituted the leading edge of defence technology, and waiters with German accents represented the chief threat to national security', David Vincent has argued.[91] In 1948, the Attlee Government ordered a civil-service purge to ensure that neither fascists nor communists worked in sensitive areas of government work.[92] In May 1949, Wallace Akers spoke out in favour of barring communists from employment in sensitive areas of the public service, in particular atomic research.[93] It was, however, not until after (and as a result of) the espionage cases of Klaus Fuchs, Bruno Pontecorvo, Guy Burgess and Donald Maclean (and owing to pressure from Washington) that Whitehall implemented the practice of 'positive vetting' in January 1952.[94] This new measure allowed security agencies to investigate the private and political background of people working in sensitive areas of government such as atomic research. 'Positive vetting' formed one of the crucial components of what David Vincent has referred to as a 'culture of secrecy'.[95] An earlier introduction of this procedure could possibly have led to an earlier exposure of Klaus Fuchs.

The immediate post-war period also saw anti-communism resurfacing. But it did not take such radical forms as in the United States. Still, an anti-communist bias affected the trial of Klaus Fuchs. At the court hearing at the Old Bailey on 1 March 1950, two key issues dominated: first, Attorney General Sir Hartley Shawcross's and Chief Justice Lord Goddard's flamboyant anti-communist rhetoric; second, and closely connected, their branding of Fuchs as a 'traitor' to the British people. 'The prisoner is a communist, and that is at once the explanation and indeed the tragedy of this case', Shawcross summarized and evaluated Fuchs's espionage activities. In a similar fashion, the Lord Chief Justice told Fuchs: 'You have betrayed the hospitality and protection given to you by the grossest treachery'. Lord Goddard's final remarks before reading out the verdict also reveal the great emotional impact of the Klaus Fuchs trial on its participants. 'Your statement which has been read shows to me the depth of self deception into which people like yourself can fall', Goddard stated, adding: 'Your crime to me is only thinly differentiated from high treason'. The Chief Justice went on to argue, 'My duty is to safeguard this country and how can I be sure that a man, whose mentality is shown in that statement you have made may not, at any other minute, allow some curious working of your mind to lead you further to betray secrets of the greatest possible value and importance to this land?'[96] Klaus Fuchs then received the maximum sentence of 14 years for his espionage activities. Since the Soviet Union had not been an enemy of the United Kingdom during the Second World War, Fuchs could not be charged with treason under the Official Secrets Act of 1911.[97]

Fuchs's split loyalties to his host country and the Soviet Union represented the second chief marker of anti-communism at his trial. As the Attorney General pointed out, Fuchs had done irreparable 'damage in breach of the loyalty that he would, one would suppose naturally, feel towards the country which had befriended him, which had enabled him to complete his training and to become a great scientist'. Shawcross emphasized the 'damage he [Fuchs] did in breach of his security undertaking, in breach of his Oath of Allegiance to the King who had granted him the privilege of British nationality'. He then concluded: 'But although these were loyalties which appeared to have meant something to him, they were, unhappily, loyalties he cast aside in favour of his loyalty to the spurious ideology of Russian Communism'. Sir Hartley Shawcross's polemical reference to the 'spurious ideology of Russian Communism' is indicative of the heated atmosphere at the court hearing, for it remains doubtful whether loyalty to a 'non-spurious' ideology would have been a defence in law. Curtis Bennett, Fuchs's defence lawyer, attempted to have his client's sentence reduced by using a kind of insanity-defence-based argument in which he elaborated on Fuchs's 'state of mind', in particular his 'controlled schizophrenia'. Lord Goddard, however, rejected this argument, saying: 'I cannot understand that metaphysical philosophy or whatever you like to call it. ... He [Fuchs] stands before me as a sane man and not relying on the disease of schizophrenia or anything else'.[98]

While the atmosphere at the trial was at times heated, the British public's response to the revelation of Fuchs's espionage was moderate, in particular by comparison with the United States. Parts of the media, however, attempted to capitalize on Fuchs's story. Immediately after the trial, some British newspapers ran headlines like 'Atomic Secrets Betrayed' that emphasized the high degree of treason that the German-born émigré physicist had committed.[99] Writing in the Beaverbrook paper the *Daily Express*, Chapman Pincher reported on the trial. 'In 90 minutes at the Old Bailey yesterday a riddle was solved: How did Russia make the atomic bomb so quickly?' The answer: 'Dr. Klaus Emil Julius Fuchs, confidant and leading member of Britain's atom team ... gave them the know-how'.[100] With this statement, Pincher helped create a myth that Fuchs was solely responsible for giving away the atom bomb. In this, his statement was reminiscent of Hollywood films like *Notorious* (1946) and *The House on 92nd Street* (1945) that depicted the atomic bomb as a weapon made up of one mathematical formula that foreign agents could easily steal.[101] Pincher continued to show an interest in Klaus Fuchs and published several articles on him over the following years, including a photo story depicting images of Fuchs in his office in Rossendorf in 1965.[102] The *Daily Express* featured a series of articles by Bernard Newman, titled 'The Spies Are Among Us', which attempted to fuel an anti-communist hysteria.[103] Strangely, the British press did not attack the British Communist Party, although several connections existed between Fuchs and other communists.[104]

For the next few years, stories of atomic espionage remained in the news thanks to journalists like Alan Moorehead and Rebecca West.[105]

Moorehead, a war correspondent and author of popular histories, worked for the *Daily Express* at the time of Fuchs's confession. In 1952, he published his book *The Traitors: The Double Life of Fuchs, Pontecorvo, and Nunn May*. Its publisher Hamish Hamilton attempted to use fears of communist infiltration to advertise it. 'The loyalties of the atomic scientist in general and the whole question of security are discussed', it read on its sleeve, 'while always in the background looms the possibility that other traitors, as dangerous as Fuchs, may still be at large'.[106] The same year also saw the release of Oliver Pilat's *The Atom Spies*.[107] Two years after the Fuchs trial, feelings still occasionally ran high. The *Sunday Express* published a series on 'Stalin's Atom Spies'.[108] And Sir John Squire concluded his review of *The Traitors* on a cynical note, commenting on the fact that Fuchs was liked by fellow inmates of Stafford Prison. 'Better to let him out now and parachute him over the lines in Korea', he suggested: 'He mightn't find any mailbags to sew there, but he could make himself useful by going out into the fields and picking up infected insects with chopsticks'.[109]

Rebecca West was a staunch anti-communist and much more extreme in her views than Moorehead and other commentators. In 1951, her series 'The Traitors' in the *Evening Standard* on 'the most sinister figures of our time' opened with a piece on Klaus Fuchs.[110] The following year saw the publication of her influential book *The Meaning of Treason*. Originally published in 1945, the London-based Reprint Society launched a second enlarged and revised edition in 1952 which contained additional chapters on atomic espionage. Viking Books published this updated edition as the *The New Meaning of Treason* in the United States in 1964.[111] West also published an article on the Rosenberg case in the *Picture Post* – Britain's leading illustrated magazine and counterpart of *Life* magazine – in 1953.[112] She argued that the Fuchs case revealed the entirely altered significance of treachery in the post-Second World War world. In her eyes, the term 'ideological espionage' did not suffice to characterize Fuchs's spying because he epitomized the archetype of the 'traitor scientist' who had abused the hospitality and trust of the British people through his dedication to communism.[113] MI5 shared this view, as a report stated: 'The history of Emil Julius Klaus Fuchs is a curious mixture of brilliant scholarship and achievement in the field of scientific research, blind devotion to the doctrines of Communism and cold-blooded treachery to the country which had done most to welcome and reward him.'[114]

The feeling that Fuchs had betrayed King and host country ran deeper than his communist background and eventually peaked in his denaturalization. On 20 December 1950, the Deprivation of Citizenship Committee convened and unanimously recommended that Fuchs be stripped of his British citizenship. 'It is just as likely and in our view probably more likely', the Committee said in its verdict, 'that the communist philosophy, in which the respondent [Klaus Fuchs] is so steeped, will again assert its ascendancy and submerge the feelings of loyalty towards the Crown, which he at present professes'.[115] Again, the issues of loyalty to state and anti-communism were closely connected in the recommendation.

Inspired by reports in the media about Fuchs's denaturalization, a film company approached the Home Office about a biopic on Klaus Fuchs. The executive producer Brock Williams of Pinnacle Production Ltd intended to jump on the anti-communist bandwagon and suggested that 'such a film produced at the present time as a first class "feature" for world distribution will prove a source of healthy enlightenment to the great cinema-going audiences and an effective antidote to the spread of Communist propaganda'.[116] By 1953, the Fuchs case had entered the realm of children's literature when Samuel Epstein and Beryl Williams published *The Real Book of Spies* in the United States. The book's first British edition came out in 1959.[117] 'The spies are all from history', commented a reviewer in the *Illustrated London News*, 'ranging from those whom Moses sent into the land of Canaan to Klaus Fuchs'.[118]

While anti-communism had been part of British politics before 1939, the political climate change which occurred in the immediate post-war period was rooted in the Second World War. During the final stages of the Second World War, the first breaches had occurred between the Western Allies and the Soviet Union. In contrast to popular belief, it was principally the British (and not the United States) government that sought confrontation with the Soviet Union in the years between 1945 and 1947.[119] At the time of Clement Attlee's two Labour Governments between 1945 and 1951, a comparatively strong anti-communist consensus emerged in the United Kingdom.[120] At the same time, revelations about the party's strategies, objectives and techniques by former, disillusioned communists provided the British public with 'insiders' views'. Works like the autobiography of Douglas Hyde, a former editor of the *Daily Worker*, as well as a collection of essays by Arthur Koestler and five other intellectuals, titled *The God That Failed*, presented 'authentic' accounts of the allegedly sinister character of communism.[121] And the *Sunday Express*, for example, dedicated its front page headline to the abandonment of the Communist Party in November 1950 by famed evolutionary biologist and geneticist John B. S. Haldane.[122]

Yet anti-communism in Britain paled by comparison with the United States and never peaked in McCarthyite, anti-communist witch hunts. After all, many refugees from McCarthyism came to Britain such as film director Joseph Losey or the classics scholar Moses I. Finley. The fact that academics like the Marxist historian Eric Hobsbawm were able to pursue steep careers in Britain, and that others such as the scientist John D. Bernal and the economist Maurice Dobb were able to express their opinions, underscores this view.

The Fuchs case also had an impact on the discourse over individual rights in a democratic society. Many contemporaries and commentators have criticized Fuchs for placing his conscience over his allegiance to his host country. Klaus Fuchs's conscience clashed with his loyalty to the United Kingdom, putting his scruples against his responsibility and accountability to his host country. Fuchs claimed that his conscience had served him as a crucial moral guide. William Skardon noted in the report of his first interrogation of Klaus Fuchs that while he 'established that Fuchs recognises that his Oath of Allegiance is a serious matter and a thing to be observed', he simultaneously 'claims freedom to act in accordance with conscience should circumstances arise in this country comparable to those which existed in Germany in 1932 and 1933, when he would act on a loyalty which he possesses to humanity generally'.[123]

Skardon's account names two of Klaus Fuchs's multiple allegiances, one to Britain and one to his conscience. Besides these two, he also had allegiances to the Soviet Union and to his friends. At times, Fuchs had faded out one or more of these multiple allegiances. Rebecca West especially challenged Fuchs's concept of loyalty to his conscience with her central line of argumentation that reduced 'conscience' to ego, arguing that 'if a state gives a citizen protection it has a claim to his allegiance'.[124] According to West, Fuchs should have subordinated the allegiance to his individual conscience under the good of his host nation which had offered him a haven from National Socialist persecution. West's argument bears striking resemblance to totalitarian beliefs. 'This is the peculiar menace of Fuchs, for if he were to propagate himself', wrote Alan Moorehead, similarly emphasizing Fuchs's egoism, 'if thousands and tens of thousands of Fuchses and their consciences were let loose on the world, they would be almost as deadly as the worst atomic bomb invented yet'.[125]

Apart from journalists and writers, former colleagues of Fuchs also expressed similar thoughts on his split loyalties. 'They were all driven by the force of the communist philosophy to take matters of life and death, perhaps for millions of people, into their own hands', Fuchs's long-time friend and boss at Harwell, Herbert Skinner, collectively judged the actions of Klaus Fuchs, Ethel and Julius Rosenberg, Alan Nunn May, David Greenglass and Harry Gold, 'and those who confessed could only in the end say, feebly, that they had simply been wrong'.[126] Many commentators have continued since to propagate views of Fuchs similar to those voiced by West, Moorehead and Skinner.[127]

In spite of his spying for the Soviet Union, and unknown to the public at the time, Fuchs intended, as he assured Michael Perrin, to fully cooperate with the British authorities in order to limit the damage done as much as possible.[128] 'I formed the impression', Perrin noted, 'that throughout the interview Fuchs was trying to remember and report all the information that he had given to the Russian agents with whom he had been in contact and that he was not withholding anything'. Perrin further observed: 'He seemed, on the contrary, to be trying his best to help me to evaluate the present position of atomic energy work in Russia in the light of the information that he had, and had not, passed to them.'[129] Fuchs's insistence that he would not talk with Skardon about any classified technical data because the latter did not have the necessary security clearance – which finally led to Michael Perrin interviewing Fuchs – shows another seemingly paradoxical and almost pedantic attitude of Klaus Fuchs.[130]

Fuchs's multiple allegiances to his adopted home country, the United States and the Soviet Union reflected this ambivalence. Not only did he work for the Manhattan Project, but he also helped the Soviet nuclear weapons programme and, at the end of his stay in Los Alamos, when it became clear that the Anglo-American wartime nuclear alliance would not be carried over into the post-war period, he even spied for Britain. Klaus Fuchs thus had a substantial part in both the atomic and hydrogen weapons programmes of the United Kingdom, the United States and the Soviet Union.[131] This has to be weighed against the (sometimes harsh) condemnation of Fuchs. Through his work for the British nuclear arms project and against the United States, Fuchs continued to try to break the American monopoly on nuclear weaponry and helped weaken the McMahon Act, which Whitehall regarded as a major obstacle in its quest to establish its own nuclear weapons project.[132] While Fuchs was never credited for his work for the British A- and H-bomb projects, it is understandable that MI5 tried to cover up his case. As a consequence, it appears almost ironic that MI5's plan failed, and that the Fuchs case marked the beginning of an enmity in MI5–FBI relations.[133]

Fuchs also showed a great deal of loyalty to his host country during the interviews: while he told Michael Perrin about the information on thermonuclear weapons which he had passed on to the Soviets, he did not mention this in his interrogation by FBI agents. The FBI, however, was to some degree aware of these omissions as MI5 had partially informed them.[134] In a letter to the State Department, J. Edgar Hoover reported on the interviews with Fuchs conducted by two FBI agents in the United Kingdom. Fuchs 'declined to furnish the details of what he had given to the Soviet Union after his return to England regarding the hydrogen bomb', he wrote, 'because of the lack of co-operation between the United States and Great Britain at the present time with regard to atomic research'.[135]

David Vincent has argued that, in the aftermath of Klaus Fuchs's confession and Bruno Pontecorvo's defection, 'The glad confidence in the value of science to the citizens of the welfare state was replaced by a less trusting and respectful attitude to those charged with discovering the hidden mysteries of nature'.[136] Alan Moorehead appropriately assessed the degree of Klaus Fuchs's deed saying that he 'had committed the crime society is least able to forgive; he made society distrust itself. And for that he was hated'.[137] Still, despite the security purge with its practice of 'positive vetting', Britain was far away from experiencing anything remotely like McCarthyism.[138] One of the major exceptions to this trend was Roy Boulting's picture *High Treason* (1951). The public uproar about an explosion that occurred at the Portsmouth naval base in July 1950 which destroyed several ships loaded with weapons and ammunition to be shipped to British forces in Korea inspired the plot. Since the incident was largely attributed to foreign-directed sabotage, it fuelled fears of a communist fifth column operating within Britain. The film opens with a re-creation of the incident and then elaborates on the idea of a communist plot to sabotage power stations in the United Kingdom, visually identifying the spies through their duffle coats. But *High Treason* represents a rare case of a British Red Scare.[139]

The Fuchs case and Anglo-American relations

While MI5 had come under severe criticism at home, the Security Service also became the target of condemnation from abroad. As implied in the third key question of the internal document circulated among staff – 'Why is it that the Americans appear to have known all about Dr. Fuchs' Communist history but not the British?' – Washington heavily criticized MI5 for its failure to detect Fuchs earlier.[140] Americans first learned about his confession on the day after President Truman had ordered a crash programme to beat the Soviet Union in the race for the hydrogen bomb.[141] His confession received considerable public attention in the United States. *Life* magazine reported in detail about the case. 'The momentous week began like any other', opened an article on Fuchs's confession that linked it with other major events, 'with everybody paying closest attention to what seemed to concern him most'. Among the week's news miscellany, it mentioned President Truman's announcement that the United States was pursuing its own thermonuclear weapons programme, the birth of Ingrid Bergman's baby and the departure of Danish-American tenor Lauritz Melchior from the Metropolitan Opera Company. 'And in Harwell, Britain's big atomic-research center', the article went on, 'an unobtrusive physicist named Klaus Emil Julius Fuchs went about his top-secret work with the satisfied feeling of a man who has recently won a promotion and a raise', before drawing an analogy to the Japanese attack on Pearl Harbor.[142] *Life* continued to cover the Fuchs case and featured, for example, an article on his court hearing including drawings. It later reported about his release and departure for East Germany in 1959 (Figure 4.1).[143]

The commanding officer of the Manhattan Project, General Leslie Groves, described Fuchs's espionage as the 'most disastrous break in security' within the Manhattan Project. Although he admitted that 'Our acceptance of Fuchs into the project was a mistake', he pointed out that 'It was a British responsibility'. With his delegation of blame towards the British security agencies that had vetted Fuchs and on whose judgement the US Army had relied, Groves's reaction exemplified a verdict commonly made by Americans. 'The United Kingdom', he complained, 'not only failed us, but herself as well'.[144]

While the Security Service has repeatedly been criticized for its supposedly lax screenings of Fuchs, these allegations appear to be by and large unjustified and often the result of hindsight.[145] As stated before, the introduction of 'positive vetting' at an earlier time might have – but not necessarily must have – led to Fuchs's exposure as a Soviet spy before 1950. Fuchs's former Los Alamos colleague John Manley rightly stressed the fact that it was relatively easy for Fuchs to pass on secrets to the Soviets. Owing to his important role in the Theoretical Division, Manley argued, Fuchs did not even have to penetrate the project.[146] Fuchs himself later made a similar point in an interview.[147] The internal MI5 memorandum thus rightly stated that it appeared doubtful whether the Americans could have provided any information 'which would have modified the British assessment of Dr. Fuchs' security record'.[148]

The news of Klaus Fuchs's confession came at a very bad time for Britain and the United States, as relations between the two countries had

Figure 4.1   Klaus Fuchs, 1960

dramatically cooled after the end of the war. Two fields – finance and atomic cooperation – suffered particularly from Washington's lack of interest and support. In the area of finance, the Truman Administration had ceased its Lend-Lease policy immediately after the end of hostilities in the Pacific theatre in August 1945, and thus left Britain without much needed American credits for its economy. What aggravated the situation further was the United States' reluctance to continue to treat Britain equally as a superpower – American government officials now rather regarded their European ally as a subordinate nation. With the passing of the MacMahon Act in 1946, which virtually cut the United Kingdom off from any American nuclear data, Britain found itself in a disadvantageous position in the early atomic age.[149]

In the United States, Klaus Fuchs's confession fuelled existing anti-communist hysteria and paranoia. Together with events which occurred in the late 1940s and early 1950s – in particular the Berlin Blockade, the 'loss' of China to Mao Zedong, the first Soviet atomic test and the Korean War – the Fuchs case added to the production of a sense of crisis. And it helped forge anti-communist consensus further in the face of the much dreaded global expansion of communism.[150] It also coincided with the rise of McCarthyism. Although it did not concern atomic espionage directly, the Alger Hiss case of 1948–9 was crucial in influencing public fears of communist fifth columnists in the United States prior to Fuchs's disclosure.[151] A week after Fuchs's confession, Senator Joseph McCarthy presented in his infamous speech in Wheeling, West Virginia, a list of over 200 alleged communists who, he claimed, had infiltrated the US State Department. This heated up Americans' anxieties of communist infiltration and expansion.

The House Committee on Un-American Activities (HUAC) under McCarthy was initially almost obsessed with investigating alleged acts of atomic espionage. HUAC even constructed the myth that communist supporters inside the Manhattan Project had given sensitive atomic-arms-related information to the Soviet Union because these cases guaranteed a lot of publicity. But when the news of Fuchs's confession broke, HUAC finally abandoned its investigation of suspected atomic spies because none of its 'witnesses' had been associated with Fuchs and, as Ellen Schrecker has observed, the committee's 'quest ... had produced many headlines, several volumes of testimony, and not a single spy'.[152]

Despite HUAC's shift of focus away from former Manhattan Project scientists, 'theoretical physicists emerged', according to David Kaiser, 'as the most consistently named whipping-boys of McCarthyism'.[153] In 1951, the chief of the FBI, J. Edgar Hoover, called Fuchs's espionage activities 'the crime of the century' in an article in *Reader's Digest*.[154] In a much more balanced way, the American *Bulletin of the Atomic Scientists* also devoted considerable attention to the Fuchs and other atomic espionage cases.[155] Public suspicion of theoretical physicists was to a large degree rooted in the openly international character and exchange of science which seemed in the eyes of many Americans to jeopardize the emerging national security state.[156]

The investigation and confession of Fuchs had eventually led to the apprehension of a number of other Soviet atom spies such as his courier Harry Gold alias 'Raymond' and David Greenglass, as well as Ethel and Julius Rosenberg. Greenglass was the younger brother of

Ethel Rosenberg, whose husband Julius had served as a courier for the Soviet Union.[157] After his arrest, David Greenglass testified in court against his brother-in-law in order to reduce his own sentence, accusing him of espionage for the Soviet Union. In spite of heavy international protests, the Rosenbergs, who represent perhaps both the most famous and tragic victims of McCarthyism, were subsequently executed in the electric chair in 1953.[158] The phobia of the Red Menace pervaded deep into American society and culture. Hollywood produced films such as *Conspirator* (1949), *The Red Menace* (1949) and *The Whip Hand* (1951), and popular singers like Carson Robinson and His Pleasant Valley Boys warned listeners in their song 'I'm No Communist' (1952) that 'Communists and spies are making monkeys out of us'.[159]

Given its strong repercussions in the United States, it comes as no surprise that Chief Justice Lord Goddard listed the negative impact that Fuchs's confession had on Anglo-American relations as among the 'gravest aspects' of Fuchs's crime in his concluding statement at the trial.[160] So grave was the effect upon the affairs between the two countries that the prime minister even shortly considered extraditing Fuchs after he had served his sentence in Britain. Under the Extradition Treaty with the United States at the time, however, this was impossible, and the idea was consequently abandoned.[161] The decline in post-war trans-Atlantic nuclear relations has to be seen as part of a larger trend of Anglophobia in the United States.[162] And, as early as 1943, there had been a good deal of distrust towards the British contributions to the Manhattan Project on the part of senior American officials such as Vannevar Bush, James B. Conant and General Leslie Groves.[163]

The fact that Fuchs was given a quick trial at the Old Bailey on 1 March 1950, just a few weeks after his confession, has to be regarded as an attempt by the British authorities, especially MI5, to limit the damage done by his confession to Anglo-American relations. 'The Fuchs trial did nothing to acquaint the public with the true nature of his offence', Rebecca West rightly observed, but 'it did something to disguise it'.[164] Officially, MI5 fully cooperated with the FBI when they forwarded, for example, a transcript of the court hearing to J. Edgar Hoover.[165] Behind the scenes, however, MI5 withheld crucial information from their American counterparts in order not to further harm the deteriorating Anglo-American relations. MI5 tried to play down the case not only because of their own security breaches and failures but also because Klaus Fuchs had gathered information on the US atomic arms programme during his time in Los Alamos.[166]

The Klaus Fuchs atomic espionage case was part of a series of events of deception by MI5 in order to limit damage to its public image. In the first instance, the Security Service led Fuchs to believe that he 'was given the chance of admitting it and staying at Harwell or of clearing out', as he stated in the dictation of his statement to William Skardon. Fuchs added: 'I was not sure enough of myself to stay at Harwell and therefore I denied the allegations and decided that I would have to leave Harwell.'[167] By early January 1950, it had already become official among members of the Security Service that Fuchs would have to leave Harwell for security reasons and, initially, it was the plan to find him a university position.[168] This suggests that MI5 deliberately misled Klaus Fuchs into the false belief that he could stay at Harwell after his confession in order to achieve his maximum cooperation and for its public image not to suffer a major blow.[169] Fuchs's defence lawyer also pointed to this fact during the court hearing.[170]

In retrospect, Fuchs described the time of his confession, trial and arrest as a 'dream' and a 'psychological state'. Initially, he had believed that he would be sentenced to death and had already finished with his life. It was only at his trial that his attorney informed him that the maximum penalty for his crime was 14 years. During the early period in custody, Fuchs suffered from the prison living conditions: he was he kept in solitary confinement, and some of the guards were also particularly tough on the 'traitor'.[171]

Not only did MI5 deceive Fuchs and lure him into his confession, while restricting the flow of information to the FBI, but it also deliberately misinformed the prime minister and the British public about the affair. As a major consequence of this campaign, Clement Attlee in a speech to the House of Commons publicly defended MI5's clearance of Klaus Fuchs for sensitive nuclear research, and helped restore the secret service's public image.[172] MI5 even drew on the publicist Alan Moorehead for their cause. Among other things, the Security Service provided Moorehead with information on Fuchs's mother, who had suffered from depression and, like her mother before and her daughter Elisabeth later, committed suicide. In this connection, the Security Service spurred speculation about 'Insanity in the family'.[173] This manoeuvre did not go unnoticed by the FBI, which regarded Moorehead's book *The Traitors* 'generally as an attempt to whitewash the previous Labour Government and the British Security Services in connection with their investigation of the spy cases involving Klaus Fuchs, Allan Nunn May and Bruno Pontecorvo'.[174]

In Britain, MI5's restored image was short-lived because later in 1950 the Italian-born émigré atomic scientist Bruno Pontecorvo, whom the Security Service had also cleared for secret work on both TA and the Manhattan Project, defected to the Soviet Union.[175] While affairs between Britain and the United States had suffered severely from the cases of Fuchs and Pontecorvo, it was the defection of Guy Burgess and Donald Maclean in 1951 that would affect relations between the two countries for years.[176] Although MI5 continued with its strategy of playing down these spy cases, it was rather unsuccessful and clashed with the press which had decided to exploit such cases to the fullest extent.[177] It therefore comes as no surprise that Fuchs featured in an internal instructional MI5 manual titled *Their Trade Is Treachery* (1964). It contained a picture of him, accompanied by the caption 'Another atomic scientist, Dr. Klaus Fuchs, also betrayed his country (14 years)', among the likes of Alan Nunn May, Guy Burgess and Donald MacLean to give an example of 'ostensibly ordinary harmless people' whose 'records of treachery stress that at no time can we relax in striving to keep our secrets secret'.[178] The British government, too, intended to overcome the embarrassment of the Fuchs case by trying to openly erase his name from the list of scientists working at Harwell, for example. The Ministry of Supply and Central Office of Information's joint booklet *Harwell* (1952), which was subsequently also published in the United States, mentioned neither Fuchs as former head of the Theoretical Physics Division nor any of his scientific papers in the respective list at its end.[179]

Conclusion

Fuchs's political radicalization in Germany had a strong impact on British nuclear culture because the revelation of his espionage emphasized the reciprocal intersections between nuclear science and political cultures in the United Kingdom, in particular with regard to national security. While it remains a difficult task to examine Fuchs's precise motivations for spying on behalf of the Soviets as well as for confessing his espionage activities in early 1950, his declaration of guilt led Britons to discuss the state of democracy in their country. It especially spurred distrust in the effectiveness of national security measures at home and abroad, especially in the United States, and had a negative impact on Anglo-American relations. So serious were the repercussions of Fuchs's confession that MI5 deliberately tried to manipulate public opinion to restore its image.

Subsequently, for many Britons, the name of Klaus Fuchs came to epitomize betrayal of their hospitality, and he was turned into a larger-than-life figure. Although the Soviet Union would have eventually developed both its own atomic and thermonuclear arms, many contemporaries believed that atomic spies such as Klaus Fuchs and Alan Nunn May had provided the Soviets almost with blueprints of these weapons. Instead, secret agents like Fuchs helped speed up the programme, probably by about one or two years.

The case of a former German POW, who had spent several years in captivity in the Soviet Union and who was now tried for espionage for the Soviets in a US court in West Berlin in 1952, illustrates that Klaus Fuchs had become a household name. *The Times* reported that the judge referred to the defendant as 'a pocket edition of the atomic spy Klaus Fuchs', but conceded that 'the evidence bore no resemblance to that in the Fuchs trial'.[180] Similarly, a 1957 biographical profile in the *New Scientist* introduced Michael Perrin, the former TA director, as 'The Man to Whom Fuchs Confessed'.[181] The Fuchs case also cast a long shadow over British foreign relations with countries other than the United States. When the British press criticized France over its atomic arms development project in 1959, in particular for the reliance on German-speaking émigré scientists, the *Paris-Presse* responded in like kind, writing: 'The British are ill-placed to accuse us of employing Germans, since their best atomic scientist was the German Klaus Fuchs'.[182]

5
Subject to Suspicion: Rudolf Peierls and the Klaus Fuchs Espionage Case

Klaus Fuchs's confession not only had a considerable impact on perceptions of national security and the efficiency of British security agencies, but it also strongly affected the community of German-speaking émigré scientists in Britain and the United States. Rudolf Peierls, who had recruited Fuchs for the TA project, and against whom allegations of being a Soviet spy have frequently been brought forward to the present day, particularly felt the repercussions of the Fuchs case. And it appeared that Fuchs's revelations represented a partial setback to his previously successful social and professional integration.

In a letter to Rudolf Peierls, Esther Simpson of the AAC/SPSL wrote that Fuchs's 'action has done a tremendous damage, but I have the feeling that British commonsense and kindliness and decency will prevent these effects from becoming exaggerated'.[1] While Britain did not experience anything remotely like Senator McCarthy's anti-communist witch-hunts, and the British public reacted by and large in a decent way, the Fuchs case affected many foreign-born scientists in Britain and the United States, including some of Klaus Fuchs's closest friends and sponsors, in particular Rudolf Peierls. Although Peierls generally shared Simpson's judgement and praised the reasonable response to the Fuchs case by the British public, he did become the target of suspicion and accusations.[2] Once again, it appeared that his German origin became a key factor in these allegations alongside Peierls's belief in values such as the freedom of science and free scientific exchange across the Iron Curtain that appeared to collide with the ideology of the emerging national security state in the Cold War.

Peierls's initial response to Fuchs's confession

Like many of Klaus Fuchs's colleagues and friends, Peierls initially refused to accept that Fuchs had spied for the Soviet Union. Immediately after the news of Fuchs's confession and the charges brought forward against him broke, Peierls requested permission to see his friend and former colleague in prison. In an interview with MI5 prior to his visit, Peierls still expressed disbelief about the allegations against him.[3] The subsequent meeting between the two men was 'slow and difficult, and both seemed a trifle embarrassed', as one Security Service report put it. Peierls, however, tried to convince Fuchs to fully cooperate with the authorities.[4]

Fuchs told Peierls during the meeting that he believed that 'knowledge of atomic research should not be the private property of any one country, but should be shared with the rest of the world for the benefit of mankind'.[5] After the meeting, Rudolf Peierls was deeply disappointed, even shocked by Fuchs's behaviour.[6] He described Fuchs's explanation of his actions and motivation as incoherent, naïve and foolish 'which [did] not suit him at all'. Peierls even declared that his 'concern [was] now not the fate of Fuchs, but the bigger issues involved'. As a consequence of the – from Peierls's point of view – deeply disturbing meeting, he reaffirmed his and his wife's full cooperation with the British authorities. As part of the couple's collaboration, Genia Peierls expressed her deep disappointment at Fuchs betraying his closest friends in a letter to Fuchs.[7] 'You were enjoying the best of the world you were trying to destroy', she wrote. 'It is not honest.'[8]

In his reply to Genia Peierls, Klaus Fuchs called the realization that he had betrayed his friends his 'greatest horror'. He complimented Peierls on her direct words: 'Funny that women see such things so much clearer than men. And that they are so much kinder by saying hard words straight out ... And don't worry if you don't see the tears. I have learned to cry again. And to love again'.[9] As an immediate consequence of Fuchs's confession, Rudolf Peierls also revoked his proposal to have Fuchs elected as a Fellow of the Royal Society.[10] In spite of the pair's disappointment about Fuchs's behaviour, Rudolf Peierls contacted Fuchs's attorney and offered assistance.[11] Peierls even acted most fairly when he sent Fuchs a letter shortly before the latter's release from prison to see if he 'need[ed] any help in getting started in life'.[12] Rudolf Peierls also received letters of solidarity from colleagues at home and abroad.[13]

Perhaps the most devastating effect that Fuchs's actions had on his colleagues was the spreading of a feeling of mutual distrust within the scientific community. As Norris Bradbury, a former Los Alamos colleague of Fuchs, put it: 'For the first time Fuchs raised the question among the scientists, "Who can you trust?" We felt as if we'd all been betrayed.'[14] Driven by a strong feeling of betrayal and disillusionment with Fuchs, Genia Peierls confronted him with the legitimate question in her aforementioned letter, asking: 'Do you realize what will be the effect of your trial on scientists here and in America?' She went on: 'Do you realize that they will be suspected not only by officials but by their own friends, because if you could, why not they?'[15]

'Perturbed Men'

While British responses to Fuchs's revelations were, in general, comparatively moderate, especially by comparison with the United States, his confession would still cast a shadow over the lives of Rudolf and Genia Peierls as well as those of other émigré scientists. In the immediate aftermath of Bruno Pontecorvo's defection to the Soviet Union, the *Sunday Express* ran the headline 'Perturbed Men: Foreign-Born Atom Experts Disturbed by Pontecorvo Case' on its front page on 29 October 1950. Alongside Peierls, the article featured pictures of the German-speaking émigrés Max Born, Otto Frisch, Nicholas Kurti and Franz Simon as well as the Polish-born scientist Joseph Rotblat. 'Many of them feel that the British Government should issue a statement testifying to their loyalty', Sidney Rodin and Joseph Garrity claimed in their article, falsely maintaining that some of the 'perturbed men' had handed their passports over to the British authorities to show their loyalty to their adopted host country.[16]

The *Sunday Express* article represented perhaps the most outspoken attack on German-speaking émigré scientists in the United Kingdom after the war. Max Born appropriately referred to the article and the accompanying pictures as a 'rogues' gallery'.[17] Immediately after the publication of the sensationalist headline, Rudolf Peierls complained strongly to the news editor of the *Sunday Express* and sent him a statement titled 'I Am Not Perturbed' which the editor promised to publish without any changes unless made by Peierls. In the letter, Peierls – on behalf of all 'perturbed men' – appealed to the spirit of fairness in British society.[18]

The *Sunday Express* printed Peierls's letter under the title 'To a Just, Fair, and Steady Britain' in its 12 November 1950 issue. 'Recently the Sunday Express printed my picture among those of a number of distinguished colleagues under the headline "Perturbed Men"', Peierls opened his statement. 'But I for one am not perturbed, and I think that goes for the others too'. He praised British society and values as 'the greatest tradition of justice and fairness and respect for human rights'.[19]

Peierls's statement received some highly positive feedback. Writing in the *Illustrated London News*, Arthur Bryant used Peierls's article as an opportunity to talk about Britain's democratic tradition of fairness. 'And since General Smuts, speaking of Britain in the "blitz" December of 1940, told his people in South Africa that he would stand by the country which, when he and his fellow-Boers were at their mercy, treated them as Christian people', wrote Bryant, 'I have read nothing about my country that has moved me so profoundly or that I have felt, with all her many and present shortcomings, to be more fundamentally true.'[20]

The 'Perturbed Men' episode was, however, not the only incident in which a tabloid reported falsely on Peierls, and he had indeed good reason to be distrustful of the British press. Immediately after Fuchs's confession, for example, he explicitly told the *Daily Mail* not to quote or refer to his opinions on this matter, which the editors ignored. As he subsequently wrote to the *Daily Mail* news editor, 'I must be grateful for the lesson you have taught me and I shall in future take care not to have anything to do with the cheap sensational section of the Press'.[21] In other instances with the *Daily Mail*, Peierls also took or threatened to take legal actions against defamations spread by the paper.[22]

Rudolf Peierls and many of the German-speaking émigré nuclear scientists, who had had a tremendous share in the manufacture of the first atomic bombs at Los Alamos now faced a paradoxical situation in which many Britons occasionally accused them of being dangers to national security. And, in a way, it appeared the Fuchs case represented a big setback to their previously successful social and professional integration into their host country's society and physics community. Their origin, it seemed, became once again the reason for their discrimination. 'I believe it is fair to say that if from the atomic energy teams in England and in America one had excluded all foreign-born scientists as well as those who in their youth had held extreme political views of one kind or other', Peierls later reflected, 'the leakage of atomic secrets would have been prevented by the fact that there would have been no atomic secrets.'[23]

Rudolf Peierls, MI5 and the FBI

In the United States, too, Peierls become the target of suspicion and criticism in the aftermath of Fuchs's confession. The case of Frederick Schlinck, a 'concerned' member of the American public, is a good example of this distrust of foreigners which was often coupled with anti-communist hysteria in the United States. Schlinck tried to actively support the anti-communist witch-hunts by alerting J. Edgar Hoover to Peierls's alleged communist ties in a letter that even found its way to the British authorities. Deeply convinced that Pelican Books, a subdivision of Penguin Books and publisher of Peierls's co-edited collection of essays *Atomic Energy*, was a communist enterprise because their books were 'sold at Communist book shops ... [and] either written by left-wingers or play[ed] a left-wing line in the text', he was highly suspicious of Peierls. 'A refugee scientist, if he were a careful and responsible person', Schlinck concluded, 'would tend perhaps more than a native Englishman to avoid the appearances of evil association'.[24] J. Edgar Hoover thanked Schlinck in a personal letter. Although the FBI director publicly denied an investigation of Peierls by Schlink's agency, he assured him that his 'courtesy of forwarding this information is indeed appreciated'.[25]

Apart from his German origin, MI5 also played up Peierls's Jewishness, linking it to communism. In 1951, for example, a Security Service report argued that Peierls used a Marxist dialectical material approach to science which he regarded as the 'panacea of mankind'. 'Another strong influence on his outlook is his Jewish origin', the report stated, 'which gives rise to his strong opposition to German re-armament'. It then concluded: 'His love of peace under all circumstances is so strong that, called upon to decide between war and Communism, he would undoubtedly choose a Communist peace as the lesser evil'.[26]

In a similar fashion, MI5 justified in part suspicions against Erna Skinner, the wife of Fuchs's long-time personal friend and colleague Herbert Skinner. Erna Skinner, whom MI5 suspected of having an affair with Klaus Fuchs and to whom Fuchs had apparently disclosed his espionage activities a week prior to his confession to William Skardon, had come to the Security Service's attention early on during the Fuchs investigation.[27] Apart from her Austrian origin, MI5 critically observed that all of her friends except for Fuchs were Jewish.[28] An MI5 report also harshly judged her as a 'woman whose unfaithfulness to her husband is a matter of common knowledge in Harwell', adding that 'She is a Jewess of Central European origin'.[29] These anti-Semitic comments in reports by the Secret Service were remainders from the Second World War era which continuously became part of the anti-communist rhetoric.[30]

Although Fuchs repeatedly stressed that none of his friends and colleagues – including Peierls, Hans Bethe, Martin Deutsch, Richard Feynman, J. Robert Oppenheimer, George Placzek, Edward Teller and Victor Weisskopf – was either communist or knew of or were involved in his espionage activities, they faced – often unknown to them – serious investigations by British and American security services.[31] To a certain degree Fuchs himself was responsible for this purge because he had indicated at several points during his interrogations by MI5 and FBI officers that his contacts during and after the war sometimes confronted him with such detailed questions and mentioned processes unknown to him. When his Soviet handlers had asked him, for example, precise questions on specific items such as the tritium bomb or electro-magnetic processes of isotope separation, he suspected that they had other informers apart from him.[32]

In the United States, FBI agents interviewed the former scientific director of the Los Alamos laboratory, J. Robert Oppenheimer. 'Fuchs impressed him at that time as a man who was carrying the "woes of the world" on his shoulders', Oppenheimer testified, 'and thought of him as a "Christian democrat" and religious man but not as a "political fanatic", or member of the Communist Party'.[33] In the aftermath of Klaus Fuchs's confession, the FBI also interviewed many of his and Rudolf Peierls's former Los Alamos colleagues.[34] Apart from Oppenheimer, Hans Bethe, Enrico Fermi, Philip Morrison and Robert Serber also came under suspicion and faced detailed investigations by the FBI.[35] In the United Kingdom, MI5 also checked former colleagues of Fuchs such as Egon Bretscher, for example.[36] 'When I heard the news I felt it to be incredible', wrote Fuchs's former mentor at Bristol University, Nevill Mott, describing his feelings and experiences after Fuchs's confession, 'but it was only too true. In fact the intelligence people descended on me and grilled me on why I had signed his naturalization certificate – not at all a pleasant experience'.[37]

The FBI did not limit its investigations solely to associates of Klaus Fuchs, but simultaneously investigated a number of Rudolf Peierls's colleagues and friends in the United States who were often the same as Fuchs', including Edward Corson, Martin Deutsch, Edward Teller and Victor Weisskopf; all of whom spoke out in favour of Peierls. Given the heated political climate of McCarthyism in the United States at the time, it appears ironic that the FBI confused Teller – who was one of the most outspoken anti-communists – with a certain Edward Teller who taught the history of the Soviet Communist Party, political economy as well as Marxism-Leninism at the Communist Workers School in New York City in 1941.[38] And, what is more, the FBI even investigated Rudolf Peierls's family and relatives in the United States.[39]

While the Fuchs case had certainly renewed MI5's interest in Rudolf Peierls, there had been earlier investigations by them into his case.

As early as 1938, Peierls had received some attention from the British authorities owing to the fact that he had many academic contacts with the Soviet Union.[40] In late 1946, MI5 launched a major investigation of Peierls which was based on three issues: his professional and private relationship with Klaus Fuchs, his wife's Russian origin, and the fact that he had previously visited the Soviet Union.[41] Because of his work in sensitive areas and his Russian-born wife, MI5 'closely investigated' Rudolf Peierls through a number of 'covert methods' in 1947 and then again and 'more intensely' in 1949 and 1950 against the background of the Fuchs case.[42]

These 'methods' included a Home Office warrant for Peierls's communications.[43] In 1949, MI5 even ordered a telephone tap for Peierls.[44] This measure was then put in place again immediately after Klaus Fuchs's arrest.[45] While it appears legitimate that, under the Home Office warrant, security staff voiced concerns about details such as the location of uranium ore deposits in publications authored by Peierls,[46] it seems rather odd that they even opened and checked letters sent from Peierls's children to their parents.[47]

MI5 also launched a major investigation into Peierls's alleged communist affiliations.[48] But Peierls was not the only foreign-born scientist on whom MI5 kept a close eye. In 1946, Lew Kowarski, who had worked on both Tube Alloys and the Manhattan Project during the war and who had returned to France after the war where he was working on the country's nuclear energy project, also aroused MI5's suspicion, for example. 'I do not put his name forward as a particular suspect', wrote John H. Marriott, a senior MI5 official, 'but he does seem to me to be a very good example of the sort of foreigner who has been closely associated with the project, and who is now working under a communist master'.[49]

As a result of MI5 investigating Rudolf and Genia Peierls, many of their close friends and associates – such as Alexander Baykov and his second wife Inna Arian-Baykov, Roy and Fania Pascal, Herbert and Erna Skinner, Max Born and even Paul Dirac – came under scrutiny by the Security Service.[50] These measures, however, failed to produce any positive evidence against Peierls.[51] As a result, MI5 closed its file on Rudolf Peierls concluding that 'In our view there is no substantial doubt about the loyalty of Professor Peierls'.[52] Given Peierls's key role in the British atomic energy project, Michael Perrin was cited in an MI5 report speaking out in favour of Peierls who had 'built the ground floor' of the British nuclear energy programme. 'Therefore, we had quite a lot to gain and little to lose in retaining his services', Perrin summed up the government's rationale for employing Peierls.[53]

American security agencies, by contrast, were highly sceptical of Peierls's innocence and suspected him of having had at least partial knowledge of Fuchs's activities and therefore viewed him as a 'bad security risk'.[54] Once hard evidence was produced against Klaus Fuchs, the FBI started to investigate Fuchs's and Peierls's movements during their time in the United States.[55] While these investigations of Peierls have to be seen against the background of the general political climate of McCarthyism at the time, the American public was also particularly suspicious of scientists' increasing political activism and engagement as well as the internationalism of the academic world.[56] These scientific ideals clashed with the emerging national security state and the anti-communist rhetoric of the day.[57] And, in particular in the United States, the government viewed the internationalist agenda of many nuclear scientists as a chief unreliability.[58]

Freedom of science and scientific internationalism

The fact that Rudolf Peierls, like many of his colleagues, believed in the freedom of science and was a strong proponent of scientific internationalism further contributed to him becoming a target of scepticism and criticism. And his leading role in the ASA, which advocated these ideals, even alerted MI5 and the FBI. The ASA was the chief organization of the British atomic scientists' movement and will be dealt with in depth in Chapters 6–7. As early as 1946, when the British Atomic Energy Act was passed, Peierls and the ASA of which he was then an executive vice-president heavily debated the secrecy clause of the new piece of legislation.[59] In June 1946, Peierls had become deeply involved in the matter when he and Harrie Massey met with the New Commonwealth Parliamentary Committee to discuss the Atomic Energy Bill. Peierls criticized the bill's 'very nebulous' wording on the issue of secrecy, which required improvement so that the secrecy provisions would not hinder basic research.[60]

In October 1946, Rudolf Peierls and Philip Moon sent a letter to the editor of *The Times* on behalf of the ASA. Speaking for 'a majority of the scientists who have been or are connected with the atomic energy project', Moon and Peierls criticized two points in the new act: the 'Lack of provision for expert advice' and, in particular, the bill's clause dealing with secrecy. 'The clause, as it stands, will tend to prevent free discussion between collaborators in fields of research bordering on the subject of atomic energy, which include a great deal of physics, chemistry, and engineering', the two ASA officials argued. 'We believe that this obstacle to scientific progress is too high a price to pay for the sake of preventing a fraction of the future discoveries from being made public', they added, 'particularly when we remember that scientists in the past have almost without exception first informed their Government of any new development of apparent military significance'.[61] Or, as James Chadwick later put it, 'When you lock the doors of a laboratory, you lock out more than you lock in.'[62]

The Association of Scientific Workers (AScW), too, regarded the clause as 'highly controversial' and shared the ASA's basic concerns.[63] This was a rare congruity between the two organizations, as the AScW normally made much more politicized statements. While the AScW, for instance, had issued an information leaflet on the Alan Nunn May atomic espionage case in which it condemned 'the harsh sentence of 10 years penal servitude' and 'demand[ed] a drastic reduction of this sentence', the ASA refrained from releasing such a political statement.[64]

Peierls's overt criticism of Western governments, especially the United States but also the United Kingdom, and their security measures which, in his view, posed a threat to the free exchange of scientific ideas brought him also to the attention of the FBI. It was in particular a summary of two talks given by Rudolf Peierls and Sir Henry Dale at an ASA conference in London in October 1948 and published in the April 1949 issue of the *Bulletin of Atomic Scientists* that aroused much suspicion on the part of the American security agency.[65] The ASA had already kept a close eye on security investigations of their American colleagues, and reported on the latest developments in the June and October 1948 issues of the *Atomic Scientists' News*.[66]

Two points made by Peierls received peculiar attention from the FBI: first, his severe criticism of the United States government and their practice of science-related security; and, second, as an FBI report phrased it, Peierls's and Dale's expressed 'desirability of scientific intercourse with Iron Curtain countries'.[67] These comments by the FBI are exemplary of views held by the United States government at the time that classified those scientists careless of obeying security rules as security risks.[68] An anonymous MI5 informant also remarked that 'Peierls frequently behaves like a silly ass in matters of security and appears to go out of his way to advertise the fact that he considers security to be nonsense'.[69]

Despite all the possible repercussions of his engagement for the freedom of science, Peierls did not abandon his views but frequently spoke out in their favour.[70] In 1950, he was, for example, involved in drafting a critical statement by the ASA Council on the recent secret purge in the British civil service.[71] 'Purges of this kind have often been understood as morality plays, with an understandable emphasis on the victimization of the innocent by the repressive state', as Richard Beyler, Alexei Kojevnikov and Jessica Wang argue.[72] As early as April 1949, the ASA had established contact with the National Council for Civil Liberties. As F. C. Champion, the ASA Hon. General Secretary, phrased it in a letter to Elizabeth Allen, the National Council for Civil Liberties General Secretary: 'we should be glad if you could keep us informed of any evidence of or material on discrimination against scientists on political grounds'.[73]

In September 1951, Peierls stated in an essay: 'If it is true that military strength is a vital factor in preventing war, then we must include in it the moral strength that depends on the beliefs in our principles.' He added: 'Free discussion, which strengthens the basis of this belief, is therefore an important military asset'.[74] In 1956, he even accepted the invitation to attend a conference in Moscow because he believed that a rekindling of scientific exchange between Soviet and Western scientists was a crucial opportunity in the post-Stalinist era.[75] While Peierls's struggle for East–West scientific cooperation made him appear suspicious to many, he was always highly critical of the Soviet Union. 'I have many good personal reasons for hating their system like poison', he wrote in a letter to William Penney, 'just as I had good personal reasons to hate the Nazis in Germany'.[76] From 1948, the official recognition of biologist-agronomist Trofim Lysenko's theories of hybridization as state science by Joseph Stalin in the Soviet Union in 1948 triggered a debate over science and politics in Britain and elsewhere. For the intervention of the Soviet state in the freedom of science appeared to confirm fears of Soviet totalitarianism, as Lysenkoism rejected the widely accepted ideas of Mendelian genetics.[77]

Peierls's liberal views were reminiscent of those propagated by two Austrian-born émigrés, the philosopher Karl Popper and the economist-philosopher Friedrich August Hayek. Popper wrote his seminal work *The Open Society and Its Enemies* in defence of liberal democracy, and directed it against any totalitarian form of government, including the Soviet Union.[78] In a similar fashion, the Austrian-born Hayek opposed socialist collectivist thought in a defence of liberalism in his 1944 book *The Road to Serfdom*. 'The guiding principle, that a policy of freedom for the individual is the only truly progressive policy', Hayek concluded, 'remains as true to-day as it was in the nineteenth century'.[79]

Articles in the journals *Nature* and *Economist* criticized especially the ASA's promotion of the idea of free scientific exchange across the Iron Curtain in connection with their proposals on the international control of nuclear energy that represented one of the organization's three main objectives, as will be discussed in the following chapter in depth. 'It is difficult to see', the *Economist* concluded, 'how anything could come of such "collaboration" except a one-way traffic to the disadvantage of the Western democracies'.[80] In a letter to the editor of the *Economist*, Rudolf Peierls in his function as the current president of the ASA defended the statement. He was particularly keen on defusing any impression that the ASA aimed at giving away secrets. '[W]e are under an obligation to guard these secrets and, whether we like it or not', he wrote, 'we are well accustomed to observe the rules in talking to those of our colleagues here, or in western Europe, or in America, to whom we have not been instructed to reveal our secrets'.[81]

As a consequence of his appraisal of the ideal of the freedom of science, Peierls faced serious problems in obtaining a US visa on two occasions in 1951: first, the American authorities refused to give him a visitor's visa to attend the International Conference on Nuclear Physics in Chicago. Peierls suspected his membership of the ASA to be the main reason for this.[82] Second, he even had severe difficulties in applying for a visa to go to the United States on official British government business to attend a conference on declassification in Washington, DC. Still in the shadow of Klaus Fuchs's confession, American security agencies brought forward numerous accusations against Peierls and cited, among other things, the latter's German origin, Genia Peierls's Russian origin, his former time in the Soviet Union and his close friendship with Klaus Fuchs as major points of objection. Most of the allegations against Peierls (and also his wife), such as his active engagement in the ASA, which he himself had identified as one of the reasons behind the Americans' behaviour and which American security services regarded as 'a Communist Front organisation', were 'pure McCarthyism', as a British official rightly labelled them. But, at the same time, the American objections to Peierls's visit to the United States indicated a deep rift in Anglo-American relations. Since the American authorities cited the McCarran Internal Security Act (a piece of internal legislation that imposed severe restrictions on the freedom of communists in the United States), they met with fierce opposition from the British government and risked a complete British withdrawal from the declassification conference.[83] In the end, the authorities granted Peierls a visa to attend this meeting.[84] He used this visa then also to go to the Nuclear Physics Conference in Chicago.[85] Peierls experienced further difficulties in obtaining a visa for a visit to the Institute of Advanced Study in Princeton, which he received only at the very last minute.[86]

The cases in which US authorities had initially denied visas to Rudolf Peierls or where they had considerably delayed the issuing process were no singular occurrences; rather they have to be seen within the context of US government policies during and after the time of McCarthyism.[87] The visa problem was apparently so aggravating for the scientific community that the *Bulletin of the Atomic Scientists* dedicated a special issue in October 1952 to 'American Visa Policy and Foreign Scientists'.[88] Prominent, foreign-born American scientists like Hans Bethe spoke out against these policies, predicting that they would 'be increasingly detrimental to the development of science in the United States'.[89] Shortly after the war, the French nuclear scientists and known communists Frédéric and Irène Joliot-Curie – the latter of whom was the daughter of Marie Curie, a pioneer in the discovery of radioactivity – were also denied visas to travel to the United States. Irène Joliot-Curie was even arrested when she travelled to the United States in 1948.[90] John Krige has argued that the denial of visas formed part of the United States government's attempts to construct American hegemony after the Second World War.[91] The United States' visa policy continued to impair science, and, in March 1955, the Federation of American Scientists (FAS) called upon readers of the *Atomic Scientists' Journal* to share their experiences so that the American organization could assess the impact of the US immigration legislation at the time on science.[92]

Apart from foreign-born scientists experiencing difficulties in obtaining US visas, the State Department also refused to issue or confiscated passports of American scientists such as Linus Pauling who allegedly held left-wing or communist views.[93] The most prominent victim of the State Department's obscure passport policy was perhaps David Bohm. As a result of this dubious practice, he left the United States for Brazil where he took up a position at the University of São Paulo and obtained Brazilian citizenship before he finally moved to the United Kingdom via Israel.[94] In Britain, the Foreign Office (FO) also intervened when Eric Burhop, who was a senior official in the ASA, was invited to visit Moscow in July 1951.[95]

Rudolf Peierls and the ASA also took sides in what was perhaps the most prominent example of how this repressive political climate affected science – the case of J. Robert Oppenheimer. While Oppenheimer had not been central to HUAC's investigations, the FBI had launched an inquiry into his alleged communist affiliations even before the Second World War.[96] To date it remains a contentious issue whether Oppenheimer joined the Communist Party or not.[97] Once the news of Klaus Fuchs's confession broke, Lewis Strauss – who served on the Atomic

Energy Commission (AEC) – started his own investigation of J. Robert Oppenheimer which finally peaked in the infamous Oppenheimer security hearings of 1954 when the AEC stripped Oppenheimer of his security clearance and thus ended his role as AEC advisor.[98] The *Atomic Scientists' Journal* dedicated considerable attention to the Oppenheimer case, including a long review article of the published transcripts and documents of the hearings by Rudolf Peierls. 'But if this case were to shake, even slightly, the enthusiasm of the educated people of a country for its way of life,' he concluded on the possible impact of the Oppenheimer security hearings, 'if it were to sow any doubt as to the survival of freedom, fairness and reason and the importance of bringing sacrifices for these principles, then the loss of moral strength which it would cause would be the equivalent of many superbombs or ships or planes'.[99] Like many of his colleagues, Rudolf Peierls was deeply disturbed about the attacks on Oppenheimer. In a personal letter to Oppenheimer, Peierls even commented on the charges brought forward against the former director of the Los Alamos laboratory because of his critical attitude towards the development of the H-bomb. 'In a Communist or Nazi state one expects as a matter of routine,' Peierls wrote, 'the holders of minority views to be regarded, and treated, as traitors, but if this happens elsewhere it makes us despondent about the prospects of survival of free society'.[100]

Throughout his life, Peierls kept on warning against the perils of an infringement of the freedom of science by security measures, as the Oppenheimer case demonstrated all too dramatically. In his autobiography Peierls later argued in favour of a critical engagement with anti-democratic ideas. 'The problem of ensuring the loyalty of people entrusted with sensitive, confidential information', he wrote, 'is sometimes confused with a desire to keep "undesirables" from entering the country or any particular profession'. This revealed in, Peierls's eyes, 'a failure to appreciate that the strength of a stable democracy rests on its citizens' understanding of the basic issues and their ability to reject simplistic extreme ideologies' because 'Marxist or fascist ideas are not like an infectious disease that can be contracted by exposure to it'. By contrast, he argued, 'familiarity with them helps to give a firmer basis to one's own convictions, and makes one's arguments against such ideas stronger'.[101]

Conclusion

The Fuchs case cast a long shadow over Rudolf and Genia Peierls's lives. In 1966, for example, the Italian public television channel Radiotelevisione

Italiana (RAI) interviewed him for a programme on Fuchs.[102] But in 1984, Peierls declined an offer to participate in a documentary on his former colleague.[103] Besides publicity on the Fuchs case, Rudolf Peierls's German and his wife's Russian origin, along with his firm belief in the free exchange of scientific ideas, also made them the target of allegations of having spied for the Soviets. And Rudolf Peierls's insistence on the freedom of science marked a site of post-war British nuclear culture where basic scientific principles clashed with the emerging national security state.

The journalist Richard Deacon created perhaps the most obscure incident concerning allegations against the Peierls when he intended to publish his book *The British Connection: Russia's Manipulation of British Individuals and Institutions* in 1979. Wrongly assuming that Rudolf Peierls was dead, the author unleashed a barrage of accusations against him. When Peierls received news about the book's impending publication, he took immediate legal action and prevented its release in the original form. The London-based publishing house Hamish Hamilton finally published Deacon's book without any accusations against Rudolf and Genia Peierls.[104]

Even after their deaths, Rudolf and Genia Peierls remained the target of defamatory allegations of spying for the Soviet Union. When the transcripts of a good number of the Venona messages intercepted in the Second World War were finally declassified and made publicly available in 1995, new accusations against Rudolf and Genia Peierls surfaced. Their names were wrongly related to those of two unidentified Soviet spies, codenamed 'Tina' and 'Fogel/Pers'.[105] In an article in the *Daily Express* in 1999, Nigel West suggested that Peierls had been one of the Soviet spies at Los Alamos. Misspelling the codename 'Fogel' as 'Vogel', which means bird in Peierls's native tongue, West brought forward a salvo of unfounded accusations. 'Tellingly,' he fuelled speculation, 'Sir Rudolf's autobiography was called Bird Of Passage, although the author gave no reason for his choice of title'.[106] Another prominent posthumous victim of such libel was Niels Bohr.[107]

While Rudolf and Genia Peierls as well as other German-speaking scientists suffered at times from the consequences of the Klaus Fuchs case and allegations of espionage, recent archival findings have also proved that MI5's suspicions were not completely unfounded. As has recently emerged, the Viennese Engelbert Broda, who worked with Hans von Halban, Lew Kowarski, Jules Guéron and Egon Bretscher on plutonium at the Cavendish, supplied secret information to the Soviet Union.[108] Besides the public repercussions the Fuchs case had in Britain and its impact on Anglo-American relations, it particularly affected the lives of other German-speaking émigré nuclear scientists, especially Rudolf Peierls, because of their origins. But Rudolf Peierls's German extraction also informed his views on the role he foresaw nuclear scientists taking after Hiroshima.

6
The Responsible Scientist: Rudolf Peierls and the Formation of the Atomic Scientists' Association

Like Klaus Fuchs, Rudolf Peierls ceased to work on nuclear weapons after the war. But unlike Fuchs, who had passed on crucial nuclear information to the Soviet Union to prevent, or better shorten, the period during which the United States held a monopoly on atomic arms, Peierls dealt differently with the social responsibility that he felt was emerging out of his previous nuclear weapons research. After his return to the United Kingdom, Peierls became a chief engine behind the British atomic scientists' movement, in particular its main organization the Atomic Scientists' Association (ASA).

While the invention and use of poison gas during the First World War had affected warfare significantly and confronted scientists such as Fritz Haber with ethical questions, the creation of the atomic bomb moved the question of science and morality to a higher and much more complex level. This development went hand in hand with an increasing turn from low to high technology. Although inventions like the Higgins Boat or Liberty Ships, penicillin, the proximity fuse and especially radar played decisive roles in the Allied war effort during the Second World War, it was the advent of nuclear weapons that not only changed the face of warfare fundamentally but also raised existential questions about human survival in general. As J. Robert Oppenheimer put it shortly after the war: 'The obvious consequence of this intimate participation of scientists is a quite new sense of responsibility and concern for what they have done and for what may come of it'.[1]

The impact of nuclear weapons, however, went beyond their creators. In the face of the recent experience of two world wars, severe economic problems and the uncertainties that atomic energy brought about, many contemporaries believed that national governments were unable to handle these new challenges. And the post-war era thus saw the emergence of internationalism as a means to solve the big issues of the day, or 'the birth of development', as Amy Staples has coined it. This internationalist agenda led to the Bretton Woods agreements that culminated in the foundation of the World Bank and the International Monetary Fund as well as the formation of further supranational organizations, especially the Food and Agriculture Organization and the World Health Organization.[2]

While a similar internationalist outlook guided the emergence of the atomic scientists' movements in Britain and the United States, the atomic scientists' motivations differed from those engaged in other organizations. Since Peierls and the majority of the nuclear scientists engaged in these movements had worked on the British and Allied nuclear weapons projects during the war, they were also driven by a very strong personal motivation to try and make sense of their wartime work and cope with the moral responsibilities that transpired from the creation of nuclear arms and energy. If Peierls's direct and indirect experience with the National Socialist regime had motivated him particularly strongly to take a leading role in nuclear arms research, his wartime work now translated into a firm social responsibility that ultimately translated into his leading position in the formation of the ASA and marked another chief facet of British nuclear culture.

American origins of the British atomic scientists' movement

The ASA's beginnings lay in the United States. As Peierls and many future ASA members worked during the Second World War as part of the British Mission at Los Alamos, they witnessed at first hand the formation of the American nuclear scientists' movement which would provide a direct model for its British counterpart. The Trinity Test of 16 July 1945 confronted Rudolf Peierls and his Los Alamos colleagues morally and ethically for the first time with the results of their work. About 40 years after the event, Peierls still recollected 'the feeling of awe at the terrible power of this weapon mixed with elation at the success of the work'.[3] Many of his colleagues shared these ambivalent feelings and were often forced, as Jon Hunner has pointed out, 'to culturally code switch' in order to comprehend the magnitude of the new weapon.[4] Klaus Fuchs recalled that while he enjoyed the sight of the explosion, he found himself confronted with unsettling questions about the future of the atom bomb, in particular its control through the military, and felt that it was the scientists' duty to prevent its use in war.[5]

Although the Trinity Test visualized in a most dramatic way the destructive force of the new weapon, Peierls and many of his colleagues faced another moral dilemma. Since they had originally set out to win the race over the atom bomb in favour of the Allies, the unconditional German surrender in early May 1945 deprived them of their justification for working on the project. Unknown to the vast majority of Los Alamos scientists at the time, the so-called ALSOS teams that the US government had sent into Germany in search of nuclear scientists and research installations had furnished conclusive evidence that Hitler had not even come close to having a working nuclear weapon at his disposal.[6]

General Leslie Groves had no interest in informing Peierls and his colleagues about these findings because this information might undermine the scientists' morale and thus jeopardize the continuation of work on the atomic arms programme.[7] As a consequence, most Los Alamos scientists were not aware of British and American intelligence reports that revealed that Germany did not possess a working nuclear device and voiced, at first, virtually no moral or political concerns about the continuation of their work on nuclear weapons.[8] At this stage, the Polish émigré physicist Joseph Rotblat was the only scientist to leave Los Alamos as a result of the German unconditional surrender.[9]

By suppressing debates on moral and political issues among the scientists through enforcing a strict regime of scheduling and time management, J. Robert Oppenheimer also contributed significantly to the project's progress.[10] And this led to what Steven Shapin has called 'the moral equivalence of the scientist'.[11] Oppenheimer's success in suppressing moral and political discussions in order to motivate scientists to resume work on the atomic bomb was largely based on his 'charismatic authority', as Charles Thorpe and Steven Shapin have argued.[12] Peierls had high confidence in Oppenheimer, in particular as a mediator between the scientists and both the military leadership and political decision-makers.[13] Through the effective use of his 'charismatic authority', Oppenheimer held back the circulation of a document that Leo Szilard had drafted on behalf of a group of scientists at the University of Chicago's Metallurgical Laboratory, the so-called Met Lab. In his petition to President Truman, Szilard urged the US president not to use the atomic bomb against Japanese cities.[14] As a result of Oppenheimer's successful repression of any dissent, Rudolf Peierls, too, was at that time not aware of the activities of his Chicago colleagues.[15]

Szilard's petition came about one month after scientists at the Met Lab around the German-born émigré James Franck had drafted an important report. The so-called Franck Report advised against the use of nuclear weapons against Japan and in favour of a demonstration of the new weapon before a United Nations (UN) delegation prior to its deployment. They passed the document on to the Secretary of War, Henry Stimson, through Arthur Holly Compton, who was the head of the Met Lab. While the Franck Report had no direct effect on the Truman Administration's decision-making process with regard to the use of the atom bomb, it would provide an important impetus for the newly formed scientists' movement after the war.[16] Despite these appeals by Chicago scientists, President Truman authorized the use of two atomic bombs, one against Hiroshima on 6 August and a second one against Nagasaki on 9 August 1945.[17] Britain's Prime Minister Winston Churchill fully backed this decision.[18]

The news of the atomic bombing of Hiroshima, which has over-shadowed the discourse over nuclear weapons to the present day, elicited mixed emotions in Rudolf Peierls and many of the Los Alamos scientists. Peierls later wrote that 'with the feeling of elation there was horror at the death and suffering that must have resulted, though we had no details yet'. His uneasiness about the consequences of the nuclear attack increased after the second atomic strike against Nagasaki just three days later.[19] Klaus Fuchs described an atmosphere of tempo-rary joy about the successful outcome of the Manhattan Project, but added that this soon gave way to serious moral and ethical quandaries.[20] Still, Rudolf Peierls, like Klaus Fuchs and many other Manhattan Project scientists, justified his participation in the creation of the atom bomb, and refrained from offering any retrospective counterfactual arguments to excuse his wartime work.[21]

When the news of Hiroshima broke and introduced the atom bomb to the world public, the atomic scientists found themselves all of a sud-den in the limelight. They became the centre of considerable public attention, and newspapers on both sides of the Atlantic ran countless headlines and articles about the development of the new weapon.[22] In Peierls's case, an awareness of the scientist's responsibility was a direct result of his work on the atomic bomb during the Second World War.[23] He believed that 'public opinion should be influenced in the right direction' once the existence of the atomic bomb had become public knowledge in the immediate aftermath of Hiroshima, 'because you can't have a reason-ably intelligent public discussion without some understanding of simple technical facts; and it was up to the scientist to explain them'.[24]

During the remainder of his stay at Los Alamos, Peierls witnessed the formation of the atomic scientists' movement in the United States, which would become a crucial prerequisite for his own role in its British counterpart. While Peierls and many of his colleagues engaged in work on the development of the atomic bomb, only a few scientists took action to stop these developments. As early as September 1942, Leo Szilard, who had previously drafted together with Albert Einstein the famous letter to FDR to call for the creation of such a weapon, now voiced serious concerns about the impact that the bomb would have on the post-war world. The Danish émigré physicist Niels Bohr had similar fears and warned of a post-war nuclear arms race. He even approached FDR and Churchill to promote the idea of international control of atomic weapons and energy. The British prime minister vehemently opposed Bohr's ideas and even wanted to have the Danish émigré arrested at one point.[25]

As a result of a growing awareness of their moral responsibility, Los Alamos scientists such as Robert Christy, Richard Feynman, Edward Teller and Victor Weisskopf started to organize themselves as the Association of Los Alamos Scientists (ALAS).[26] The group intended, as it declared in its statement of purpose, 'to promote the attainment and use of scientific and technical advances in the best interests of humanity'.[27] It formed at a time when the Los Alamos laboratory, including the British Mission members Klaus Fuchs and Egon Bretscher, started preparations for Operation Crossroads, the first post-war series of nuclear tests at the Bikini Atoll.[28] Similar organizations emerged at other Manhattan Project installations: at Oak Ridge, Tennessee, the Oak Ridge Association of Scientists and Engineers formed shortly after the war.[29] Here, Lothar Nordheim engaged in the organization's efforts.[30] In Chicago, the Committee on Social and Political Implications, which scientists had established under the chairmanship of James Franck at the Met Lab shortly after Hiroshima and Nagasaki, evolved into the Atomic Scientists of Chicago, Inc. in early September 1945.[31]

In an attempt to disseminate information about the nature of the new weapon to the wider public, the ALAS drafted and distributed a statement that was also published in the *New York Times*. The document emphasized six main points: first, the firm belief that other nations – apart from Britain, Canada and the United States – were capable of developing nuclear arms; second, the United States' high vulnerability to atomic attack, especially on its industrial centres; third, the unlikelihood and difficulty of initiating effective countermeasures; fourth, the new weapon gave the aggressor an advantage as a nuclear attack could destroy any means for retaliation; fifth, the threat that the atomic bomb represented to human civilization as such; and, finally, the call for international control of atomic energy.[32]

Members of the British Mission showed a particularly strong interest in these ideas. Owing to their immigration status as alien subjects, however, they refrained from signing the ALAS memorandum that was addressed to the US government and press. Instead, several British scientists, including Klaus Fuchs, drafted a letter to Sir James Chadwick to which they attached a copy of the ALAS document. They assured Chadwick that they supported the points expressed by their American colleagues, and indicated their interest in issuing a similar statement to the British government and media.[33] Consequently, the ALAS statement marked the starting point of a concerted effort by British scientists to express their thoughts on nuclear weapons and energy univocally.

Despite concerns over their status as 'guests' in the United States, Peierls and several members of the British Mission drafted a 'Memorandum from British Scientists at the Los Alamos Laboratory, New Mexico' in the autumn of 1945. Its signatories included – apart from Peierls – Klaus Fuchs, Egon Bretscher, Anthony French, James Hughes, William Marley, Donald Marshall, Philip Moon, William Penney, Tony Skyrme, Ernest Titterton and James Tuck. In the document, the undersigned atomic scientists revealed great concern about 'the implications of this completely new weapon'. They went on to explain the purpose of their statement: 'We feel it our duty to bring our knowledge and ideas on this subject to the attention of those responsible for British policy'. The memorandum repeated, by and large, the principal points made in the ALAS statement, especially the new quality of the atom bomb's destructive force, the insufficiency of an appropriate defence against nuclear weaponry, and the need for the international control of military and civilian applications of atomic energy, as well as the necessity of the 'Free movement of scientific personnel and information'.[34]

The British scientists at Los Alamos aimed at giving more weight to their statement when they followed the example of their American colleagues and used fear as a tool in conveying their message. 'A single bomb of the present type can completely cripple the life and resources of a city the size of Bristol or Coventry', they stated, adding: 'There is no specific defence against atomic bombs'. The members of the British Mission further stressed the defencelessness against nuclear arms in their memorandum, arguing: 'The prospects of preventing their delivery, or intercepting a large fraction of them, seem extremely remote, particularly since they could be delivered in a variety of ways, for example by rockets of the V. 2 type'.[35]

On 1 November 1945, American scientists made another crucial move with far-reaching consequences for the British atomic scientists' movement when the groups that had organized at the Manhattan Project installations at Chicago, Los Alamos, Oak Ridge and New York merged in the Federation of Atomic Scientists. Shortly after its inauguration, they renamed the organization the FAS. The newly formed organization arranged the National Committee on Atomic Information (NCAI) to provide public education of atomic matters. The launch of the *Bulletin of the Atomic Scientists of Chicago* by Eugene Rabinowitch and Hyman Goldsmith in December of the same year further strengthened the young federation. Officially not an FAS publication, the journal, which was a short time later renamed the *Bulletin of the Atomic Scientists*, became the scientists' movement's chief organ at home and abroad. As a fundraising organization, the FAS set up the Emergency Committee of Atomic Scientists (ECAS) in May 1946. Chaired by Albert Einstein, it included the German-speaking émigrés Hans Bethe, Leo Szilard and Victor Weisskopf.[36]

Initially, the FAS followed an ideal of political 'objectivity', refraining from any 'expression of political opportunism'.[37] It had four main objectives: the realization that the atomic bomb was 'a revolutionary weapon'; the view that there existed no proper defence against the new weapon; the firm belief that scientists in other countries were already working on other national atomic arms research programmes; and the necessity – that arose from the third point – to implement a system of international control of nuclear energy.[38]

The formation of the Atomic Scientists' Association

Rudolf Peierls – like many members of the British Mission to Los Alamos – brought these influences and ideas back with him to the United Kingdom. For Peierls, the scientist's new moral responsibility translated into his strong personal engagement with the ASA. The Association was the British counterpart of the FAS and had similar goals to its American sister organization.[39] The ASA emerged out of the Atomic Scientists' Committee (ASC), which a group of scientists including Peierls had founded under the auspices of the AScW in 1946.[40] ASC members included – apart from Peierls – Patrick Blackett, John D. Bernal, Harrie Massey, Nevill Mott, Marcus Oliphant, Eric Burhop, John Moore, Alan Nunn May and Joseph Rotblat.[41] The ASC listed three chief objectives in its mission statement that were highly reminiscent of those held by its successor, the ASA: educating both the public and fellow experts in atomic-energy-related matters; advising political decision-makers about nuclear power, and promoting its international control of nuclear power; and collaborating with sister organizations abroad.[42]

Shortly after the ACS's creation, several of its members had already called for its complete independence from the AscW, and put forward first ideas for 'forming an all-embracing Federation of British Scientists'.[43] TA scientists at the Clarendon Laboratory in Oxford and at the University of Liverpool were very sympathetic to these proposals.[44] And on 28 February 1946, the ASC decided to disband the committee and launch the ASA instead.[45] Initially, the new organization was called the Association of Atomic Scientists, but shortly afterwards was renamed the ASA.[46]

The decision to form an independent organization of British atomic scientists lay, to a large extent, rooted in the AScW's political minded-ness. After all, the AScW was a trade union that commented on political developments on a regular basis. Unlike many of his colleagues who became politically active and did not refrain from making political statements in public, Rudolf Peierls followed a different path and advo-cated the ideology of 'unpolitical' science. In February 1946, he thus informed Sir James Chadwick about 'a somewhat complicated situation with the Committee of Atomic Scientists'. Peierls expressed serious con-cern about the ASC's dependence on the left-wing AScW because, in his eyes, 'such a connection would only antagonize certain people'. What Peierls disliked in particular about the umbrella organization was that the AScW intended 'to act as a trade union for scientists, and to express the general view of scientists as unbiased experts'. In his opinion, the AScW combined two areas that 'do not mix and should be carried out by separate bodies'. As a consequence, Peierls suggested the formation of an independent committee. But he was aware of possible problems that such an independent council in the style of the FAS could face. For its formation required the acceptance of members 'from all branches of science outside of the project' which generated 'much controversy' in the current discussions of the group.[47] Here, Peierls referred to the fact that the ASC struggled with the broad political spectrum of its members, which often affected the committee's decision-making process.[48]

Chadwick shared Peierls's concerns and criticized political involve-ment by the FAS (Figure 6.1).[49] Other scientists had similar concerns. Joseph Rotblat, for example, was critical of the close association between the ASC and the AScW.[50] Peierls, however, not only distrusted political activism from the left but harboured a general suspicion towards scientists' involvement in politics. He completely disagreed, for example, with some of Lord Cherwell's ideas and, at a conference in Chicago in 1951, these opposing views led to Peierls falling out with Cherwell.[51] Peierls tremendously influenced the policies of the ASA in

Figure 6.1   From left to right; Rudolf Peierls, Sir James Chadwick (centre) and Geoffrey Ingram Taylor, Cambridge, date unknown

this realm. Although the ASA cooperated in some instances with the AScW, the organization remained critical of the union. In a similar fashion, the ASA Council instantly rejected a proposal by the World Federation of Scientific Workers to organize a joint 'Conference on the Social Implications of Atomic Energy' in November 1954.[52]

While the ASA followed in many ways the example of the FAS, Peierls envisaged some essential differences to the American organization, especially with regard to political statements, 'by insisting from the outset that a proper division should be made between statements of scientific facts and opinions held by scientists'. Peierls stressed that 'on the latter type of question the association should not express any definite views or advocate any definite policies'. Instead, he believed that the ASA should only make public statements if these were agreed upon by all Council members.[53] Peierls's faith in the Association's capability of reaching unanimous decisions through a grass-roots democratic decision-making process indicates a good deal of idealism on his part. In retrospect, he admitted, 'There was also a friction because some of the vice-presidents objected to statements being made without their approval'.[54] When the London Staff Correspondent of the International News Service, for example, sent Peierls a questionnaire to learn about the ASA's views on nuclear energy, the Council discussed the answers.[55] In another case, the ASA Council even had to approve an article that introduced the organization to the readers of the *Bulletin of the Atomic Scientists*, delaying its publication.[56]

On 8 March 1946, Peierls, Blackett, Burhop, Gwyn Owain Jones, Massey, Oliphant, Rotblat, Simon and eight other scientists met in London to form the new association.[57] At the beginning, only a Provisional Committee of the ASA existed whose members included – apart from Peierls – Rotblat, S. Devons and Gwyn Owain Jones. Harrie Massey chaired the committee, and W. J. Arrol and Eric Burhop were its joint honourable secretaries. The ASA's Provisional Committee's purpose was to form the ASA as a full organization.[58] The Oxford branch soon started drafting a constitution for the new association.[59] The ASA listed three primary objectives in its mission statement: to educate Britons about nuclear energy, its promises and perils; to promote the international control of atomic power; and to advise the government on nuclear issues.[60] By May 1946, the ASA already had over 60 members and 95 by July 1946.[61]

Starting in April 1946, the Provisional Committee arranged for the election of the ASA Council to take place on 1 June 1946, and scheduled its first general meeting for 15 June at the University of Birmingham.[62] At the meeting, delegates elected Rudolf Peierls along with Massey, Oliphant, Blackett, Rotblat, Burhop, Kurti, Gerry Pickavance, Arrol, Mott, Skinner, William Marley, K. E. Chackett, S. Devons, Gwyn Owain Jones and Nikolai Kemmer as members of the ASA Council.[63] Peierls then served besides Maurice Pryce as one of the ASA's two executive vice-presidents.[64] At the General Meeting in Oxford in late June 1948, Peierls was elected president of the ASA.[65] He was re-elected twice, and stayed in office until October 1950 when he became one of the organization's several vice-presidents.[66] So significant was Peierls's role as a leading functionary in the Association that it prompted the Central Committee of the Soviet Communist Party to gather and compile information on him.[67]

The ASA was finally incorporated 'as a company limited by guarantee' on 15 April 1947.[68] A national Council – which ASA members directly elected – governed the Association.[69] The Council consisted of a president and a secretary and several vice presidents.[70] While the ASA restricted full membership to atomic scientists so that it comprised experts in nuclear matters, all members of the public could become associate members.[71] At its peak, the ASA had about 140 full members.[72] By 1950, the ASA had around 500 associate memberships.[73] Early on, the ASA actively approached 'useful people' who were attached to universities, unions, churches or were in other important positions in society to become associate members.[74]

From the start, the Association attempted to establish contact with other organizations, including in Britain (despite significant programmatic reservations) the AScW, the Social Relations of Science Committee of the British Association, the New Commonwealth Committee, the National Women Citizens Association and the Council of the Churches' Atomic Energy Committee. Internationally, the ASA forged links with the Atomic Scientists of Chicago, the FAS and the NCAI, as well as with the Australian Association of Scientific Workers and the Dutch International Council of Scientific Unions.[75]

In July 1946, the ASA held a conference in Oxford as its first major event. The meeting focused on three key issues: proposals for an international organization of nuclear scientists; the international control of atomic energy; and uses of nuclear energy.[76] Otto Frisch, William Marley, Gwyn Owain Jones, Nicholas Kurti, Philip Moon, Joseph Rotblat and Franz Simon served on the sub-committee in charge of organizing the conference.[77] Foreign visitors included George Placzek and Lise Meitner (both USA), Hans Halban (France), Homi Bhaba (India) and other scientists from the United States, France, the Netherlands, Sweden, Switzerland and Norway. British guests included Roy Innes of the AScW, Sir James Chadwick, Sir John Cockcroft, Lord Brabazon, Sir Henry Dale, Sir Henry Tizard and Kurt Mendelssohn. Rudolf Peierls chaired the first session on 'National Legislation and International Control of Atomic Energy'.[78]

Public outreach and education

Public outreach and education represented a key concern of the newly formed Association. The ASA followed the examples of the FAS and the ALAS whose activities many ASA members had directly experienced during their time at Los Alamos. The ALAS had published the weekly *Los Alamos Newsletter* and Robert Serber's eyewitness account about the situation in Hiroshima and Nagasaki shortly after the atomic bombings.[79] While ASA members participated in public discussions on the political implications of nuclear power, Peierls insisted upon the ASA refraining from 'offering and advocating a patent remedy'. By contrast, he underlined the organization's commitment to 'provide the non-expert reader with a basis on which to think out his attitude to the problem' of nuclear energy.[80] With its focus on 'the layman' as the target audience, the ASA followed a general understanding in the 1940s and 1950s that imagined the public on the whole as 'informed laymen'.[81] As informed citizens, they believed, these 'laymen' would consequently form their own opinions about the subject.

In April 1947, Peierls published an important article in the *Endeavour* magazine that explained the necessity for public nuclear education because 'In public discussion of the prospects of atomic energy, the threat to the future of civilization which the atomic bomb represents tends to overshadow the promise of benefits from the constructive applications of atomic power'.[82] Peierls's motivation to write his article has to be seen against the ambivalent public opinion towards atomic power at the time. As in the United States, the news of Hiroshima generated an ambivalent response in Britain.[83] By 1946, many Britons felt that the negative consequences of nuclear energy would outweigh its benefits in the long run. The radio broadcast of John Hersey's book *Hiroshima* (1946) served as a catalyst to reinforce these pessimistic feelings, and a majority of British people believed that only strong global organizations could cope with the crisis produced by nuclear weaponry.[84] In a Gallup poll of May 1946, 46 per cent of the respondents agreed that nuclear power would do 'more harm than good' in the long run, while only 28 per cent saw a more positive outcome of atomic energy in the future and 26 per cent were undecided. And these figures did not change considerably over the next two years.[85]

On the eve of the first US post-war atomic tests in the Bikini Atoll in 1946, Whitehall's publication of *The Effects of the Atomic Bombs at Hiroshima and Nagasaki*, a report by a British team that had visited the cities of Hiroshima and Nagasaki in November 1945, informed the British public in graphic detail about the nuclear bombings of the two cities.[86] Ambivalent feelings towards nuclear energy even pervaded the highest circles of government, and arguably delayed Clement Attlee's decision to pursue an independent British nuclear weapons project until 1947.[87]

As a consequence, Peierls set out in his article to present a more balanced picture of nuclear energy. He dedicated only a short section to the atom bomb, which he described as 'the only important practical application' of nuclear energy so far and whose 'comparative cheapness' he identified as its 'essential new feature'. The 'comparative cheapness' of atomic weapons, especially in the long run in comparison to large conventional militaries, represented indeed one of the major reasons why the United Kingdom took the decision to pursue its own nuclear weapons project before the Marshall Plan and the North Atlantic Treaty Organization (NATO) revealed a stronger American dedication to Europe.[88]

With his *Endeavour* article, Peierls also set out against a nuclear euphoria and nuclear fantasies that were sweeping the country in order to give a more 'realistic' picture of the peaceful uses of nuclear power. Originating in the United States, much speculation went on about the benefits of nuclear power, including atomic planes, trains, cars and ships, and medical science also appeared to be a major benefactor of the new energy source.[89] Among the first and most influential books to deal with these atomic hopes was *Atomic Energy in the Coming Era* by the popular-science writer David Dietz.[90] It was first published in the United Kingdom in 1946 under the title *Atomic Energy Now and Tomorrow*.[91] Peierls, by contrast, was realistic about the impact of atomic energy on the electricity price, and delivered a blow to such predictions as atomic-powered cars.[92] The same year as his *Endeavour* article appeared, Peierls co-edited with John Enogat the second volume of the Penguin Books *Science News*.[93]

It was in this idealistic spirit that Peierls and the ASA stepped up as the major forum in the United Kingdom to educate members of the general public about the benefits and perils of atomic power in what they claimed was a politically 'objective' way. With their nuclear education campaign, the ASA offered a third way, an alternative to public information campaigns launched by the churches, the British Council of Churches and the Anglican Church, and Whitehall. Along with pacifists, members of the clergy had been the first and foremost critics of nuclear weapons after the news of Hiroshima had reached Britain.[94] Unlike American churchmen, the British clergy acted almost unanimously.[95] While the ASA spoke with the voice of scientific experts who had been involved in the creation of the new weapons, the British Council of Churches intended to assume the role of a guardian of ethics and morality. In 1946, it published the report *The Era of Atomic Power*.[96] The report was the result of the work of a commission, the first ecumenical one of its kind, which had been appointed by the British Council of Churches.[97] But *The Era of Atomic Power* failed to provide answers. As a result, the Church lost credibility as a chief guardian of morality.[98] In the immediate aftermath of Hiroshima and Nagasaki, the ASA thus had a decisive programmatic advantage over the British Council of Churches because it never intended to offer any 'patent remedy', as Peierls had put it.[99]

Alongside the British Council of Churches, the British government also tried to educate its citizens about the new era of nuclear power. Apart from *The Effects of the Atomic Bombs at Hiroshima and Nagasaki*, Whitehall published its version of the story of the making of the atom bomb shortly after the war. The 40-page booklet *Statements Relating to the Atomic Bomb*, however, was unable to compete with its American counterpart – Henry DeWolf Smyth's *Atomic Energy for Military Purposes*.[100]

Following the example of the American scientists' movement, the ASA started to publish a monthly newsletter. The ASA's Birmingham branch launched it in August 1946.[101] Since the Birmingham group was soon overwhelmed by the ever-increasing numbers of the publication, the ASA collaborated with the New Commonwealth Society in producing the newsletter.[102] By June 1947 the ASA newsletter had a circulation of about 600 copies per issue.[103] As a result of these increasing numbers, the ASA decided to publish a journal in July 1947 as a major form of outreach.[104] Compared to the *Bulletin of the Atomic Scientists*, the *Atomic Scientists' News* was, as Rudolf Peierls rightly observed, 'In size, scope, and circulation ... very modest'.[105] By November 1950, it had a print run of 1,200 per issue.[106] Like its American counterpart, which occasionally featured articles from the *Atomic Scientists' News*,[107] the ASA's journal quite frequently included essays that had previously appeared in the *Bulletin of the Atomic Scientists*.[108] In September 1951, the ASA started with a new series of the *Atomic Scientists' News* which was published bi-monthly by Taylor & Francis.[109] In an attempt to reach a wider readership, the ASA renamed the *Atomic Scientists' News. New Series* the *Atomic Scientists' Journal* in September 1953.[110] With its July 1956 issue, the *Atomic Scientists' Journal* folded because, as it stated in its final editorial, 'the Association's aims were not now being fully met by a publication which, by its very nature, could never reach a wide public'.[111] Initially, the ASA considered plans to relaunch the *Atomic Scientists' Journal* under the name the *Atomic Age* with the publisher Pergamon, which had close connections with other institutions, organizations and businesses in the nuclear-energy sector.[112] But in the end, these plans failed to materialize largely due to an incident involving the publisher advertising the

Atomic Age in *The Times'* 'Scientific Supplement' without permission from the ASA and inaccurately naming Rotblat as its editor.[113]

Although the Association was now left without its own published organ, the ASA cooperated with the editors of the newly launched weekly magazine the *New Scientist*. The Association prepared a monthly 'Atomic Science' and a news review section. Rudolf Peierls was among the contributors to the first 'Atomic Science' section in the *New Scientist*'s 3 January 1957 issue.[114] In terms of outreach to 'interested laymen', the *New Scientist* seemed to provide a promising medium because the new periodical was, as its editors declared in its first issue, 'published for all those men and women who are interested in scientific discovery and in its industrial, commercial and social consequences'.[115]

Besides publications, the ASA planned and realized its most spectacular outreach event – the travelling Atom Train exhibition, which it launched in early November 1947 at Liverpool's Central Station.[116] With 146,000 visitors and travelling the United Kingdom for 168 days, the Atom Train was a huge success for the ASA. The accompanying exhibition catalogue, which *Nature* termed 'the most popular and widespread publication on atomic energy in Britain', sold 46,000 copies and supplied an ample financial stock for the young organization.[117] The Atom Train later even toured Scandinavia, visited Paris and Cairo, and participated in a conference organized by the UN Educational, Scientific and Cultural Organization in the Lebanese capital of Beirut.[118]

Rudolf Peierls played a chief role in organizing the Atom Train. He served on a special committee which the ASA's Council had appointed for this purpose. Besides Peierls, it included F. C. Champion, William Marley, Philip Moon as well as John Curry, Michael Moore and Joseph Rotblat from the University of Liverpool's Physics Department where the exhibition was put together.[119] While the ASA initially faced some difficulty in mobilizing sponsors and partners for the travelling exhibition, Peierls suggested early on that the ASA actively seek support from, and closely work with, British government departments in planning and carrying out the exhibition.[120] The Association thus turned to Michael Perrin, whom many ASA members knew through their wartime work for TA and who was now the deputy controller of the Ministry of Supply's Department of Atomic Energy, for support.[121]

This led to collaboration between the Association and government offices, especially the Ministry of Supply's Department of Atomic Energy and the AERE Harwell, in bringing the exhibition home to people all across the United Kingdom. Above all, the Treasury loaned the Association a total of £6,000 against possible losses and to prepare the exhibition.[122] The Department of Atomic Energy and the AERE provided most instrumentation and radioactive sources for the Atom Train, while the Ministry of Supply's Directorate of Information provided support in organizational matters. In addition, the DSIR, the Ministry of Works, the British Museum and the Science Museum, the AEC and the United States Navy Department as well as private corporations such as British Thomson-Houston Co. Ltd., Gaumont British Equipments Ltd. and Gaumont British Instructional, Metropolitan-Vickers Electrical Co. Ltd. and the Press Association supported the exhibition.[123] Harrie Massey drafted the exhibition script that the ASA then sent to the Ministry of Supply for approval.[124] And Peierls subsequently prepared the text in the exhibition guide.[125] Thanks to the support by government offices, the Atom Train was comparatively well equipped at a time of severe shortages and rationing of goods shortly after the war.[126]

In most locations where the Atom Train called, Atomic Energy Weeks were held simultaneously during its stopover. These were often organized by the local UN Association chapter. At these events, films such as the Gaumont-British Instructional's animated short film *Atomic Physics* (1947) enjoyed great popularity.[127] Moreover, these Atomic Energy Weeks featured meetings (an ASA press release advertised them as events) 'to discuss what the ordinary citizen can do to ensure that atomic power is used for our benefit and not for our destruction'.[128] Nearby universities usually provided students as tour guides, and Rudolf Peierls delivered public lectures among the likes of Sir James Chadwick, Marcus Oliphant and Sir John Cockcroft.[129]

The Atom Train consisted of two railway carriages that each housed one of the exhibition's two sections. Part I considered the 'Fundamental Facts' of nuclear energy, whereas Part II concerned its 'Practical Applications'.[130] Following the ideal of political impartiality and a similar educational purpose as outlined in detail by Rudolf Peierls in his 1947 *Endeavour* article, the organizers informed visitors in the introduction to the accompanying exhibition catalogue that: 'The Atom Train Exhibition has been brought to your town to help you understand the facts about atomic energy. Everyone knows that this new power can be used for destruction; much less is known as yet of its possibilities for good'. They stressed: 'One can get a balanced view only by understanding a little of what is behind it', and thus underlined the ASA's role as an objective educator.[131]

Yet several panels in the exhibition were clearly intended to provoke a strong audience response. Although the exhibition focused by and large on civilian uses of atomic energy, the organizers framed it within a chiaroscuro of two opposing choices – nuclear utopia or nuclear obliteration – and thus were by no means free of political bias. The first panel that visitors faced constituted one part of this discursive frame. 'Atomic Energy for Good or Evil' showed a human hand and a skeleton hand which both pointed their index fingers from opposite sides at an idealized atom in the centre.[132]

When visitors had reached about half way through the second carriage, they encountered the section on the atom bomb that proved to be equally problematic with regard to the ASA's mission of 'objective' public education. 'The Atomic Bomb' established a link with the opening panel 'Atomic Energy for Good or Evil'. The section provided a technical overview of the manufacture of nuclear weapons and included photographs of atomic explosions in the Trinity Test, Hiroshima, Nagasaki and the Baker Test in the Bikini Atoll to demonstrate the force of the atom bomb.[133] And, what is more, 'The Atomic Bomb' showed a map of Central London, detailing the effects of a Hiroshima-type nuclear attack on the city.[134] To bring the peril of the effects of nuclear war further home, the exhibition also contained a map of the town where the train was currently calling that specified the effects of such an attack on the local area.[135]

At the end of their tour of the Atom Train, visitors viewed a window titled 'The Choice' that constructed a frame with 'Atomic Energy for Good or Evil' and presented them with what the organizers believed to be the two choices that lay ahead of them – a bright atomic future or nuclear annihilation. In order to visualize this dichotomy, a screen displayed two continually changing pictures. The first image depicted children peacefully playing, whereas the second one showed a crying baby among the rubble of Hiroshima.[136] Underneath, two texts further explained these images to get the message across to the public. 'There is no defence against the atomic bomb', a text under the title 'Destruction' emphasized, warning that 'In five to ten years, any industrial nation could discover these secrets and make the bomb'. In the second text, 'Construction', the ASA promoted the international control of atomic energy as a solution 'to allow the full, constructive advance of atomic energy research'.[137]

With the presentation of these facts to sharpen public awareness of the perils of nuclear warfare, the ASA abandoned its 'objective' approach and made a clearly political statement. The Association borrowed this technique in particular from the American scientists' movement. Early on, members of the British Mission to Los Alamos, including Peierls and several other future ASA members, had used fear in their 'Memorandum from British Scientists at Los Alamos, New Mexico'.[138] Perhaps this deviation from the ASA's concept of 'objective' nuclear education ensured that the Atom Train was such a big success.

The international control of atomic energy

Besides creating a heightened public awareness of nuclear energy and educating Britons about it, the ideal of scientific 'impartiality' in political matters also governed the ASA's second main objective – the international control of nuclear energy. Shortly after the war, Rudolf Peierls and many of his colleagues started to advocate this idea 'to help in the solution of this most pressing problem', as the ASA declared in its mission statement.[139] The organization devoted considerable time and energy to discussing this issue.[140] An article published in *Nature* in June 1946 reveals how significant many contemporaries deemed the idea of international control. 'Until the individual nations are prepared to renounce national sovereignty to that limited extent', the article concluded in a bleak tone on the current state of affairs, 'atomic energy will continue to represent the great menace of our age, and its potentialities for good will remain an unsubstantial shadow'.[141]

During the final stages of the Manhattan Project, British atomic scientists had given first thoughts to this concept. In their 'Memorandum from British Scientists at the Los Alamos Laboratory, New Mexico', Rudolf Peierls and other members of the British team had already spoken out in favour of international control in the autumn of 1945.[142] In a memorandum to Sir John Anderson, which built on the Los Alamos missive, Peierls had made further recommendations, especially on the imposition of a system of inspection as a means of achieving such a control scheme, as early as November 1945.[143]

The issuing of a statement on international control was one of the ASA Provisional Committee's first chief tasks. Its members intended to submit a communiqué to the United Nations Atomic Energy Commission (UNAEC).[144] Rudolf Peierls, Philip Moon and Marcus Oliphant drafted its first version before the Provisional Committee discussed and amended it.[145] The large array of members from all areas of the political spectrum, and the release of the Acheson-Lilienthal Report, complicated its composition process considerably. Despite all varying views, 'it was felt very strongly that the time had come when British atomic scientists should take some action along the lines similar to those taken by their American colleagues to try to ensure that enlightened politics are followed in connection with the control of atomic energy', the ASA Joint Honourable Secretary put it, 'so that this outstanding scientific achievement will not prove an unparalleled disaster to the human race'.[146]

Shortly after its formation as an independent organization, the ASA adopted an official line on the international control of atomic energy that closely followed the American Lilienthal-Acheson Plan and later the Baruch Plan as it was presented before the UN.[147] Originally, US Secretary of War Henry Stimson and Undersecretary of State Dean Acheson had proposed a plan under which all future nuclear arms were to be placed under international control. In early 1946, Acheson was appointed as head of a committee in charge of drafting a plan on the international control of atomic energy to be presented before the AEC by Secretary of State James Byrnes. This led to the so-called Acheson-Lilienthal Report which promoted international control across a broad range of areas from atom bombs to uranium ore. Subsequently, President Harry Truman appointed Bernard Baruch to present the proposals to the UNAEC's first session on 14 June 1946. Baruch amended the plan substantially so that it allowed the United States to continue its nuclear arms research and granted the Truman Administration the right to veto the plan if it had not been properly implemented. For the Soviet Union, Andrei Gromyko delivered the official reply to Baruch's proposals, insisting that all nations, including the United States, should ratify a moratorium on the development and use of nuclear arms before any accord on the issue of the international control of atomic power could be reached. Unsurprisingly, the Soviet Union vetoed the 'Baruch Plan' in the UN Security Council.[148] While 'the Acheson report contains all the proposals for the control of atomic energy that our committee considered essential', the ASA memorandum to the UNAEC stated, 'it still seemed worthwhile for us to issue a statement embodying the findings of our committee and to indicate the type of reasoning which had led us to make our recommendations'.[149] Here, the ASA statement sounded like a justification of its right to existence rather than a statement by experts offering advice.

Alerted by an article in *The Times* on the ASA memorandum, the FO gave some attention to the ASA statement. In a letter to Sir John Anderson's personal assistant, Denis Rickett (later Sir), J. G. Ward of the FO rejected in particular the ASA's view of the UN as 'a genuine international super-national body with the will and capacity to act as a sort of super state and to override the Governments of sovereign States'. Ward criticized this outlook as 'the common weakness when scientists launch out as prophets in the field of international politics'. What surprised

Ward in particular was the fact that Rudolf Peierls, in his position on the division of nuclear energy research into safe and dangerous areas, had flip-flopped from opposing it to backing the Acheson-Lilienthal Report that advocated in particular this separation.[150] Rickett shared Ward's critique.[151] The fact that Franklin A. Lindsay, the Executive Officer of the US Representative to the UNAEC, thanked Philip Moon on behalf of Bernard Baruch for the ASA statement and expressed his gratitude 'that on nearly all the basic points your Association is in agreement with the United States policy' is indicative of how far the ASA in fact moved into the political arena – despite its official, 'unpolitical' line.[152]

Prime Minister Clement Attlee took an ambivalent stance: while his Government officially backed American plans during 1946–47, Whitehall was, at the same time, deeply concerned about the implications these proposals would have on British nuclear research. Although the international control of atomic energy had initially seemed to provide solutions to many problems created by the atomic bomb, growing distrust of the Soviet Union finally led Attlee to decide that Britain should pursue its own nuclear weapons and energy research project.[153] In hindsight, Rudolf Peierls viewed the feasibility of the Baruch Plan very sceptically, admitting that 'There never was a chance of getting such a plan accepted by the Soviet Union'.[154]

At the time, however, Rudolf Peierls and the ASA, by contrast, were still enthusiastic about the idea of instituting a system of international control. But not all ASA members agreed with the ASA's line on the matter. As a result, Rudolf Peierls called for an ASA Council meeting to discuss and settle the issue. At the meeting, the Council members agreed that Rudolf Peierls should draft a statement and circulate it to ASA Council members for feedback.[155] Peierls drafted the memorandum with the help of Blackett, Burhop, Devons, Jones, Kemmer and using proposals made by the Cambridge branch. Following its democratic decision-making process, various drafts of the document were circulated among members of the Council before its publication.[156] The Liverpool branch and Herbert Skinner as well as Patrick Blackett objected to the draft statement so that it had to undergo modification, which delayed its publication considerably.[157] In the end, 12 of 14 Council Members (Patrick Blackett and Eric Burhop objected to it) agreed with the ASA statement on international control, and it was finally published.[158] On 20 January 1947, the ASA Council ultimately released it to the press, and the *Bulletin of the Atomic Scientists* reprinted it.[159] The statement was widely received. Even the Soviet media covered it, and Andrei Gromyko made references to it in his speech to the UN Security Council. As a result, the ASA Council held a special meeting to discuss the latest developments in the matter of international control of atomic energy.[160]

In spite of the failure to reach an agreement on the terms under which an international control scheme could be implemented in the UN Security Council, the ASA was still optimistic about the idea in its statement in February 1947, declaring that 'efforts must be continued to find a workable scheme acceptable to all countries'. Although it officially still followed its 'impartial' approach to politics, the ASA openly argued, as the 'most important objection to the Baruch Plan from the point of view of other nations', that 'technically Congress could withhold ratification at any rate'. Following similar proposals to those made by Rudolf Peierls to Sir John Anderson in November 1945, the ASA recommended inspection as a key feature to achieve a working system of international control.[161]

In its conclusion, the statement acknowledged the fact that 'an effective system of control acceptable to all concerned is a very doubtful proposition in the present state of distrust between nations, since it must contain, at least in embryonic form, a measure of world government'.[162] The ALAS had already promoted the idea of a One World government in charge of all nuclear matters – military and civilian.[163] The engine behind the promotion of this principle remained in the United States where many scientists, including Albert Einstein, saw the concept of a world government as an adequate answer to the vast problems caused by atomic weaponry in the immediate post-war period.[164] The idea of a world government was also popular with the British public: in a Gallup poll conducted in September 1946, 50 per cent of the respondents stated that the United Kingdom should follow suit if other leading countries placed their conventional and atomic forces under the command of a world parliament.[165]

While the ASA intended to stay away from any partial judgments, the FAS even contributed a statement to Dexter Master's and Katharine Way's edited collection of essays *One World or None*, a key text in galvanizing nuclear fears in order to promote the world government idea.[166] Where the FAS and Albert Einstein publicly sided with the One World movement, Peierls sympathized with the more abstract and less polemic proposals by Niels Bohr, who had put forward a 'Memorandum on the Open World' shortly after the war.[167] Despite his claims to act 'unpolitically', Peierls once again acted ambivalently when his name was among the signatories of a statement published in *The Times* in March 1947 calling on the prime minister and the foreign secretary to advocate the international control of atomic energy and weapons.[168]

The same year, Peierls shared his thoughts on international control with the readers of the *Bulletin of the Atomic Scientists* on the occasion of the second anniversary of the bombing of Hiroshima. He argued that British scientists showed a similar degree of awareness to American ones about the necessity of achieving a regime of international control for nuclear energy. He pointed to Britain's role as an intermediary between the superpowers, accentuating that 'this function, however, can hardly be assisted by public action and propaganda – it is essentially a question of diplomacy'.[169]

Although the chances of reaching significant progress in the area of international control of atomic power became increasingly slimmer, the ASA Council decided to set up a study group concerned with the issue in July 1947. Alongside Rudolf Peierls, the group comprised Nevill Mott, Patrick Blackett, Marcus Oliphant and Herbert Skinner.[170] It continued to investigate and report on the issue as well as draft and disseminate 'non-controversial' articles on the subject.[171] In July 1948, the ASA Council issued a further statement.[172] Recipients included the British and Soviet representatives to the UNAEC, Sir Alexander Cadogan and Andrei Gromyko respectively, as well as the AScW, the ministers of defence and supply, the foreign secretary, Prime Minster Clement Attlee, the UN secretary-general and the editor of *Nature*.[173]

The statement acknowledged the fact that the ASA's second objective had basically failed by the summer of 1948 because of growing tensions between the Soviet Union and the United States.[174] That the ASA was lobbying for an increasingly unpopular cause was, for example, revealed in a Gallup poll of October 1948 where only 3 per cent of the respondents specified the control of atomic arms as 'the most urgent international problem at the present time' after British–Soviet relations (25 per cent), preserving peace (24 per cent), the Berlin question (22 per cent), miscellaneous issues (14 per cent) and undecided interviewees (12 per cent).[175]

In spite of these figures, the ASA continued to pay considerable attention to this issue. With the coming of thermonuclear weapons, the ASA and the FAS held a joint meeting in Oxford on 14 and 15 September 1950 where international control was at the top of the agenda.[176] Besides Rudolf Peierls, participants in the two-day meeting included from the ASA J. I. Michiels, Kathleen Lonsdale, Nevill Mott and Herbert Skinner, and from the FAS its chairman William Higinbotham, A. Roberts, Samuel Allison, D. L. Hill, Carson Mark, George Placzek and M. Shapiro as well as Lew Kowarski of the French Atomic Energy Commission.[177]

In light of the H-bomb, the ASA adapted its line on this issue to the changed realties which had occurred over time. Since international control could not be achieved and an ever-increasing number of countries worldwide sought to acquire atomic power for either military or civilian purposes or both, nuclear proliferation had replaced the implementation of a global control scheme as top issue on the agenda. At an ASA Council meeting in February 1957, Joseph Rotblat stressed the continued danger that the increasing proliferation of nuclear technology in reactors posed to the issue of international control, with many smaller countries potentially in a position to manufacture their own atomic arms.[178] As a consequence, Rudolf Peierls, together with L. F. Bates, Norman Feather, Gwyn Owain Jones, Kathleen Lonsdale, Philip Moon, Maurice Pryce, Joseph Rotblat and George Thomson, called attention to this new danger in a letter to the editor of the *The Times* in November 1957.[179]

Advising the government

While both public nuclear education and proposals for international control had often conflicted with Peierls's and the ASA's ideal of political 'objectivity', the association's third objective – 'To help to shape the policy of this country in all matters relating to Atomic Energy' – proved even more problematic.[180] For this area inevitably caused Peierls and other ASA members to become caught up in British politics. Peierls and the ASA pursued an ambiguous course towards shaping government policy. 'I hope you will feel sure that at any rate all British scientists in this country are aware of the difficulties that confront the statesmen in this situation', Peierls assured Sir John Anderson in November 1945, 'and desire nothing more than to assist them in this task'.[181] Although Peierls underlined once again his idealistic intention, he became (perhaps unconsciously) more deeply involved in politics.

From December 1945, Rudolf Peierls officially engaged in advising the British government on nuclear energy matters. On 20 August 1945, the prime minister formed an Advisory Committee on Atomic Energy to determine the feasibility of military and civilian applications of nuclear power in the United Kingdom and to explore possible ways to achieve the international control of atomic energy in close coordination with Henry Stimson and Dean Acheson. Chaired by Sir John Anderson, members of the Advisory Committee on Atomic Energy included Sir Edward Appleton, Patrick Blackett, Sir Henry Dale, Sir James Chadwick as well as Sir George Thomson.[182] Rudolf Peierls – along with Blackett, Sir John Cockcroft, Sir Charles Galton Darwin, Norman Feather,

Maurice Pryce and Thomson – served on its technical Subcommittee on Nuclear Physics, which the government set up in December 1945 under Chadwick's chairmanship.[183]

Within the context of the Atom Train exhibition, Peierls suggested early on that, in order to secure support for the project and not to jeopardize its realization, it would be wise 'to make contact with some people on a high level in government circles and convince them that the activities of the Association [of Atomic Scientists] are going to be constructive and that our aims are the same as those of the Government'. In Peierls's view, Sir John Anderson (a former member of Churchill's War Cabinet and chancellor of the exchequer, who now chaired the Advisory Committee on Atomic Energy), Lord Portal (who served as the Ministry of Supply's controller of production of atomic energy), and the minister of Supply were crucial contacts in this endeavour. Peierls proposed that he should use his personal contacts with Anderson to make the ASA's agenda known to key players in the government.[184]

In May 1946, Peierls consequently wrote to Anderson. He defined the ASA's role as an 'impartial' educator of the public in nuclear-energy-related matters. 'At the same time', he tried to diffuse any fear on the part of Anderson that the ASA would work against the government, saying, 'we realize our obligations not to infringe the limitations about disclosure to which we have to adhere and, in general, not to make irresponsible statements that would embarrass the Government'. Peierls also attached to this letter the text of a memorandum on international control which the ASA was about to submit to the UNAEC, and asked him for a personal meeting.[185] Anderson replied immediately agreeing.[186]

This openly non-confrontational course towards the government also had to do with the fact that many ASA Council members were at the same time civil servants working at government research facilities. In cases where UKAEA staff served on the ASA Council, R. M. Fishenden explained in 1958: '[W]e have in the past taken the view that the Committee as such cannot issue a statement on a contraversial [*sic*] matter of political importance which is contrary to official Government line'. To illustrate this point, he referred to a statement on fallout dangers from thermonuclear weapons testing that the ASA's Committee on Radiation Hazards had issued the previous year.[187] Since several ASA Council members were AERE staff members, the ASA sent a copy of the draft statement to the UKAEA. As it 'may cause embarrassment to the Government', Fishenden had justified this move at the time and even suggested 'the statement ... be redrafted in such a way that the controversial part of the statement is attributed to the expert committee set up by the A.S.A. rather than to Council itself'.[188] But this was not the only case where a clash of interests occurred. Cockcroft, for example, refrained from any potentially political entanglement when he declined the offer to be on an ASA advisory committee to the Pugwash movement because of his senior position within the UKAEA.[189]

Initially, Rudolf Peierls intended to cooperate with the government in order not 'to embarrass' Whitehall, as he wrote to Anderson, but did not always support government action. In a letter to the editor of the *The Times* in January 1950, he and F. C. Champion on behalf of the ASA openly criticized Whitehall's information policy regarding nuclear energy. 'A document of the type of a White Paper issued every few years', they complained, 'is not an adequate means of keeping the public informed both of what has been achieved in this country and of the plans and possibilities for the future development of atomic energy'.[190]

In addition, Peierls's argument that there was no working defence against nuclear weapons, which he and Otto Frisch had first made in the 'Frisch-Peierls Memorandum', also undermined official British government policy on civil defence.[191] In a 1946 article in *Nature*, Peierls reiterated this point on behalf of the ASA.[192] Other leading atomic scientists such as Sir James Chadwick shared this view.[193] Before the Soviets detonated their first nuclear bomb, the ASA adopted this stance and published a report on the effects of atomic arms and civil defence in July 1949. In spite of the ASA's usual avoidance of giving qualitative comments, its authors underlined the impossibility of defence against atomic weaponry.[194]

By contrast, the government played down the effects of nuclear war in its first official civil defence pamphlet on the subject that was based on data gathered at Hiroshima and Nagasaki.[195] A review of the handbook in the *Atomic Scientists' News* criticized the fact 'that it has not included enough general information on the "large-scale" picture of a city after an atomic bomb attack, and that consequently it fails to give a sufficient indication of the magnitude of the task facing the Civil Defence forces'.[196] The government booklet revealed that perceptions of nuclear war were still deeply rooted in the experience of the Second World War. Even after the so-called Strath Report – a secret document that assessed the damage of a hypothetical thermonuclear attack on the United Kingdom – confirmed that civil defence had become nothing more than wishful thinking after the arrival of the H-bomb, Whitehall continued to promote it.[197] And Matthew Grant has exposed the British civil defence programme as a 'façade'.[198] These observations support Peierls's argument that no defence existed against (thermo)nuclear weapons.

Conclusion

By the late 1940s, the ASA had established itself as the chief organization of the British atomic scientists' movement. Because of his involvement in making the first atomic bombs during the Second World War, Rudolf Peierls – like many of his colleagues – felt a strong social responsibility after Hiroshima. He confronted the atom bomb by assuming a decisive role in founding the ASA and shaping its agenda, especially educating the public and advising political decision-makers about the dangers and benefits of nuclear power as well as promoting the international control of atomic energy. His direct and indirect experiences with National Socialism, which had caused him to become an integral part of TA and MED work, now also had an impact on his engagement with post-war British nuclear culture; for his wartime work translated into his strong involvement with the ASA.

Like many other aspects of British nuclear culture, the ASA had American origins, as several of its future members, including Peierls, witnessed the beginnings of the American atomic scientists' movement during their stay at Los Alamos. The ASA had its heyday in 1947 and 1948 with its Atom Train exhibition. While nuclear education and science advising continued to be key aims of the association, its third goal – the international control of atomic energy – had basically failed by 1946. But this was not the only area where the ASA lost credibility.

7
The 'Unpolitical' Scientist: Rudolf Peierls, the Concept of 'Objective' Science and the End of the Atomic Scientists' Association

If Rudolf Peierls helped create the Atomic Scientists' Association (ASA) and shape its objectives tremendously, he also had a significant share in its demise. Shortly after its formation, the ASA faced problems regarding its future existence, especially its approach to politics and recruitment. Peierls substantially contributed to these issues through his insistence on the concept of 'objective' science. His socialization in the academic milieu of inter-war Germany informed this ambivalent concept that ultimately led to the dissolution of the ASA. The ideal of political 'objectivity' not only generated internal problems for the ASA, but it also had a serious impact on its mission. What seemed to work in the immediate post-war period soon became outdated and was, as Matt Price has argued in the context of the United States, 'a fiction'.[1] But Peierls himself was inconsistent in adhering to his concept of political 'objectivity', even after the coming of the H-bomb.

Peierls's concept of the 'unpolitical' scientist reconsidered

Peierls's concept of political 'objectivity' was highly complex and informed by three key factors: his socialization in the German intellectual milieu, his exposure to British scientific cultures and his acknowledgement of universal scientific principles. Like many of his colleagues around the world, he was a strong proponent of the ideology of 'apolitical' science. Although its proponents present it as being objective, this concept is in fact highly subjective. Since the norms which form the underlying basis of this ideology depend both on their particular context and the consensus of a specific group of scientists, in Peierls's case the ASA, they cannot be objective as he and other scientists have often claimed. As a result of this subjectivity of their respective

151

scientific norms, scientists who follow this ideology act ambiguously. Claiming that science is 'objective' and 'unprejudiced', they define any conduct or statement which is in line with the established norms and as such based on consensus, in this case the ASA Council, as 'apolitical'. By contrast, they classify any behaviour that violates these norms and has not been reached by consensus as 'political'.[2] This ideology partly explains why Peierls made what he thought were 'unpolitical' statements but which outsiders often perceived as 'political' ones when he actively campaigned for the freedom of science during the early Cold War, as Chapter 5 has shown. When the ASA discussed, for example, the secrecy clause of the 1946 Atomic Energy Act, Peierls followed the association's line of argumentation. While the ASA criticized the government, Peierls stressed in a letter to the ASA Council that 'Scientists do not object to the Official Secrets Act' but were 'anxious that in its application to fundamental scientific knowledge it should be administered in an enlightened and generous spirit'.[3]

This ideology, however, has a further consequence that links the area of politics with that of morality: a scientist like Peierls, who follows the ASA's rules, regulations and argumentation – for example, on the division of pure and applied science – can claim that he is 'apolitical', regardless of the practical applications of his work. Peierls addressed this distinction in an article in the *Atomic Scientists' News* in September 1951.[4] He warned that hysteria among scientists about an imminent war could translate into 'turning over whole laboratories or their senior staff from academic work to ... war research' and thus harm basic research significantly. Such short-sighted decisions, Peierls feared, would imperil 'the long-term interests of science' and 'endanger the future development by neglecting the training of younger people'.[5]

Many Manhattan Project scientists, who often apologetically referred to the making of the atomic bomb retrospectively as an engineering task, used the division between pure and applied research – at least in part – to cope with their involvement in nuclear weapons research.[6] In 1946, J. Robert Oppenheimer wrote that 'knowledge is a good in itself'.[7] Still, Oppenheimer – like other Los Alamos scientists such as Hans Bethe, Richard Feynman and Edward Teller – was aware that pure knowledge could find both peaceful and destructive applications.[8] 'At present the research in nuclear physics in the universities', wrote Rudolf Peierls, 'is similarly directed to a better understanding of the basic laws of physics, and is part of our search for the truth about nature'. He added: 'It is most unlikely that any particular part of this work will in the near future find any practical application'. While this statement can certainly be read as an attempt by Peierls to come to terms with his past involvement in atomic-weapons-related work, it is important to remember that he ceased working on nuclear weapons after the war and called on his colleagues' 'duty to science' not to jeopardize the future of independent pure research and thus the advance of theoretical physics in the long run.[9] And this belief in basic research even applied to former enemies: Peierls and Klaus Fuchs condemned, for example, the destruction of Japanese cyclotrons by US occupation troops in late 1945.[10]

At the same time, however, Peierls admitted that it was often a difficult task 'to define the exact boundary between pure science and its practical application', especially 'where science has application to warfare, the best-known (though not the only) example being atomic energy'.[11] In a statement issued in 1947, the ASA had already expressed this view, declaring: 'We do not believe that it is necessary to class research on atomic explosives as a dangerous activity; the danger does not come from research as such, but from the application of the results.'[12] Shortly after Peierls's article had appeared in the *Atomic Scientists' News*, Albert Einstein formulated these thoughts as a question, asking readers of the *Bulletin of the Atomic Scientists*: '[S]hould we consider the search for truth – or, more modestly expressed, our efforts to understand the knowable universe through constructive logical thought – as an autonomous objective of our work?' He went on: 'Or should our search for truth be subordinated to some other objective, for example to a "practical" one?'[13] Accordingly, Einstein clearly foresaw the peril of scientists becoming tools of government policies and using their knowledge for their respective governments' sake. Michael J. Neufeld has called this phenomenon 'technocratic amorality', and cited the German-born rocket pioneer Wernher von Braun as perhaps its most notorious example because of the latter's 'single-minded obsession with his technical dreams' which differentiated him from thousands of other Nazi 'fellow travelers'.[14]

Besides Peierls's statements on the division of pure and applied knowledge, his comments on the morality of nuclear weapons were especially – although perhaps most often subconsciously – political. In response to a Conservative MP's proposition made in Parliament that supported President Truman's threat to use nuclear weapons in the Korean War, Peierls wrote a letter to the editor of the *The Times* on behalf of the ASA, arguing that the atomic bomb 'was most unsuitable as a "police weapon" to enforce order in local disputes'.[15] Since Peierls followed one of the ASA's main lines of argumentation which had declared in 1946 that it 'does not commit itself to regarding the atomic bomb as a desirable or suitable weapon for police functions', his statement was, from the Association's subjective point of view, 'unpolitical'.[16] What frightened Peierls in particular about nuclear weaponry was the ease with which it could be used and the disproportionality of its use in a conventional war because, as he wrote in 1950, 'after all, the police in this country still go unarmed in the face of risks because of the feeling that if they carry weapons they might use them under stress where violence would not be necessary, and that this would be immoral'.[17]

While Peierls clearly condemned any use of atomic arms as 'police weapons' alongside conventional warfare in regional conflicts, he firmly believed in the atom bomb's potential as a deterrent on the global scale. He and Otto Frisch had come up with the concept of nuclear deterrence in their seminal memorandum in early 1940, as discussed in Chapter 2. In 1947, Peierls reiterated this idea because 'the destructive nature of atomic weapons may itself be of ultimate benefit if it helps to bring home to everyone the lesson that the use of military force for aggression does not pay'.[18] Two years later, he made the point that global war was in his view unlikely, stating: 'The more we believe in the imminence of war, the more we shall be concerned with the effect of our policy on the course of the war and its outcome'. He continued: 'If we believe war is unavoidable, we must only think of minimizing the loss it will bring us, even at the risk of thereby precipitating it at an earlier date'.[19] And Peierls remained a proponent of nuclear deterrence well after the arrival of the H-bomb.[20]

Alongside these universal factors, Peierls's socialization in Germany played a key role in the formation of his political awareness. Born into a middle-class *Bildungsbürgertum* milieu in the period of the German Empire, he received his higher education during the subsequent era of the Weimar Republic. It thus seems plausible that Peierls, like many Germans of his generation, often unconsciously made political statements while believing themselves to be acting 'unpolitically'.[21] The writer Thomas Mann popularized this view of the 'unpolitical German' in his polemical treatise *Reflections of an Unpolitical Man (Betrachtungen eines Unpolitischen)* in 1918.[22]

Furthermore, Peierls was sceptical of a strong political influence in science as his rejection of the AScW's role as both a trade union and impartial informer of the public indicated. While scientists worldwide commonly followed an anti-political ideology to protect their own interests – such as authority, standing, reputation and rank of members of their respective scientific community – from any political interference, this principle underwent a significant modification in Weimar Germany. The middle and upper classes who usually provided the scientific staff and scholars at the time, and who in many cases were still rooted politically in the monarchical and anti-democratic ideology of the former *Kaiserreich*, utilized this more universal ideology for their nationalistic purposes. Out of an aversion to the democratic Weimar constitution, they believed, as Paul Forman has shown, that 'The bureaucratic authoritarianism of the old regime, basing policy not on "politics" but on objective, impartial judgments, served the true interests of the nation, while every policy of the parliamentary-democratic Weimar regime was *ipso facto* "political", unobjective'.[23]

Although Rudolf Peierls had first-hand experience of this German ideology and was sceptical of strong government influence in science, he held a deep belief in the democratic values and institutions of the United Kingdom and the United States. The structure of the decision-making process in the ASA Council, which he had decisively influenced and advocated, reflected this. Arguably, his exposure to British political and scientific cultures had resulted in Peierls's partial Anglicization. Here, a comparison between Rudolf Peierls and the post-war political involvement of those scientists such as Werner Heisenberg and Carl Friedrich von Weizsäcker, who had stayed inside Germany during the war and occupied leading positions in the National Socialist nuclear weapons project, is instructive (Figure 7.1).[24]

Figure 7.1 Peierls on a visit to the West German capital of Bonn in 1979. Carl-Friedrich von Weizsäcker (far left), Rudolf Peierls (second from right), Victor Weisskopf (far right)

After the war, Heisenberg and von Weizsäcker were among the most outspoken critics of nuclear weaponry in the Federal Republic of Germany (FRG). They acted more politically than scientists had been permitted to during the inter-war years and commented on numerous issues on the political agenda such as the hotly debated nuclear armament of the newly founded West German Army, the Bundeswehr.[25] Because of their opposition to the proposals by the West German chancellor, Konrad Adenauer, and the minister of defence, Franz Josef Strauß, who called for a nuclear-armed Bundeswehr, their names appeared among those of the so-called 'Göttingen 18', the signatories of the famous 'Declaration of 18 Atomic Scientists' ('*Erklärung der 18 Atomwissenschaftler*', commonly known as the 'Göttingen Manifesto') in 1957.[26] Max Born, who had re-emigrated to the FRG in 1954, also signed this manifesto and frequently warned the public about the dangers of nuclear weapons.[27] Arne Schirrmacher has appropriately referred to the West German physicists' political activism as '*politisches Grenzgängertum*', as scientists' engagement with politics.[28]

In their political mindedness, Heisenberg and von Weizsäcker intended to pick up (and at times even reinforce) a German academic tradition, which had existed since the late nineteenth century.[29] With the exception of the Weimar Republic and the immediate post-war years before the foundation of the two German states, scientists had traditionally viewed support for the government as 'apolitical' and opposition to it as 'political'.[30] In Germany, physicists had regarded themselves traditionally as both the country's scientific and cultural elite.[31] Physics, as a form of education (*Bildung*), was intrinsically linked with culture.[32] Fritz Ringer termed this group of German academics 'mandarins' who represented 'a social and cultural elite which owes its status primarily to educational qualities'. These omnipotent 'mandarin intellectuals' were found across all disciplines ranging from chemistry and physics to the humanities and the social sciences.[33] In particular, the liberal idea of *Überparteilichkeit*, of having no political affiliation, saw a comeback in the FRG.[34] In Heisenberg's and von Weizsäcker's case, their past involvement with the German atomic bomb project also partly motivated their open engagement in post-war politics in the FRG.[35]

Rudolf Peierls never assumed the role of a 'mandarin', but shared a sense of *Überparteilichkeit* and a similar concept of 'objectivity' as Heisenberg, who 'over his career ... found himself weakening the notion of the scientist's objectivity in sometimes self-conscious ways', as Cathryn Carson observes, and 'ended up with a liberal pluralism of perspectives held together by increasingly tenuous forms of discursive coherence'.[36] In addition, Peierls became partly Anglicized in his view of science and politics. His constant calling upon scientists and the ASA to assume the position of information suppliers was how many of his British colleagues defined their role. It was thus a major marker of his Anglicization, and differentiated him from West German scientists, in particular Heisenberg and von Weizsäcker, who acted as political and moral observers.[37] In spite of the inevitable involvement with politics, especially in the realm of science advising, Peierls and the ASA were less political than science advisors, especially those in the United States.[38] Here, other German-speaking émigré scientists who had played important roles in the development of the atom bomb – such as Hans Bethe, Albert Einstein, Leo Szilard and Victor Weisskopf – made public statements against the development of atomic arms, especially the hydrogen bomb.[39] At the other end of the spectrum, Edward Teller became almost over-motivated in his determination to develop the H-bomb when he issued his call 'Back to the Laboratories' to fellow scientists to commence work on the next generation of atomic arms.[40]

Rudolf Peierls remained a supporter of the ideology of political 'objectivity' in science until the ASA's demise and beyond. On the occasion of the second anniversary of the nuclear bombing of Hiroshima, he postulated his idea of the scientist's role in politics. 'The scientists' job is ... an unspectacular one', he wrote in the *Bulletin of the Atomic Scientists*, '... to make and keep public opinion alive to the importance of the problem and ready to support or criticize any scheme that in the future may reach a stage where one can realistically talk about its practical implications'.[41] But Peierls's favoured concept was not unanimously accepted in the ASA, and generated some controversy among its members from the beginning. As early as October 1947, Kathleen Lonsdale voiced the criticism that 'not only has the Atomic Scientists' Association not agreed views but that ... its members have not given sufficient consideration to the serious questions at all' because of its policy of 'apolitical' mindedness.[42] Yet Rudolf Peierls insisted on this principle as, for example, at the Association's annual conference in October 1948.[43]

Peierls reiterated his view in the context of the ASA when he published a lengthy discussion of the current international situation in the *Atomic Scientists' News* in 1949. 'The policy of the A.S.A.', his essay opened, 'is to make pronouncements only on matters on which Council (or perhaps on some occasions all members) are unanimous or very nearly unanimous'. Peierls went on to say that 'This may be expected to happen in questions of technical and scientific fact, but it would be surprising in questions in which we are influenced by our political outlook', adding: 'On such problems the Association wants to encourage objective discussion so as to clear up the issues and help others to form their views with least prejudice'.[44] That Peierls and the ASA still refrained from commenting on political questions, but intended 'to encourage objective discussion' as late as 1949, reveals a good deal of idealism on their part.

On several occasions, ASA members engaged in debates over their organization's purpose and future course.[45] As early as 1946, Marcus Oliphant had argued against the Association's 'unpolitical' approach to science.[46] In order to receive input by members, the ASA had also published a questionnaire in the spring of 1948.[47] In the light of serious recruitment problems, F. C. Champion wrote to Peierls in 1950: 'I feel that the whole question of a continuance of the Association should be discussed at the next Council meeting together with a serious consideration of the possible future officers if the Association is in fact to continue'.[48]

In July 1951, Herbert Skinner commented in the *Atomic Scientists' News* on the issue. His article showed the growing discontent within the Association. It was in particular the ASA's 'policy ... in regard to the general political question of the international control of atomic energy' as expressed in a letter to the editor of *The Times* by Kathleen Lonsdale and J.L. Michiels that 'worried' Skinner, and the organization's approach to politics that worried him in general. The letter to the editor, he complained, revealed 'woolliness of thought in its advocacy of "compromise" between conflicting international views' and was symptomatic of the ASA. Since neither Lonsdale nor Michiels had engaged in work on the atom bomb, Skinner questioned whether 'such publications represent the views of genuine atomic scientists'. Skinner went on to complain about the lack of more pronounced statements, which he largely blamed on the Association's diverse make-up, leading to 'a wishy-washy compromise between the diametrically opposed opinions of various members of the council'. The *Bulletin of the Atomic Scientists*, by contrast, was a positive example of 'how it is possible to maintain a lively and interesting periodical on these matters'. In the case of the ASA, however, Skinner suggested that 'the Association should abstain altogether from political subjects' because 'its views on political matters are simply worth nothing at all'.[49]

It appears remarkable that the ASA still followed its 'unpolitical' course despite this internal criticism. Outsiders also viewed the ASA's 'apolitical' approach to science as problematic. As early as July 1948, a *Nature* article called a statement by the organization on the international control of nuclear energy 'lamentably weak' and criticized that,

'In its anxiety to avoid taking sides ... the Council of the Association has become vague'. The review added, 'The commendable attempt to avoid political bias is carried too far when it avoids passing judgment of facts.'[50] In a similar fashion, the *Economist* described the statement as 'commendably free from political bias' and called the ASA's argument that war was simply to be avoided in the atomic age 'platitudinous'.[51] These judgements reveal that many commentators regarded the ASA's concept of 'unpolitical' science outdated fairly shortly after the Association's creation. But Rudolf Peierls and the ASA officially continued to advocate this policy and adhere to it.

Enter the hydrogen bomb

If Rudolf Peierls and many ASA officers contributed to the Association's decline by refraining from making politicized statements, exterior factors put additional strain on the organization – in particular the coming of the hydrogen bomb. President Harry S. Truman's announcement that his administration had ordered the development of the H-bomb generated public interest, and Rudolf Peierls – in his function as ASA President – answered in the *Sunday Express* '11 Vital Questions on the Big, Big Bomb', ranging from its design, development and yield to possible civilian applications.[52] But, at the same time, internal dissent over an ASA statement on the H-bomb revealed a deep rift within the organization. Here, the H-bomb served as a catalyst to bring internal ASA problems to the forefront. That the Association comprised members from across the political spectrum further complicated the situation.

In spite of the fact that Peierls and other ASC members had originally founded the ASA as an independent body, it faced many problems owing to its members' diverse political views from the start. On the political left, Patrick Blackett, who had served as the AScW's president from 1943 to 1947, and Eric Burhop, another leading figure in the union, were members of the ASA's Council from the beginning. Blackett was among the sharpest internal critics of the Association.[53] While he was a member of the government's scientific advisory committee, he argued against an independent British atomic bomb.[54] In 1948, Patrick Blackett published his book *Military and Political Consequences of Atomic Energy* that had grown out of his frustration with British nuclear policy at the time, especially the failure to implement a system of international control of atomic energy.[55] Two years later, he contributed the foreword to the AScW pamphlet *Atomic Attack: Can Britain Be Defended?* that heavily undermined official British government policy on civil defence.[56] Sir

John Anderson, as the former chairman of the Advisory Committee on Atomic Energy to the government, criticized Blackett and the AScW on the grounds that they 'may have shown undue haste and unjustified self-confidence in putting forward their own views at this juncture'.[57]

In another instance, an article by Eric Burhop published in April 1949 in the *Atomic Scientists' News* on the civil service purge in Britain generated dissent among many ASA members.[58] On the political Right, Lord Cherwell or Sir George Thomson stood for conservative values.[59] Thomson had reservations about the views expressed by some ASA Council members, especially left-wing ones, and therefore declined Peierls's offer to run for ASA president in 1950.[60] Cherwell was very critical of a public declaration on the civil service purge and opposed a statement suggested to be made by the ASA Council on the hydrogen bomb.[61]

Peierls himself was 'very disturbed' by the proposed statement on the hydrogen bomb, writing to F. C. Champion: 'I appreciate, of course, the idea behind it but I take strong exception to the words "aggravates the arms race"'. He feared especially its negative impact on both the ASA's public reputation and Anglo-American relations. Peierls added: 'I am certainly not prepared to sign this and I feel that this must not be allowed to come out from [*sic*] the office of the Association even if it bears only a few signatures ... as long as it does not come out with the authority of Council.'[62]

In the end, the ASA Council never sanctioned the proposed statement on thermonuclear weapons, and it was not released to the press as a Council statement but as a declaration by individual ASA members.[63] The ASA sent it to the *The Times* for consideration. Despite his initial reservation, Rudolf Peierls's name was among those ASA members who had undersigned the memorandum, including Eric Burhop, F. C. Champion, Kathleen Lonsdale, Joseph Rotblat and Sir George Thomson. 'The recent decision of the USA to develop the hydrogen bomb shows that an atomic arms race is in progress', the missive opened, 'and emphasises the dangerous direction in which humanity is moving'. Peierls and his colleagues stressed that 'We believe that if a disastrous atomic war is to be avoided, the utmost attempts must be made now to eliminate atomic warfare either by a new effort to solve the problem of the effective control of atomic energy or by a new contribution to the wider problem of international relations'. In fairly broad terms, they suggested: 'Any solution must be acceptable to all nations though all nations would have to be prepared to sacrifice some of their national interests for a realist hope of continued peace'.[64]

The debacle over reaching a common line on the issue raised crucial questions about the Association's future. 'Not unconnected with all these difficulties is the question whether our Association can continue to function', wrote the ASA's Honorary General Secretary, F. C. Champion, to Rudolf Peierls. The H-bomb thus became the litmus test to determine the Association's future because, as Champion pointed out, 'the continuance of our Association depends on whether or not we have anything to say about the H-bomb'.[65]

By May 1950, Sir George Thomson circulated another draft statement on the hydrogen bomb to Council members and vice-presidents that concerned its technical features. Thomson emphasized three points about the hydrogen bomb which he deemed of particular importance: first, a 'conventional' plutonium or uranium device was required to trigger a thermonuclear reaction; second, the upper limit of the H-bomb's explosive yield was open; and finally, a thermonuclear explosion produced a large number of neutrons with severe effects on life and health. Thomson stressed the H-bomb's special nature as opposed to 'the original atomic bomb' which 'was a weapon like other weapons, more powerful indeed but of the same kind'. In conclusion, he argued: 'So it seems that hydrogen bombs, in exchange for the risk of making the world uninhabitable, cannot offer any improvement in the waging of war. Nor is it even likely to change appreciably the balance between the East and the West'. Thomson viewed proposals that called for mutual inspections as unfeasible, and sympathized instead with 12 US scientists around Hans Bethe and Victor Weisskopf who urged President Truman to declare that he would not use the H-bomb.[66]

Unlike their 12 American colleagues, who received considerable media coverage with their public statement demanding that the Truman Administration renounce the pre-emptive use of the hydrogen bomb, ASA members failed to attract similar attention from the press on the issue.[67] Shortly after President Truman's announcement, however, the *Picture Post* featured short statements by Rudolf Peierls and other British scientists including the ASA members Kathleen Lonsdale, Harrie Massey and Eric Burhop.[68]

The clergy, by contrast, took a much stronger stance against the H-bomb and appeared to be much more political than shortly after the end of the Second World War.[69] The archbishop of York demanded the H-Bomb be outlawed and the archdeacon of London called it a decisive issue in the 1950 elections.[70] In April 1950, the British Council of Churches supported a statement made by the Executive Committee of the World Council of Churches[71] which called on governments to engage in peaceful cooperation, warning against the danger of future thermonuclear war. 'The hydrogen bomb is the latest and most terrible step in the crescendo of warfare which has changed war from a fight between men and nations to a mass murder of human life'.[72]

Although the arrival of thermonuclear weapons received considerable attention from the ASA, the Association failed once again to make a clear statement on the issue. In March 1950, the ASA dedicated a special issue of the *Atomic Scientists' News* to the H-bomb that had evolved out of a symposium organized by the Association on the topic.[73] The *Daily Mirror* devoted an article on its front page to the special number, specifically mentioning Rudolf Peierls's contribution.[74] In his essay, Peierls commented on global security and thermonuclear weapons. He clearly differentiated between work on the atom bomb in which he had been involved himself during the war and which had implied the capacity 'to gain new constructive powers', on the one hand, and work on the H-bomb with its purely destructive potential, on the other. In a very ambiguous way, Peierls emphasized the hydrogen bomb's function as a deterrent, while simultaneously commenting on the role of scientists in this progress, warning that 'there is responsibility in inaction as well as in action, and if the scientists in America or anywhere else collectively boycotted this project their responsibility might be very hard to bear if later on their country was attacked with this or a similar weapon'. He asked readers: 'Are we, then, to blame our American colleagues for being optimistic about the working, in the long run, of democratic institutions?'[75] Yet the debate over political and moral issues involved in the development of the H-bomb was more complex and went beyond the discussions revolving around the making of the atomic bomb during the war, with scientists taking different sides.[76]

Given the gravity of the debate over the hydrogen bomb, Peierls's 'apolitical' approach appeared to be outdated and a remnant of the immediate post-war period. Yet the ASA continued to cover the hydrogen bomb because, the editorial of the following issue stated, 'The importance of adequate public discussion of this subject cannot be over-estimated'.[77] The *Atomic Scientists' News'* May 1950 issue featured a ten-page section 'The Hydrogen Bomb – American Reactions' entirely made up of reprinted articles from the *Bulletin of the Atomic Scientists'* March number. This can perhaps be seen as a compromise on the behalf of the ASA to present more opinionated articles in its journal without releasing any such statement in its name.[78] In 1957, the *New Scientist*'s 'Atomic Science' section featured two highly biased articles on the H-bomb, one by Bertrand Russell and another by Angus Maude, a Conservative MP. Russell, who had played a significant role in galvanizing public opinion against the H-bomb when he had co-authored the famous Einstein-Russell Manifesto in July 1955,[79] emphasized its perils. In his article, he called for a test stop and urged Whitehall to cease its H-bomb development programme, as he did 'not wish to be an accomplice in a vast atrocity which threatens the world with overwhelming disaster'.[80] Angus Maude, by contrast, argued in favour of a continuation of the British hydrogen bomb tests, emphasizing the H-bomb's 'peculiar effectiveness ... as a deterrent, ... a total destroyer'. Maude concluded, 'After all, the risk of possible mutations in a hundred years has to be compared with the risk of universal destruction in ten', He added: 'I know which I would choose, even if I did not believe that we may soon have the chance to prevent both'.[81]

While the presentation of these polarized views surely represented a departure from the ASA's fairly moderate approach to politics, Rudolf Peierls continued to advocate his 'unpolitical' view of science, even after the tragic *Lucky Dragon* incident. In the aftermath of the United States' Castle Bravo Test in the South Pacific on 1 March 1954, the crew of the Japanese fishing vessel *Lucky Dragon*, which was at the time of the detonation outside the designated danger zone, was heavily contaminated by fallout from the thermonuclear device. And, extensive US and Soviet thermonuclear weapons testing programmes drastically revealed that fallout was an international problem that could not be limited to a country or test site by the mid-1950s.[82]

When the ASA Council debated a memorandum issued by the *Bulletin of the Atomic Scientists* around Eugene Rabinowitch, in which he and his colleagues called for an international conference in the light of these recent events, Peierls's insistence on 'objectivity' resurfaced once more.[83] 'Evidently we cannot expect unanimity for a statement supporting American foreign policy, or policy in Atomic Energy matters', Rudolf Peierls strongly objected to the proposal. 'If we make a statement which is critical of American policy, and thereby stress the distinction between the views of American scientists and those of their government', he argued 'the results will only make things worse, since in the present mood of American opinion any foreign intervention in their internal disagreements can do nothing but harm'. In addition, Peierls generally disliked the idea of an international conference on the subject.[84]

Rabinowitch then drafted a second, revised memorandum for an international conference that stressed the need for an international meeting on the subject and proposed to set up study groups to prepare the conference.[85] The ASA set up a special sub-committee that consisted of K. Lonsdale, G. N. Walton, Hodgson and Rotblat to discuss the memorandum and ask for further details from Rabinowitch.[86] Eventually, the

ASA Council agreed in principle to support an international conference, especially by preparing it through the work of study groups. In line with its 'unpolitical' approach, the ASA suggested it should work closely with other bodies, including the British Association for the Advancement of Science, the Royal Society and the Royal Institute of International Affairs. The Council further proposed to change the names of the 'Science versus Society' and 'Threat of Science to Mankind' groups into the less polemic 'Scientists and Society' and 'Atomic Science and Mankind' groups respectively, as well as to include study groups on the environmental impact of nuclear waste, 'the psychological factor' and civil defence.[87] The ASA actively lobbied for the establishment of study groups with other British organizations and institutions such as Chatham House and the House of Commons.[88] By February 1957, ASA members contributed to four study groups: Study Group I under the directorship of Rotblat examined radiation hazards; Study Group II under Hodgson looked into inspection and control; famed scientist and broadcaster Jakob Bronowski headed Study Group III on social responsibility'; and Haddow's Study Group IV investigated the formation of an international organization.[89]

The Atomic Scientists' Association's last battle

Joseph Rotblat's group on radiation hazards was behind the ASA's final publicity stunt, a statement on strontium-90 hazards in 1957. The radioactive isotope strontium-90 is a fission product that started to occur in increasing doses in the atmosphere as a consequence of atmospheric nuclear testing in the 1950s. If consumed with food or liquids, especially milk, strontium-90 is stored in human teeth and bones and can act as a carcinogenic agent. The statement has to be seen within the context of United States, Soviet and the impending first British H-bomb tests as well as a report by the British government's Medical Research Council (MRC) that dismissed any fallout-related radiation risks. In early 1957, Rotblat raised the question of following the FAS's model and issuing a statement on thermonuclear weapons tests before an ASA Council meeting, asking its members for their views.[90]

Sir George Thomson felt that he 'must protest most strongly against' the planned public statement. He cited a report on radiation hazards from nuclear tests by the government's MRC as evidence that this risk 'was negligible'.[91] At a subsequent meeting, the ASA Council then decided to ask vice-presidents and Council members 'about the proposal that the A.S.A. should express publicly its opposition to H-bomb tests'.[92] Sir John Cockcroft took a Peierlsian line, arguing 'the A.S.A.

should confine itself to scientific matters and publicising the facts'. Like Thomson, he cited the MRC report on the subject as evidence against the need for such a statement.[93]

Since there was an overwhelming majority for issuing 'a statement on the scientific problems concerning radiation hazards, but not on the political aspect', the ASA Committee on Radiation Hazards drafted the statement based on its findings. Alongside Joseph Rotblat, who chaired the committee, it comprised J. W. Boag, C. E. Ford, A. Haddow, W. M. Levitt, Patricia Lindop, S. B. Osborn, L. S. Penrose, P. A. Sheppard, G. Simon and L. A. Salmon. In line with ASA policy, they circulated their draft statement among vice-presidents and Council members.[94] Since ASA Council members disagreed about the exact wording of the statement, they called a special Council meeting to discuss the proposed statement.[95]

The report warned of the health hazards posed by strontium-90, if the isotope were ingested into the human body where it was stored in bones. Rotblat and his group dismissed the findings of the MRC report that had assessed the additional irradiation of human reproductive organs at 1 per cent of the naturally occurring radiation. The government missive based this estimate on the assumption that thermonuclear testing remained at the current rate. Rotblat and his colleagues, by contrast, calculated that 'by the year 1970 the radiation dose to bone from all tests carried out so far will amount to about 40% of the dose from all natural sources, including the radium which is normally contained in bone'.[96] And, what is more, they even attempted to establish a correlation between individual H-bomb tests and the cases of bone cancers that occurred worldwide. Rotblat's group concluded that thermonuclear weapons such as the ones tested at the Bikini Atoll in 1954 'if exploded high in the atmosphere, may eventually produce bone cancers in 1,000 people for every million tons of T.N.T. of equivalent explosive power. (It has been stated that bombs equivalent to a total of 50 million tons have already been exploded.)'[97]

Rudolf Peierls criticized the draft statement, doubting that the 'proportionality between the dose and the number of casualties should extend to very small doses'.[98] Sir John Cockcroft objected to the statement being issued in the name of the ASA Council and found that 'In particular the assumption about the relationship of bone cancer incidence to the dose is a very dubious one'. Instead, he suggested it should be published under the name of the ASA Committee on Radiation Hazards 'so that the scientific world can see who it is who is responsible'.[99] Herbert Skinner was 'strongly against the A.S.A. issuing an official statement based on so shaky an assumption as the proportionality of carcinomas and leukemias down to zero doseage, the constant of proportionality being based on a high dose'. He added: 'I don't think the A.S.A. should lend its name to publicise a completely unproven scare'.[100] Others were in favour of the wording of the draft statement. Bates and Pryce approved of it, and Nevill Mott called it a 'useful document'.[101] The ASA Council agreed to issue the statement in the name of the Committee on Radiation Hazards, and with a different opening paragraph that stressed the committee's role in crafting the public statement.[102]

Once the authorship issue had been settled, the ASA sent the statement to government offices and the press.[103] Given that many ASA Council members were government employees, Cockcroft distanced himself and other AERE staff members from the statement vis-à-vis the UKEAE Chairman, Sir Edwin Plowden.[104] The British government asked the MRC for their comments on the statement.[105] Sir Harold Himsworth, the MRC chairman, rejected it 'very strongly' and wrongly associated Sir John Cockcroft with it so that he considered discussing 'the question of possible resignation from the A.S.A.' with the latter.[106] As a result, Cockcroft informed Massey, the ASA president: 'The recent statement has been somewhat embarrassing to me since the *Manchester Guardian* and other papers have formally associated my name with the statement in spite of the fact that it was issued as a statement of the Committee'.[107] Massey, too, was 'somewhat disturbed by the recent statement'.[108]

Despite Cockcroft's embarrassment, the statement on strontium-90 hazards was another major publicity stunt and represented the ASA's last battle before its demise. It received considerable attention at home and abroad.[109] The governments of Belgium, Britain, Japan, West Germany and organizations such as the UN and the FAS as well as institutions, including the Max-Planck Institute for Biophysics in the FRG, the United Free Church and the Rockefeller Institute for Medical Research, expressed great interest in the report.[110]

As a result of the considerable disquiet that the statement had generated within the Association, the ASA Council was more careful regarding future public declarations. Once again, the ASA moved away from making political announcements. This return to its 'unpolitical' approach to science and politics that Peierls had so crucially helped to forge found expression in the process of drafting a letter to the editor of *The Times* in the autumn of 1957. This addressed the issue of nuclear proliferation as a consequence of the international community's failure to agree on and institute an international system of control of nuclear energy.[111] While Sir John Cockcroft welcomed the fact that the letter was 'non-party and non-political', he was unable to sign it because of his role at the UKAEA.[112] Herbert Skinner 'was glad to get [the] letter stating that the proposed letter to the Times is not to be an A.S.A. statement of any kind'. He went on: 'On re-reading the letter, I still think that it gives the impression that it would be quite easy for various nations to make A-bombs.'[113] In a letter to Sir John Cockcroft, Skinner even stated, 'I thought we had an understanding that they would stop producing half-baked statements of this kind'.[114] And the new ASA President Sir George Thomson assured Cockcroft 'that there would be no further letters of this kind, and on that understanding I agreed to hold up my resignation as Vice-President'.[115] *The Times* finally published it on 29 November 1957 undersigned by Rudolf Peierls, L. F. Bates, Norman Feather, Gwyn Owain Jones, Kathleen Lonsdale, Philip Moon, Maurice Pryce, Joseph Rotblat and George Thomson.[116]

The Atomic Scientists Association's final act

Although the memorandum on strontium-90 represented the Association's last major public statement, Peierls's and the organization's insistence on refraining from making political statements had become increasingly obsolete with time. By comparison, American scientists around Barry Commoner, Edward U. Condon and Linus Pauling issued an appeal to governments and people all around the world to stop nuclear testing in September 1957.[117] The ideology of 'apolitical' science was thus a major factor in bringing about the ASA's disbandment. Rudolf Peierls wrote in his autobiography that the ASA 'ran out of steam' during the 1950s. He retrospectively blamed this development primarily on two factors: first, the ASA failed to attract the adequate number of younger people needed to keep it up and running. Second, and more significant, he attributed the ASA's demise to a general feeling that 'however important the international problems, there was little that a British organisation could do' because '[t]he future depended essentially on what the United States and the Soviet Union were doing'.[118]

While in 1947 Peierls had defined Britain's role euphemistically as a mediator between the superpowers in the struggle for achieving an international system for the control of atomic power, he revised his position shortly afterwards when the Soviets vetoed the Baruch Plan and detonated their own atomic bomb. In a 1950 article in the *Bulletin of Atomic Scientists*, he had already revealed a good deal of pessimism about the realization of the ASA's mission, admitting: 'We have not so far seen a solution which could be worked for by organizations in Great

Britain and which offered a realistic chance of success.'[119] In 1970, he elaborated on Britain's dwindling role as a world power as a major reason for the ASA's decline and final disbandment in the late 1950s. Peierls argued that 'the danger of world war looks from here almost as it looks from the "have-not" viewpoint of the non-nuclear countries'. Hence, British groups like the ASA faced more difficulty in mobilizing members than their American counterparts. 'The prospect of giving abstract thought to possible solutions which might commend themselves to others', he concluded, 'does not attract wide support'.[120]

As Peierls rightly observed, the changed international situation certainly had a major impact on the ASA's significance. The failure to implement a scheme for the international control of nuclear energy in the late 1940s played a substantial role in the Association's slow but steady decline. Since the creation of such a system ranked among the its three chief objectives, its failure posed a serious problem for justifying both the ASA's existence and purpose, as some of its members had already realized along the way.[121] In the United States, similar developments occurred: the NCAI collapsed and membership in the FAS declined significantly.[122]

Rudolf Peierls's comments about the ASA's loss of significance in the face of Britain's decline as a world power reveal the general limitations of groups like the ASA. Holger Nehring has classified them as 'nationalist internationalists'. These movements have objectives that are connected to global issues like the international control of atomic energy, but they simultaneously view them from a national perspective.[123] In the case of the ASA, the scientific practice of atomic scientists, in particular physicists, who had traditionally been highly international in their professional conduct collided with a more nationalist political agenda.[124] This clash was often a source of conflict between governments and individual scientists, as has been shown in the case of Rudolf Peierls in Chapter 5. Although German scientists had represented a special case and had not always been as fully integrated into international networks as their American or British colleagues, Rudolf Peierls, together with Hans Bethe, Klaus Fuchs, Werner Heisenberg, Pascual Jordan, Wolfgang Pauli and Victor Weisskopf, had been among the first generation of physicists based in Germany to participate more regularly in international events and networks, even before the forced departure of the many émigrés among this group.[125]

While Peierls had rightly assumed that the United Kingdom's decreasing weight on the international stage was a crucial factor in bringing about the end of the ASA, his explanation as to why it failed to attract more young members, so desperately needed to operate it, appeared to be somewhat idealistic. The ASA faced severe recruitment problems quite early on. By 1950, dwindling numbers of full members started to pose a serious problem to the organization and to threaten it financially.[126] Here, apart from the changing geo-strategic realities and its lack of political mindedness, it was especially the question of membership that was discussed. As early as April 1948, the ASA considered the introduction of so-called Graduate Memberships.[127] In early 1949, it also contemplated the idea of hosting an annual dinner for key political decisionmakers to generate publicity.[128]

A significant part of the problem lay rooted in the fact that the ASA comprised an elitist circle of full members, with non-atomic-scientists only being granted associate membership status. 'We have considered the question of widening this, as the Americans have done, into an organisation including all scientists in this country', Peierls had remarked on membership and the ASA's position among the existing organizations as early as 1946. The ASA Council soon discarded this idea because 'it was felt that this would be invading territory which at present other such as the A.Sc.W. and the British Association [for the Advancement of Science] regard as their own'. And again, Peierls was concerned that such a move towards widening the membership could entail 'getting definitely labelled in any political direction, as would be the case under the A.Sc.W.'[129]

The Atom Train epitomized the ASA's heyday when its objectives still seemed achievable, and the ASA's departure from its ideology of 'objective' science made the exhibition a big success. Like their American colleagues, British scientists would never again feature so prominently in the public eye as during the 1940s.[130] Later attempts to revive the association failed. These included a series of public lectures. 'It is a long time now since the Atomic Scientists' Association have organised any public lecture on atomic energy', Joseph Rotblat observed in March 1953, 'and we feel that the subject has crystallised enough during the last few years to merit an up to date review'.[131] In 1954, the ASA thus collaborated with the University of London's Department of Extra-Mural Studies in organizing six public lectures on nuclear energy. Speakers and topics were: 'Atomic Research at Harwell' by Sir John Cockcroft, 'Atomic Weapons' by Otto Frisch, 'Power from Atomic Energy' by Franz Simon, 'Radiation Hazards of Atomic Energy' by J. F. Loutit, 'Medical Uses of Atomic Energy' by E. E. Pochin and 'Atomic Energy and Moral Issues' by Kathleen Lonsdale and Sir George Thomson.[132]

Other projects to revitalize the ASA included renaming the *Atomic Scientists' News. New Series* as the *Atomic Scientists' Journal*. This move went hand in hand with the Association's reassessment of its strategy and goals as well as possible ways of attracting more members and readers of its journal in September 1953.[133] In a similar vein, the ASA Council briefly considered but finally abandoned plans to rename the ASA the Nuclear Age Society and the *Atomic Scientists' News* the *Nuclear Age* by the spring of 1955. '"Atomic Scientists' Association" and "Atomic Scientists' Journal" means to the layman, intelligent or otherwise, "highly technical and therefore not my concern"', Ford argued in an internal ASA Council memorandum in 1955, adding, 'the very opposite of what we want'.[134]

By September 1955, the ASA Council had made no progress with regard to the ASA's reformulation of its objectives.[135] Paradoxically, the ASA continued to follow the very same ideal of political 'objectivity' that had played a considerable part in the cancellation of its own journal. 'As before, every effort will be made to present all sides of a controversial issue while at the same time being particularly careful that scientific facts should be distinguished from opinion', emphasized the ASA president Harrie Massey in the first 'Atomic Science' section in the *New Scientist*.[136]

In June 1957, Harrie Massey called a special meeting of vice-presidents and officers to discuss the future of the organization.[137] After the meeting, Sir John Cockcroft suggested to Massey that the relatively high number of vice-presidents be reduced.[138] In the aftermath of an ASA Council meeting, a disappointed Sir George Thomson wrote to Sir John Cockcroft in early November 1957: 'Massey has resigned the Presidency and I think the Association will shortly come to an end, but there is a prospect of its continuing in modified form as part of the British Association [for the Advancement of Science; BA] organisation, though I think this prospect is a pretty nebulous one'.[139]

By February 1958, plans for a merger with the BA took more shape. At a meeting of the ASA Council with their BA counterpart, the ASA Council members were asked to sign an agreement so that six ASA members and six BA members could serve on a Joint Executive Committee.[140] Under this new collaboration, the ASA provided, for example, speakers for public talks that the BA organized.[141] At the ASA's annual general meeting on 12 July 1958, full members were to vote on the Council's decision to continue the successful cooperation with the BA. If it proved successful again and 'if there is no longer any need for the continued existence of the ASA as an independent body' in a year's time, H. R. Allen wrote to all full ASA members, the ASA Council would then ask attendees of the 1959 general meeting for the ASA's disbandment. He recommended that those members who were still concerned about atomic issues join the BA and engage in the work of its Division for Social and International

Relations.[142] At its 103rd meeting on 14 March 1959, the ASA Council decided to officially wind up the ASA at its final annual general meeting on 11 July 1959 when it officially folded.[143]

Conclusion

While the ASA was still operating, the formation of a broad anti-nuclear movement had taken place. A highly opinionated article by J. B. Priestley in the 2 November 1957 issue of the *New Statesman* was crucial for mobilizing this mass movement, with the Campaign for Nuclear Disarmament (CND) becoming the chief organization.[144] After its creation in 1958, an increasing politicization Rudolf Peierls had always warned against in the ASA took place in the CND. It seemed his concept of the 'apolitical' or 'objective' scientist was dead once and for all. As the new left won considerable influence over the CND in the early 1960s, the organization took – by comparison with the ASA – highly politicized views when it demanded from Whitehall that Britain leave NATO, for example.[145] To put the ASA's disbandment into perspective, it is important to bear in mind that not all attempts to establish national organizations with internationalist or international agencies were fruitful. In what was perhaps the most famous case, the international community failed to create the International Trade Organization in 1947–8.[146]

It was perhaps against the background of his experience with the ASA and the emergence of the highly political anti-nuclear movement that Peierls showed signs of resignation during the late 1950s. After the ASA's demise in 1959, he, along with many former ASA members, became involved in the international Pugwash Conferences that started at about the same time.[147] When it reached its final stages, the ASA also collaborated with the Pugwash movement.[148] But Peierls did not attend any of their conferences until the 1960 meeting in Moscow.[149] The Pugwash movement, which exists to the present day, survived the Bretton Woods system, which collapsed in the early 1970s.[150]

Despite the eventual failure of his concept of the 'unpolitical' scientist and the disbandment of the ASA, Peierls's involvement in the British atomic scientists' movement had paved the way for an anti-nuclear mass movement, most notably the CND. In what can perhaps best be termed 'the CND factor', this organization has dominated public debates over nuclear disarmament until the present day. And the awareness of nuclear issues clearly entered the public arena thanks to a large degree to Peierls's tireless efforts.

Conclusions and Afterthoughts

Klaus Fuchs and Rudolf Peierls played significant roles in the making of British nuclear culture in the years from 1939 until 1959. They shaped two chief components of atomic culture in their host country: the practice of nuclear science and the social, political and cultural implications of their work. Fuchs's and Peierls's ethnicity, socialization and schooling in Germany as well as their exposure to German culture before emigrating to Britain had a strong impact on their involvement with British nuclear culture. Their experiences with National Socialism and their personal knowledge of some of the key scientists who were believed to be behind the German nuclear weapons project led to a strong determination in both of them to become involved in nuclear arms research and to beat Germany in the race for the atomic bomb. At the same time, however, as Chapters 1 and 2 have shown, their German origins made them 'enemy aliens' and did not allow them to work on important war projects such as radar so that they were – almost accidentally – pushed into the direction of nuclear weapons research, which was not deemed as crucial to the war effort in 1939. And because of his extraction, Fuchs was even interned in Canada for several months.

Klaus Fuchs and Rudolf Peierls shaped 'scientific culture' in Britain and the United States significantly. Peierls became a key player in the early British nuclear weapons project. It was, in particular, the seminal memorandum, which he co-authored with Otto Frisch in early 1940, that galvanized Whitehall's efforts to pursue its own atomic arms programme. Not only did it lead to the two Maud Reports and the establishment of TA, but it also had a strong impact on the formation of the Allied Manhattan Project. Peierls also made further crucial contributions to establishing Anglo-American nuclear cooperation through a visit to the United States early in the war, for example. In what would later have serious implications for Peierls in the aftermath of Klaus Fuchs's confession of espionage for the Soviet Union, he also recruited Fuchs into TA work.

In 1943, Fuchs and Peierls joined the Manhattan Project and worked, at first, briefly in New York City and then at the secret laboratory at Los Alamos, New Mexico. At Los Alamos, Fuchs and Peierls came across many German-speaking émigré atomic scientists, as Chapter 3 has demonstrated, including Hans Bethe, Egon Bretscher, Otto Frisch, George Placzek, Edward Teller and Victor Weisskopf. Peierls and Fuchs, together with the other German-speaking émigré scientists, as a cohort, helped tremendously to shape a new approach to nuclear science that consisted of a close cooperation between theoretical and experimental scientists and built on large-scale government funding. Here, many German-speaking émigré scientists profited from their education in German universities with their focus on the theoretical side of science and their experience with state-funded universities in Continental Europe. With their input into the creation of the first atomic bombs at Los Alamos, Fuchs and Peierls thus stepped up the establishment of the emerging culture of Big Science. As a result of their wartime work in the MED, Peierls and Fuchs also contributed to the United States' emergence as the dominant scientific power after the Second World War.

After the war, Fuchs and Peierls returned to Britain where they took different paths. While Peierls resumed his professorship at the University of Birmingham, Fuchs became head of the theoretical physics division at the newly founded AERE Harwell. It was during this period that Fuchs's and Peierls's achievements in 'scientific culture' started to show definite results in the form of 'cultures of insecurity'. In 1950, Fuchs confessed that he had spied on his host country for the Soviet Union since the day he had become engaged in TA work. As Chapter 4 has demonstrated, Fuchs's experience with National Socialism, especially during his time at Kiel University, not only led to a strong motivation to work on the British and Allied nuclear weapons projects but also radicalized him politically. The Fuchs case affected public opinion in Britain where it caused members of the public and political decision-makers to question the efficiency of national security agencies, especially MI5, in their defence of the democratic order. But the espionage affair had further repercussions for Anglo-American relations, and even led MI5 to deceive both the British public as well as the prime minister and other key political decision-makers to restore its image. While Fuchs's confession in general had a strong effect on public opinion, it had particularly serious implications for Fuchs's former mentor. Chapter 5 showed that

Rudolf Peierls and other German-speaking émigré atomic scientists in Britain and the United States became the target of suspicion owing to their German origins. The effects of Fuchs's radicalization in Germany could thus long be felt, and marked a partial setback to their successful social integration that had previously taken place.

Peierls also engaged with another area of British nuclear culture, namely the social responsibility that arose from scientists' work on nuclear weaponry. After his return from the United States, he became a crucial figure in the emerging British atomic scientists' movement through his involvement in the ASA. Chapters 6 and 7 have revealed that Peierls's socialization in and exposure to research cultures in Germany informed his understanding of the relationship between science and politics and the role he envisaged for science in public education and advising political decision-makers. The resulting concept of the 'unpolitical' scientist was decisive in both the formation and end of the ASA. While it represented an integral part of the ASA's approach to politics from the start and worked well in the immediate post-war years, it became increasingly outdated once the Cold War advanced and realistic chances for implementing a system of international control had failed.

While the time period under investigation ends in 1959, Rudolf Peierls did not withdraw from nuclear culture. After the ASA's demise, he, along with many former ASA members, became involved in the international Pugwash Conferences as the first chairman of their British group.[1] Peierls continued to adhere to his belief in the ideology of 'objective' science. 'We believe that there is a need, particularly in the scientific community, for unbiased information, and for a forum for objective discussion', he stated in September 1976 in his function as chairman of the British Pugwash Group.[2] By the 1980s, however, Peierls started to openly advocate nuclear disarmament. He joined the British branch of the Nuclear Weapons Freeze Campaign, the so-called Nuclear Freeze movement, which demanded Britain's unilateral abandonment of its nuclear weapons programme, and he even assumed its directorship the following year.[3] While he continued to back the concept of nuclear deterrence, he rejected NATO's doctrine of limited nuclear war, in particular the use of tactical nuclear weapons to stop a large-scale conventional attack by the Warsaw Pact on Western Europe.[4] In 1995, Peierls co-authored a report by the British Pugwash Group that made a case against an independent British nuclear deterrent.[5]

Peierls also carried on with his work as science advisor. As early as February 1957, he had accepted an invitation by the British government to serve as a representative of universities on the Governing Board of the newly founded National Institute of Research in Nuclear Science.[6] In late December 1964, he accepted the FO invitation to serve on a 'Consultation Panel on Disarmament'.[7] The following year, he even presented a paper on 'Disarmament – the Answer to Nuclear Stalemate: A Scientific View' at a UN conference.[8] In 1965, Peierls also resumed his consultancy for the AERE Harwell.[9]

While Peierls's host country honoured his legacy with symposia dedicated to his scientific achievements and, above all, his knighthood in 1968, Klaus Fuchs's confession overshadowed the latter's standing.[10] Media coverage of the Fuchs case – such as the Radio 4 play *Atomic Lunch* (25 July 2005) and the second episode, 'Superspy' (22 January 2007), of the five-part BBC 2 miniseries *Nuclear Secrets* – demonstrates this. And, apart from his espionage, Britons know little about his scientific work and life after 1959 today.

In 1959 Klaus Fuchs moved to the GDR where he settled in Dresden and became deputy director of the nearby Rossendorf atomic energy research installation.[11] He was a member of the Central Committee of the Socialist Unity Party (SED), the Academy of Science (Akademie der Wissenschaften) and the GDR Research Council (Forschungsrat der DDR), and received the Order of Patriotic Merit (Vaterländischer Verdienstorden) in silver (1962) and gold (1971).[12] By the mid-1980s, he ceased to take an active part in its work due to his age and deteriorating health, but remained an active member of the German–Soviet Friendship Society.[13]

In spite of his service to the Soviet Union, Fuchs kept a very low profile in the GDR.[14] He published a few articles on the importance of science for the socialist state, and spoke out occasionally when he accused the West German government, for example, of pursuing the development of nuclear weapons.[15] Even his courier, the anonymous woman with whom he had met several times in Banbury, alias Ruth Werner, did not mention him at all in the first edition of her autobiography *Sonjas Rapport* (*Sonya's Report*).[16] It was only after the end of the Cold War that an expanded edition was published first in English in 1991 and then in 2006 in German; this comprises several pages of previously withheld material on Fuchs.[17] But the Soviet Union apparently held Fuchs in such high esteem that, for medical attention, he and his wife spent some time in 1968 in the Central Committee's resort of Barhiva near Moscow, for instance.[18] Ironically, it was Fuchs's boss at Rossendorf, Heinz Barwich, who attracted some publicity when he defected to the United States in 1964 and it transpired that he had spied for the Central Intelligence Agency (CIA).[19]

Despite their contributions to the making of British nuclear culture, the achievements of Fuchs – apart from his spying – and Peierls have remained fairly unknown to the British public. Here, Albert Einstein represents a rare exception to the fate which Fuchs, Peierls and the overwhelming majority of German-speaking émigré scientists in Britain and the United States share. 'This has even proved very convenient to physicists like myself when asked in social contact to explain one's profession', Rudolf Peierls remarked on Einstein's fame. 'The occasion usually does not warrant a dissertation on the nature of physics; the answer "Einstein was a theoretical physicist" generally satisfies, even if it does not enlighten, the questioner'.[20]

What contributed to this lack of public awareness of Fuchs's and Peierls's scientific achievements was the (often forced) recruitment of German rocket scientists like Wernher von Braun and Ernst Stuhlinger by the United States and the Soviet Union after the war.[21] Paradoxically, scientists such as von Braun who had served under the National Socialist regime found themselves in the limelight. As they were celebrated in the United States for their success in the space race, they seemed to overshadow the legacy of the émigré atomic scientists who had come to Britain and the United States after Hitler's accession to power and supported the Allied war effort. A Gallup poll conducted in the United Kingdom in January 1959 demonstrates this phenomenon: 39 per cent of the respondents declared that 'Former German scientists' were the engine behind the United States' advancement in rocketry and satellites, while only 20 per cent accredited the progress in space technology to American scientists and 40 per cent did not give a clear answer. When asked about the Soviet space and missile programmes, a similar picture emerged (34 per cent for the 'German scientists'; 28 per cent for 'Russian scientists' and 38 per cent did not know the answer).[22]

So strong was the impact of German or 'former German' scientists like von Braun on public awareness that popular culture capitalized on them, too. The remark by one of the characters in the espionage film *Ice Station Zebra* (1968) reveals this. After an Anglo-American rescue party have recovered a special film developed by American scientists which together with a secret British-made camera had been abducted by Soviet agents, the British security officer David Jones (Patrick McGoohan) explains to US Navy Commander James Ferraday (Rock Hudson) how the film found its way not only into Russian hands but also onboard one of their satellites, saying 'then the Russians put our camera, made by our German scientists, and your film, made by your German scientists, into their satellite, made by their German scientists'. While the Gallup poll indicated that many Britons were in general aware of the presence of German scientists in Britain, the United States and the Soviet Union, Jones's statement reveals a self-conscious cynicism about the way in which the two superpowers used German science and scientists in the Cold War context.

But it was then in the character of Dr Strangelove (Peter Sellers), the protagonist of Stanley Kubrick's 1964 film of the same title, that this cynical view of 'former German' scientists such as Wernher von Braun was blended with negative stereotypes of German-speaking émigré nuclear scientists. In the film, Dr Strangelove embodies the nuclear threat in a supposedly 'German' body, which comprises an ambivalent and contradictory hodgepodge of traits drawn from various scientists from Leo Szilard to Edward Teller to Wernher von Braun. The protagonist of Kubrick's picture has emerged as perhaps the most dominant epitome of the German-speaking émigré nuclear scientist. It is against such stereotypes that this book set out to examine the elemental roles played by Rudolf Peierls and Klaus Fuchs in the making of British nuclear culture.

Notes and References

Introduction

1. Francis Nicosia, 'Nazi Persecution in Germany and Austria, 1933–1939', in *The Holocaust: Introductory Essays*, ed. by David Scrase and Wolfgang Mieder (Burlington, 1996), pp. 51–64.
2. 'Gesetz zur Wiederherstellung des Berufsbeamtentums vom 7. April 1933', *Reichsgesetzblatt*, 1. 34 (8 April 1933), 175–7.
3. Claus-Dieter Krohn, 'Vereinigte Staaten von Amerika', in *Handbuch der deutschsprachigen Emigration 1933–1945*, ed. by Claus-Dieter Krohn and others (Darmstadt, 1998), pp. 446–66 (p. 446); Herbert Strauss, 'The Movement of People in a Time of Crisis', in *The Muses Flee Hitler: Cultural Transfer and Adaptation 1930–1945*, ed. by Jarrell Jackman and Carla Borden (Washington, DC, 1983), pp. 45–59 (p. 47).
4. Gerhard Hirschfeld, 'German Refugee Scholars in Great Britain, 1933–1945', in *Refugees in the Age of Total War*, ed. by Anna Bramwell (London, 1988), pp. 152–63 (pp. 152–3); Louise London, *Whitehall and the Jews, 1933–1948: British Immigration Policy and the Holocaust* (Cambridge, 2000), pp. 11–12.
5. David Cassidy, 'Understanding the History of Special Relativity: Bibliographical Essay', *Historical Studies in the Physical and Biological Sciences* (hereafter *HSPS*), 16. 1 (1986), 177–95 (pp. 182–3).
6. *The Bethe-Peierls Correspondence*, ed. by Sabine Lee (Singapore, 2007); *Sir Rudolf Peierls: Selected Private and Scientific Correspondence*, ed. by Sabine Lee, 2 vols (Singapore, 2007–9).
7. Rudolf Peierls, *Atomic Histories* (Woodbury; New York, 1997); Rudolf Peierls, *Bird of Passage: Recollections of a Physicist* (Princeton, 1985); Richard Dalitz, 'Peierls, Sir Rudolf Ernst (1907–1995)', in *Oxford Dictionary of National Biography*, Oxford University Press, Sept 2004; online edn, Jan 2008 <http://www.oxforddnb.com/view/article/60076> [accessed 10 August 2011]; Sabine Lee, 'Rudolf Ernst Peierls, 5 June 1907–19 September 1995', *Biographical Memoirs of Fellows of the Royal Society* (hereafter *BMFRS*), 53 (December 2007), 265–84.
8. Here two short biographical sketches are perhaps the only exceptions: 'Fuchs, Klaus Emil Julius', in *Biographisches Handbuch der deutschsprachigen Emigration nach 1933*, 3 vols, ed. by Werner Röder and Herbert Strauss (Munich, 1980–3), I (1980), p. 206; Mary Flowers, 'Fuchs, (Emil Julius) Klaus (1911–1988)', in *Oxford Dictionary of National Biography*, Oxford University Press, 2004; online edn, May 2008 <http://www.oxforddnb.com/view/article/40698> [accessed 10 August 2011].
9. Alan Moorehead, *The Traitors: The Double Life of Fuchs, Pontecorvo and Nunn May* (London, 1952; repr. New York, 1963); Oliver Pilat, *The Atom Spies* (New York, 1952); Rebecca West, *The Meaning of Treason*, 2nd enlarged and rev. edn (London, 1952).

10. Montgomery Hyde, *The Atom Bomb Spies* (London; New York, 1980); Norman Moss, *Klaus Fuchs: The Man Who Stole the Atom Bomb* (New York, 1987); Robert Chadwell Williams, *Klaus Fuchs, Atom Spy* (Cambridge, MA, 1987).

11. Günter Flach, 'Klaus Fuchs – Sein Erbe bewahren', *Sitzungsberichte der Akademie der Wissenschaften der DDR, Mathematik – Naturwissenschaften – Technik*, 2/N (1990), 5–10; Gert Lange and Joachim Mörke, *Wissenschaft im Interview: Gespräche mit Akademiemitgliedern über ihr Leben und Werk* (Leipzig, 1979), pp. 33–44. Here, his father's autobiography represents an exception and offers some interesting views on Klaus Fuchs; Emil Fuchs, *Mein Leben*, 2 vols (Leipzig, 1957–9), II (1959).

12. Ronald Friedmann, *Der Mann, der kein Spion war: Das Leben des Kommunisten und Wissenschaftlers Klaus Fuchs* (Rostock, 2005); Eberhard Panitz, *Treffpunkt Banbury oder wie die Atombombe zu den Russen kam: Klaus Fuchs, Ruth Werner und der größte Spionagefall der Geschichte* (Berlin, 2003).

13. See *Changing Countries: The Experience and Achievement of German-Speaking Exiles from Hitler to Britain, from 1933 to Today*, ed. by Marian Malet and Anthony Grenville (London, 2002); *Forced Migration and Scientific Change: Émigré German-Speaking Scientists and Scholars after 1933*, ed. by Mitchell Ash and Alfons Söllner (Washington; New York, 1996); *German-Speaking Exiles in Great Britain*, ed. by Ian Wallace (Amsterdam, 1999); Jan-Christopher Horak, 'On the Road to Hollywood: German-Speaking Filmmakers in Exile 1933–1950', in *Kulturelle Wechselbeziehungen im Exil – Exile Across Cultures*, ed. by Helmut F. Pfanner (Bonn, 1986), pp. 240–8.

14. John Cornwell, *Hitler's Scientists: Science, War, and the Devil's Pact* (New York, 2004), pp. 38–40; Jean Medawar and David Pyke, *Hitler's Gift: Scientists Who Fled Nazi Germany* (London, 2001), p. 3.

15. I borrow this approach from Jutta Vinzent, *Identity and Image: Refugee Artists from Nazi Germany in Britain 1933–1945* (Weimar, 2006), pp. 23–8.

16. *Atomic Culture: How We Learned to Stop Worrying and Love the Bomb*, ed. by Scott Zeman and Michael Amundson (Boulder, 2004); Paul Boyer, *By the Bomb's Early Light: American Thought and Culture at the Dawn of the Atomic Age* (New York, 1985; repr. Chapel Hill, 1994); Allan Winkler, *Life under a Cloud: American Anxiety about the Atom* (New York, 1993; repr. Urbana, 1999).

17. Kirk Willis, 'The Origins of British Nuclear Culture, 1895–1939', *Journal of British Studies*, 34. 1 (1995), 59–89 (p. 60).

18. Clifford Geertz, *The Interpretation of Cultures* (New York, 1973; repr. London, 1975), p. 29.

19. See on the making of the atomic bomb: Margaret Gowing, *Britain and Atomic Energy, 1939–1945* (London, 1964); Lillian Hoddeson and others, *Critical Assembly: A Technical History of Los Alamos During the Oppenheimer Years, 1943–1945* (Cambridge, 1993); and on its consequences: *Atomic Culture*; Boyer, *By the Bomb's Early Light*; Winkler. A notable exception is *The Atomic Bomb and American Society: New Perspectives*, ed. by Rosemary Mariner and Kurt Piehler (Knoxville, 2009).

20. Andrew Pickering, *The Mangle of Practice: Time, Agency and Science* (Chicago, 1995), p. 3.

21. Jutta Weldes and others, 'Introduction: Constructing Insecurity', in *Cultures of Insecurity: States, Communities, and the Production of Danger*, ed. by Weldes and others (Minneapolis, 1999), pp. 1–33 (p. 2).

22. See Ulrich Beck, *Risikogesellschaft: Auf dem Weg in eine andere Moderne* (Frankfurt a.M., 1986).
23. David Nye, *American Technological Sublime* (Cambridge, MA, 1994), pp. 225–56.
24. Arnold, *Windscale 1957*, p. xxii.
25. See Linda Colley, *Britons: Forging the Nation 1707–1837*, new edn (London, 2003); Sonya Rose, *Which People's War? National Identity and Citizenship in Wartime Britain 1939–1945* (Oxford, 2003).
26. Spencer Weart, *Nuclear Fear: A History of Images* (Cambridge, MA, 1988), pp. 3–74.
27. See Bernhard Rieger, *Technology and the Culture of Modernity in Britain and Germany 1890–1945* (Cambridge, 2005).
28. Bernhard Kellermann, *The Tunnel* (London, 1915); H. G. Wells, *The World Set Free* (London, 1914).
29. See *Meanings of Modernity: Britain from the Late-Victorian Era to World War II*, ed. by Martin Daunton and Bernhard Rieger (Oxford, 2001).
30. Becky Conekin, Frank Mort and Chris Waters, 'Introduction', in *Moments of Modernity: Reconstructing Britain 1945–1964*, ed. by Conekin, Mort and Waters (London, 1999), pp. 1–21 (p. 3).
31. John Baylis, *Ambiguity and Deterrence: British Nuclear Strategy 1945–1964* (Oxford, 1995), pp. 34–66; Margaret Gowing, 'Britain and the Bomb: The Origins of Britain's Determination to Be a Nuclear Power', *Contemporary Record*, 2. 2 (1988), 36–40.
32. Gabrielle Hecht, *The Radiance of France: Nuclear Power and National Identity after World War II* (Cambridge, MA, 1998), p. 15.
33. Dennis Kavanagh and Peter Morris, *Consensus Politics: From Attlee to Major*, 2nd edn (Oxford, 1994), pp. 93–4.
34. Jim Tomlinson, 'Reconstructing Britain: Labour in Power 1945–1951', in *From Blitz to Blair: A New History of Britain Since 1939*, ed. by Nick Tiratsoo (London, 1998), pp. 77–101.
35. Central Office of Information, *Nuclear Energy in Britain*, 2nd end (London, 1960), pp. 9, 11–13.
36. Lorna Arnold, *Windscale 1957: Anatomy of a Nuclear Accident*, 2nd edn (Basingstoke, 1995), pp. xxi–xxii, 21–6; David Holloway, *Stalin and the Bomb: The Soviet Union and Atomic Energy, 1939–1956* (New Haven, CT, 1994), pp. 347–8.
37. Bernhard Rieger, 'Envisioning the Future: British and German Reactions to the Paris World Fair in 1900', in *Meanings of Modernity: Britain from the Late-Victorian Era to World War II*, ed. by Martin Daunton and Bernhard Rieger (Oxford, 2001), pp. 145–64.
38. Margot Henriksen, *Dr. Strangelove's America: Society and Culture in the Atomic Age* (Berkeley, 1997), p. xxiii.
39. Rieger, *Technology and the Culture of Modernity in Britain and Germany*, p. 233.
40. Donald Howard Avery, 'Atomic Scientific Co-Operation and Rivalry Among Allies: The Anglo-Canadian Montreal Laboratory and the Manhattan Project, 1943–1946', *War in History*, 2. 3 (1995), 274–305; Ross Galbreath, 'The Ruherford Connection: New Zealand Scientists and the Manhattan and Montreal Projects', *War in History*, 2. 3 (1995), 306–19; M. M. R. Williams, 'The Development of Nuclear Reactor Theory in the Montreal Laboratory

of the National Research Council of Canada (Division of Atomic Energy) 1943–1946', *Progress in Nuclear Energy*, 36. 3 (2000), 239–322.

41. See Lorna Arnold and Mark Smith, *Britain, Australia and the Bomb: The Nuclear Tests and Their Aftermath*, rev. edn (Basingstoke, 2006); John Crawford, '"A Political H-Bomb": New Zealand and the British Thermonuclear Weapon Test of 1957–58', *Journal of Imperial and Commonwealth History*, 26. 1 (1998), 127–50.

42. S. J. Ball, 'Military Nuclear Relations between the United States and Great Britain under the Terms of the McMahon Act, 1946–1958', *Historical Journal*, 38. 2 (1995), 439–54; Septimus Paul, *Nuclear Rivals: Anglo-American Atomic Relations, 1941–1952* (Columbus, 2000), pp. 94–108.

43. See John Krige, *American Hegemony and the Postwar Reconstruction of Science in Europe* (Cambridge, MA, 2006).

44. 'Atom village is agog – over duckling', *Daily Mirror*, 5 August 1946, p. 5.

45. Cmd. 9075, 'Statement on Defence 1954' (London, 1954), p. 4.

46. 'Atomic Energy Study Group: The Atomic Problem. Comments on AE/171 (Revised version), from Prof. M. L. Oliphant', May 1948, the Papers and Correspondence of Sir Rudolf Peierls, 1907–1995, Department of Western Manuscripts, Bodleian Library, University of Oxford, Oxford, United Kingdom (hereafter Peierls Papers), MS Eng. Misc. b. 223, F 5, p. 1.

47. Matthew Jones, 'Anglo-American Relations after Suez: The Rise and Decline of the Working Group Experiment and the French Challenge to NATO, 1957–59', *Diplomacy & Statecraft*, 14. 1 (2003), 49–78; Dilwyn Porter, '"Never-Never Land": Britain under the Conservatives 1951–1954', in *From Blitz to Blair* (see Tomlinson, above), pp. 102–31 (pp. 113–16).

48. Margaret Gowing, *Independence and Deterrence: Britain and Atomic Energy, 1945–1952*, 2 vols (London, 1974), I, 1–2.

49. See Ian Clark and Nicholas Wheeler, *The British Origins of Nuclear Strategy 1945–1955* (Oxford, 1989).

1 Difficult Beginnings: Social Integration between Survival and Internment

1. Hans Bethe, 'The Happy Thirties', in *Nuclear Physics in Retrospect: Proceedings of a Symposium on the 1930s*, ed. by Roger Stuewer (Minneapolis, 1977), pp. 11–31.

2. Fanchon Fröhlich, 'Biographical Notes', in *Cooperative Phenomena*, ed. by Hermann Haken and Max Wagner (Berlin, 1973), pp. 420–1 (p. 421).

3. Strauss, 'Movement of People', pp. 50–1.

4. Jan-Christopher Horak, 'Filmkünstler im Exil: Ein Weg nach Hollywood', in *Die Künste und die Wissenschaften im Exil 1933–1945*, ed. by Edith Böhme and Wolfgang Motzkau-Valeton (Gerlingen, 1992), pp. 231–54 (p. 231).

5. 'Conditional Landing. Immigration Officers Report, 25 September 1933', The National Archives, Kew, Richmond, Surrey, United Kingdom (hereafter TNA), KV 2/1245.

6. Peierls, *Bird of Passage*, pp. 56–81.

7. *Refugee Scholars: Conversations with Tess Simpson*, ed. by Ray M. Cooper (Leeds, 1992), p. 58; Peierls, *Bird of Passage*, pp. 90–8.

8. Lenz to Peierls, 9 March 1933, Peierls Papers, MS Eng. Misc. b. 197, A 2, fols 28^r–30^r; Rudolf Peierls, Interview by Charles Weiner, 11–13 August 1969, Oral History Collections, Niels Bohr Library and Archives, American Institute of Physics, College Park, Maryland, United States (hereafter AIP), p. 3.
9. Peierls, Interview by Weiner, p. 2.
10. Otto R. Frisch, Interview by Charles Weiner, 3 May 1967, AIP, pp. 12–13.
11. Clausen to Frisch, 19 June 1933, Society for the Protection of Science and Learning Papers, Department of Western Manuscripts, Bodleian Library, University of Oxford, Oxford, United Kingdom (hereafter MS S.P.S.L.) MS S.P.S.L. 327/10, fol. 460^r.
12. Peierls, Interview by Weiner, p. 2.
13. Peierls, *Bird of Passage*, pp. 140–1, 151. See also the correspondence between Rudolf Peierls and his father and stepmother in Germany as well as between Rudolf and Genia Peierls and their children, Peierls Papers, supplementary catalogue (hereafter sup. cat.), A.120.
14. Rudolf E. Peierls, 'Our Relations with German Scientists', n.d., TNA, AB3/94, pp. 1–2, 9.
15. Irene Dittrich, 'Die "Revolutionäre Studentengruppe" an der Christian-Albrechts-Universität zu Kiel (1930–1933)', *Demokratische Geschichte*, 4 (1989), 175–84 (pp. 180–2); Emil Fuchs, *Mein Leben*, II, 200–1, 228–38, 245–68; Peter Wulf, 'Die Stadt auf der Suche nach ihrer neuen Bestimmung (1918 bis 1933)', in *Geschichte der Stadt Kiel*, ed. by Jürgen Jensen and Peter Wulf (Neumünster, 1991), pp. 303–58 (p. 358); *Vertriebene Wissenschaftler der Christian-Albrechts-Universität zu Kiel (CAU) nach 1933: Zur Geschichte der CAU im Nationalsozialismus – Eine Dokumentation bearbeitet von Uta Cornelia Schmatzler und Matthias Wieben*, ed. by Ralph Uhlig (Frankfurt a.M., 1991), pp. 48–9.
16. Emil Fuchs, *Christ in Catastrophe* (Wallingford, 1949).
17. Marion Berghahn, *Continental Britons: German-Jewish Refugees from Nazi Germany* (Oxford, 1988), p. 138.
18. Thomas Elsaesser, 'Ethnicity, Authenticity, and Exile: A Counterfeit Trade? German Filmmakers and Hollywood', in *Home, Exile, Homeland: Film, Media, and the Politics of Place*, ed. by Hamid Naficy (New York, 1999), pp. 97–123 (p. 113).
19. Max Born, *My Life: Recollections of a Nobel Laureate* (London, 1978), p. 281.
20. Berghahn, p. 82.
21. In the German version of his autobiography, Max Born uses this term; *Mein Leben: Die Erinnerungen des Nobelpreisträgers* (Munich, 1975), p. 377.
22. Thomas Lekan has stressed the significance of the German landscape and the relationship between German communities abroad and their homelands; 'German Landscape: Local Promotion of the *Heimat* Abroad', in *The Heimat Abroad: The Boundaries of Germanness*, ed. by Krista O'Donnell, Renate Bridenthal and Nancy Reagin (Ann Arbor, 2005), pp. 141–66.
23. Daniel Snowman, *The Hitler Émigrés: The Cultural Impact on Britain of Refugees from Nazism* (London, 2002), p. 60.
24. Peierls, *Atomic Histories*, p. 362.
25. James M. Ritchie, *German Exiles: British Perspectives* (New York, 1997), p. 9.
26. This is even the title of Berghahn's study.
27. Otto R. Frisch, *What Little I Remember* (Cambridge, 1979), p. 71.

28. Frisch, Interview by Weiner, p. 19.
29. Kurt Mendelssohn, 'The Coming of the Refugee Scientist', *New Scientist*, 26 May 1960, pp. 1343–44 (p. 1343).
30. Peierls, *Bird of Passage*, p. 149.
31. Frisch to Peierls, 13 May 1941, TNA, AB 1/574.
32. Frisch, Interview by Weiner, p. 40.
33. Angus Calder, *The People's War: Britain 1939–1945* (London, 1969; repr. London, 1997), pp. 118–26.
34. 'Alien Registration Form: Fuchs, Emil Julius Klaus', n.d., TNA, KV 2/1259.
35. The Under Secretary of State, Aliens Department, Home Office, London to Klaus Fuchs, 30 August 1937, TNA, KV 2/1259.
36. Paul Hoch, 'The Reception of Central European Refugee Physicists of the 1930s: U.S.S.R., U.K., U.S.A.', *Annals of Science*, 40. 3 (1983), 217–46 (p. 222).
37. Robin Rider, 'Alarm and Opportunity: Emigration of Mathematicians and Physicists to Britain and the United States, 1933–1945', *HSPS*, 15 (1985), 101–76 (p. 131).
38. Hoch, 'Reception of Central European Refugee Physicists', pp. 224–5, 230.
39. Note that because the Academic Assistance Council (AAC) was renamed the Society for the Protection of Science and Learning (SPSL) in 1936, I refer to the organization hereafter jointly as AAC/SPSL. Rudolf Peierls, personal information form, 11 October 1934, MS S.P.S.L. 335/9, fol. 471^r.
40. Hans Bethe, Interview by Lillian Hoddeson, 29 April 1981, AIP, p. 29.
41. Walter Moberly, 'Ministry of Labour, Employment and Training Department. Application for Permission to Employ an Alien or Aliens Not Now in the United Kingdom: Peierls, Rudolf', 3 October 1933, MS S.P.S.L. 438/2, fols 266^r–266^v; Moberly to Gibson, 3 October 1933, MS S.P.S.L. 438/2, fol. 267^r; Peierls, *Bird of Passage*, p. 96; Bill Williams, '"Displaced Scholars": Refugees at the University of Manchester', *Melilah: Manchester Journal of Jewish Studies*, 3 (2005) <http://www.mucjs.org/MELILAH/2005/3.pdf> [accessed 10 August 2011], 1–29 (p. 6 note 31).
42. *Refugee Scholars*, p. 58. David Zimmerman offers a concise overview of the history of the AAC/SPSL in 'The Society for the Protection of Science and Learning and the Politicization of British Science in the 1930s', *Minerva*, 44. 1 (2006), 25–45.
43. Bragg to AAC/SPSL, 1 February 1935, MS S.P.S.L. 335/9, fol. 476^r.
44. Peierls, *Bird of Passage*, pp. 114, 127.
45. Rudolf Peierls, personal information form, 11 October 1934, MS S.P.S.L. 335/9, fols 471^r–475^r.
46. Note that in the rare cases where German-speaking, émigré scientists anglicized the spelling of their names (usually quite soon after their arrival in Britain), this form is used throughout the thesis. The original spelling is given in brackets the first time a particular name is mentioned. The case of Franz Simon represents the only exception to this rule because he changed his name in 1946, upon the award of the CBE, Later to Sir Francis Simon.
47. AAC/SPSL, 'Displaced Scientists Resident in Great Britain', n.d., attached to letter, Simpson to Mills, 28 February 1936, MS S.P.S.L. 330/1, fols 89^r–92^r.
48. Hans Bethe, personal file, MS S.P.S.L. 324/4, fols 125^r–138^r; Max Born, personal file, MS S.P.S.L. 325/3, fols 43^r–198^r; Otto R. Frisch, personal file, MS S.P.S.L. 327/10, fols 441^r–524^r; Nikolai Kemmer, personal file, MS S.P.S.L.

509/4, fols 478^r–515^r; Heinrich Kuhn, personal file, MS S.P.S.L. 333/3, fols 45^r–102^r; Nicholas Kurti, personal file, MS S.P.S.L., fols 136^r–163^r; Fritz London, personal file, MS S.P.S.L. 334/4, fols 260^r–340^r; Heinz London, personal file, MS S.P.S.L. 334/5, fols 342^r–391^r; Lothar Nordheim, personal information form, n.d., MS S.P.S.L. 335/7, fols 324^r–332^r; Erwin Schrödinger, personal file, MS S.P.S.L. 339/4, fols 289^r–374^r; Franz Simon, personal file, MS S.P.S.L. 339/8, fols 450^r–541^r.

49. Bernard Wasserstein, 'Intellectual Émigrés in Britain, 1933–1939', in *The Muses Flee Hitler* (see Strauss, 'Movement of People', above), pp. 251–2.

50. Peierls to Adams, 10 June 1937, MS S.P.S.L. 175/1, fol. 528^r; Peierls to Adams, 19 June 1937, MS S.P.S.L. 175/1, fol. 74^r.

51. Hirschfeld, '"A High Tradition of Eagerness..."', p. 603.

52. Hoch, 'Reception of Central European Refugee Physicists', pp. 224–5, 230; Murrow to Adams, 18 June 1934, MS S.P.S.L. 324/4, fol. 128^r.

53. Peierls to Adams, 31 January 1935, MS S.P.S.L. 335/9, fol. 507^r.

54. Skepper to Kuhn, 19 October 1934; Kuhn to Adams, 26 October 1934, fols 78^r–80^r; Secretary [of the AAC] to Kuhn, 26 January 1935, MS S.P.S.L. 333/3, fol. 82^r.

55. Hoch, 'Reception of Central European Refugee Physicists', pp. 222–3.

56. Skepper to Nordheim, 6 September 1934, MS S.P.S.L. 335/7, fol. 347^r.

57. Berghahn, p. 122.

58. Charles Weiner, 'A New Site for the Seminar: The Refugees and American Physics in the 1930s', in *The Intellectual Migration: Europe and America, 1930–1960*, ed. by Donald H. Fleming and Bernard Baylin (Cambridge, MA, 1969), pp. 190–233 (pp. 190–1).

59. Helmut G. Asper, 'Film', in *Handbuch der deutschsprachigen Emigration* (see Krohn, above), pp. 957–70 (pp. 957, 964); Hoch, 'Reception of Central European Refugee Physicists', pp. 231–4; Strauss, 'Movement of People', p. 50.

60. Born to Simpson, 27 October 1937; Simpson to Born, 29 October 1937; Born to Simpson, 5 November 1937; Simpson to Born, 6 November 1937, MS S.P.S.L. 328/1, fols 149^r–152^r.

61. Klaus Fuchs, personal information form, 8 November 1937, MS S.P.S.L. 328/1, fols 141^r–146^r.

62. Adams to Born, 18 January 1938; Born to Adams, 19 January 1938; Adams to Fuchs, 21 January 1938, MS S.P.S.L. 328/1, fols 153^r–155^r.

63. Fuchs to Adams, 24 January 1938, MS S.P.S.L. 328/1, fol. 156^r.

64. Demuth to Simpson, 15 July 1938; Simpson to Demuth, 16 July 1938; Adams to Demuth, 22 July 1938, MS S.P.S.L. 328/1, fols 159^r–161^r.

65. Herbert Fröhlich, 'Moving On: Experiences of a Scientist in Exile', *University of Liverpool Recorder*, 93 (October 1983), 222–28 (p. 226); Heitler to Adams, 28 February 1936, MS S.P.S.L. 330/1, fol. 93^r.

66. See, for example, Peierls to Simpson, 11 October 1938; Simpson to Cooper, 11 October 1938, MS S.P.S.L. 438/2, fols 276^r–277^r.

67. The AAC/SPSL kept copies of the Home Office files of the following scientists: Rudolf Peierls (MS S.P.S.L. 438/2, fols 266^r–283^r); Klaus Fuchs (MS S.P.S.L. 430/2, fols 286^r–351^r); Herbert Fröhlich (MS S.P.S.L. 430/2, fols 249^r–285^r); Egon Orowan (MS S.P.S.L. 428/2, fols 222^r–239^r).

68. Hartley-Hodder to Fuchs, 23 October 1934, MS S.P.S.L. 430/2, fol. 287^r; Mott to The Secretary, AAC, 25 October 1934, MS S.P.S.L. 430/2, fol. 288^r.

69. Skepper to Mott, 26 October 1934; Skepper to The Secretary, The High Commission, 26 October 1934, MS S.P.S.L. 430/2, fols 289^r–291^r.

70. Fuchs to Thomson, 15 September 1939; Thomson to Fuchs, 15 September 1939 (included the questionnaire which Fuchs completed and returned to Thomson with his letter (fol. 162^r), MS S.P.S.L. 328/1, fols 162^r–163^r; Simpson to London, November 1940, MS S.P.S.L. 334/5, fol. 357^r).

71. Ronald Stent, 'Jewish Refugee Organisations', in *Second Chance: Two Centuries of German-Speaking Jews in the United Kingdom*, ed. by Werner E. Mosse and others (Tübingen, 1991), pp. 579–98.

72. William H. Beveridge, *A Defence of Free Learning* (London, 1959), pp. 8–22, 24–6; Gerhard Hirschfeld, 'A High Tradition of Eagerness...', in *Second Chance* (see Stent, 'Jewish Refugee Organisations', above), pp. 599–610.

73. Rider, pp. 146–150.

74. Hoch, 'Reception of Central European Refugee Physicists', p. 226.

75. William Farren and George P. Thomson, 'Frederick Alexander Lindemann, Viscount Cherwell. 1886–1957', *BMFRS*, 4 (November 1958), 45–71 (p. 45); Adrian Fort, *Prof: The Life of Frederick Lindemann* (London, 2004), pp. 15–40, 115–27, 182; Mendelssohn, p. 1343.

76. Stefan L. Wolff, 'Frederick Lindemanns Rolle bei der Emigration der aus Deutschland vertriebenen Physiker', in *German-Speaking Exiles in Great Britain*, ed. by Anthony Grenville (Amsterdam, 2000), pp. 25–58.

77. Fuchs to Thomson, 15 September 1939, MS S.P.S.L. 328/1, fol. 162^r.

78. Born to Simpson, 5 July 1940; Born to Simpson, 24 October 1940; Simpson to Born, 29 October 1940, MS S.P.S.L. 328/1, fols 167^r–168^r, 170^r; Mott to Simpson, 26 October 1940, MS S.P.S.L. 328/1, fol. 169^r.

79. Peierls to Born, 5 November 1940; Peierls to Born, 27 November 1940, TNA, AB 1/572.

80. Born to Peierls, 11 March 1941; Peierls to Born, 12 March 1941; Peierls to Born, 22 March 1941; Peierls to Born, 10 May 1941; Peierls to Born, 16 May 1941; Born to Peierls, n.d. (reply to a letter sent by Peierls to Born dated 26 May 1941), TNA, AB 1/572.

81. Bethe, Interview by Hoddeson, p. 4; Frisch, *What Little I Remember*, p. 130; Secretary [of the AAC/SPSL] to Fröhlich, 30 July 1935, MS S.P.S.L. 430/2, fol. 253^r; W. J. Skardon, 'Emil Julius Klaus Fuchs', 22 December 1949, TNA, KV 2/1249, p. 2.

82. Secretary [of the AAC/SPSL] to Peierls, 28 January 1935, MS S.P.S.L., 328/1, fol. 41^r; Peierls to Adams, 31 January 1935, MS S.P.S.L. 335/9, fol. 507^r. Max Born also helped; Born to Adams, n.d.; Adams to Born, 11 June 1935, fols 42^r–43^r; Born to Adams, 26 June 1935, MS S.P.S.L. 328/1, fols 45^r–45^v.

83. Skepper, note, 'Herbert Fröhlich, Physics', 24 January 1935, MS S.P.S.L. 328/1, fol. 14^r; Peierls to Adams, 6 July 1935; General Secretary [Adams] to Peierls, 10 July 1935, MS S.P.S.L. 328/1, fols 48^r–49^r.

84. Berghahn, p. 139.

85. Robert Vansittart, *Black Record: Germans Past and Present* (London, 1941).

86. Born to Peierls, April 1942, TNA, AB 1/572; Victor Gollancz, *Shall Our Children Live or Die?* (London, 1942).

87. Berghahn, pp. 140–2; Stent, 'Jewish Refugee Organisations', p. 588.

88. Peierls, *Bird of Passage*, p. 145.

89. Mendelssohn, p. 1344.

90. Louise London, 'British Immigration Control Procedures and Jewish Refugees 1933–1939', in *Second Chance* (see Stent, 'Jewish Refugee Organisations', above), pp. 485–517 (p. 500).
91. Chief Constable, Central Police Office, Bristol to Sir Vernon, RE: Emil Julius Klaus Fuchs, 5 November 1934, TNA, KV 2/1245.
92. Translation of letter, Police Chief, Kiel, to German Consulate, Bristol, 16 October 1934, TNA, KV 2/1245.
93. Translation of letter, Klaus Fuchs to Registration Office, Kiel, Germany, RE: Certificate of no objections against the new issuing of a passport, 6 October 1934, TNA, KV 2/1245.Reference to this letter is also made, for example, in J. C. Robertson, 'Emil Julius Klaus Fuchs', 23 November 1949, TNA, KV 2/1248, p. 5.
94. German Embassy, London, to German Consulate, Bristol, 7 August 1934; Hartley-Hodder, Konsul, to Fuchs, 9 August 1934; Hartley-Hodder, Konsul, to Fuchs, 23 October 1934, TNA, KV 2/1245.
95. Charmian Brinson, 'The Gestapo and the German Political Exiles in Britain During the 1930s: The Case of Hans Wesemann – and Others', *German Life and Letters*, 51.1 (1998), 43–64 (pp. 43–7). See Müller to Leiter der Staatspolizeistellen oder Vertreter im Amt, RE: Flüchtige Kommunisten, Landesarchiv Schleswig-Holstein, Schleswig, Germany, (hereafter LASH), Abt. 455, Nr. 21. The list for the city of Kiel, however, dates from 1936 and does not contain the name of Klaus Fuchs; 'Emigranten-Liste. Nachweisung über emigrierte Personen, die im Bezirk der Stapo Kiel bekannt georden sind', n.d., LASH, Abt. 455, Nr. 9.
96. 'Merkblatt zur Beachtung bei der terminmäßigen Berichterstattung an Gestapa', attached to the 'Emigranten-Liste. Nachweisung über emigrierte Personen, die im Bezirk der Stapo Kiel bekannt geworden sind', n.d., LASH, Abt. 455, Nr. 9.
97. 'Attachment. Received 28.11.49', TNA, KV 2/1248, p. 1.
98. Gerhard Paul, 'Die Gestapozentrale in der Düppelstraße 23: Die Zentrale des NS-Terrors in Schleswig-Holstein', in *Täter und Opfer unter dem Hakenkreuz: Eine Landespolizei stellt sich der Geschichte*, ed. by Förderverein 'Freundeskreis zur Unterstützung der Polizei Schleswig-Holstein e.V.' (Kiel, 2001), pp. 43–50 (p. 46).
99. London, *Whitehall and the Jews*, p. 12.
100. Hirschfeld, '"A High Tradition of Eagerness ..."', p. 604; Wasserstein, pp. 250–1.
101. Peierls, *Bird of Passage*, p. 145; Peierls, Interview by Weiner, pp. 93–4.
102. Fletcher to Ladd, RE: Emil Julius Klaus Fuchs, 21 September 1949, Ferenc Szasz Papers, Center for Southwest Research, University of New Mexico, Albuquerque, New Mexico, United States, MSS 552 BC (hereafter Klaus Fuchs FBI File), 65-58805-7, vol. 1, serials 1–26, p. 27.
103. Frisch, *What Little I Remember*, pp. 127–8.
104. Rudolf Peierls, *Bird of Passage*, p. 151.
105. 'Fuchs, Klaus', p. 206.
106. The Under Secretary of State, Home Office (Aliens Department), London to Fuchs, 17 August 1938; The Chief Constable, City Police, Edinburgh to the Superintendent, Central Register of Aliens, Home Office, RE: Fuchs, Emil Julius Klaus, 23 August 1938, TNA, KV 2/1259.

107. Berghahn, p. 139.
108. Frisch, Interview by Weiner, pp. 40–1; Frisch, *What Little I Remember*, p. 127; Peierls, *Bird of Passage*, p. 145.
109. Klaus Fuchs's 'Application by an Alien for Permission to Reside in a Protected Area, 24 April 1940', TNA, KV 2/1259.
110. Fuchs to the Chief Constable, the City of Edinburgh, 26 May 1942; Assistant Chief Constable, City of Edinburgh, 28 May 1942; The Chief Constable, City of Birmingham to the Chief Constable, City Police, Edinburgh, RE: Aliens (Movement Restriction) Order, 1940, 30 May 1942, TNA, KV 2/1259.
111. Peierls, *Bird of Passage*, pp. 145, 147–8.
112. Panikos Panayi, 'An Intolerant Act by an Intolerant Society: The Internment of Germans in Britain During the First World War', in *The Internment of Aliens in Twentieth Century Britain*, ed. by David Cesarani and Tony Kushner (London, 1993), pp. 53–75.
113. Peter Gillman and Leni Gillman, *'Collar the Lot!' How Britain Interned and Expelled Its Wartime Refugees* (London, 1980), p. 5.
114. David Cesarani, 'An Alien Concept? The Continuity of Anti-Alienism in British Society Before 1940', in *The Internment of Aliens in Twentieth Century Britain* (see Panayi, above), pp. 25–52.
115. *The Gallup International Public Opinion Pools: Great Britain 1937–1975*, ed. by George H. Gallup, 2 vols (New York, 1976), I, 22, 33, 34.
116. Tony Kushner and David Cesarani, 'Alien Internment in Britain during the Twentieth Century: An Introduction', in *The Internment of Aliens in Twentieth Century Britain* (see Panayi, above), pp. 1–22 (p. 1).
117. François Lafitte, *The Internment of Aliens*, new edn (London, 1988), p. vii.
118. Ibid., pp. 62–5.
119. Chief Constable, City Police, Edinburgh to the Chief Constable, City Police, Birmingham, 9 June 1941, TNA, KV 2/1259; Peierls, *Bird of Passage*, pp. 145–6.
120. Aliens Registration Department, City Police Chambers, Edinburgh to Emil J. K. Fuchs, 25 October 1939; A. Macauley to Chief Constable, Birmingham City Police, RE: Control of Aliens. Emil Julius Klaus Fuchs – German, 16 August 1941, TNA, KV 2/1259.
121. 'Male Enemy Alien – Exemption from Internment – Refugee: Fuchs, Emil Julius Klaus, 2 November 1939', TNA, KV 2/1259.
122. Born [to Aliens Tribunal], 30 October 1939; Sturge to the Chairman, Aliens Tribunal, Edinburgh, 30 October 1939, TNA, KV 2/1259.
123. Gillman and Gillman, p. 45.
124. London, *Whitehall and the Jews*, p. 170.
125. Cited in Gillman and Gillman, p. 153.
126. Tony Kushner, 'Clubland, Cricket Tests and Alien Internment, 1939–40', in *The Internment of Aliens in Twentieth Century Britain* (see Panayi, above), pp. 79–101 (pp. 87–8).
127. 'Alien Registration Form: Fuchs, Emil Julius Klaus', n.d., TNA, KV 2/1259; Assistant Chief Constable to the Manager, Employment Exchange, Edinburgh, 1 November 1940, TNA, KV 2/1259; Robert Williams, *Klaus Fuchs*, pp. 32–3.
128. Born to Sir Thomas, 29 May 1940, TNA, KV 2/1246.

129. Max Born, *My Life*, p. 286.
130. Born to Simpson, 22 May 1940, MS S.P.S.L. 328/1, fols 164^r–165^r (fol. 164^r).
131. Simpson to Born, 25 May 1940, MS S.P.S.L 328/1, fol. 166^r.
132. Heitler and London to Simpson, 1 July 1940, MS S.P.S.L. 328/1, fol. 104^r; Hoch, 'Reception of Central European Refugee Physicists', p. 228; Leonard Rotherham, 'Hans Kronberger, 1920–1970', *BMFRS*, 18 (November 1972), 412–26 (p. 414); Tyndall to Heitler, 17 June 1940; 'University of Bristol: Alien Scientists and Research Workers in Internment', 7 August 1940, attached to letter, Tyndall to Under Secretary of State, Aliens Department, 7 August 1940, Departmental Archives of the H.H. Wills Physics Laboratory, University of Bristol, Bristol, United Kingdom (hereafter H.H. Wills Physics Laboratory).
133. Tyndall to Egerton, 26 June 1940, MS S.P.S.L. 328/1, fols 103^r–103^v (fol. 103^r). Arthur M. Tyndall became involved in the matter of his interned colleagues, Tyndall to [Simpson], 10 July 1940, MS S.P.S.L. 328/1, fol. 105^r.
134. See, for example, Tyndall to Thomson, 12 June 1940; Tyndall, 'Memorandum on Nuclear Physics Work at Wills Physical Laboratory, University, Bristol', n.d., attached to letter, Tyndall to Pye, 24 June 1940, H.H. Wills Physics Laboratory. The memorandum also mentions another German-speaking émigré, K. Sternschluss (p. 2).
135. Tyndall to Thomson, 18 June 1940, H.H. Wills Physics Laboratory, p. 3.
136. Thomson to Tyndall, 17 June 1940, H.H. Wills Physics Laboratory, p. 1.
137. 'Kronberger, Hans', in *Biographisches Handbuch der deutschsprachigen Emigration nach 1933* (see 'Fuchs, Klaus Emil Julius', above), II (1983), p. 668.
138. Bünemann to The Secretary, Society for Protection of Science [and Learning], 20 January 1941, MS S.P.S.L. 474/3, fols 359^r–361^r.
139. Oscar [*sic*] Bünemann, 'Curriculum Vitae', May 1939, MS S.P.S.L. 474/3, fol. 353^r.
140. Perrin to Gorrell Barnes, 'T.A. Staff for U.S.A.', 31 January 1944, TNA, CAB 126/331, p. 3; Chadwick to Bünemann, 1 June 1945; Chadwick to Massey, 25 July 1945, the Papers of Sir James Chadwick, 1914–1974, Churchill Archives Centre, Churchill College, University of Cambridge, Cambridge, United Kingdom (hereafter CHAD), CHAD IV/3/7; 'A.E.R.E. Programme No. 3, September 1947', pp. 1–5.
141. Wasserstein, p. 254.
142. Moorehead, *The Traitors*, pp. 78–81; Michael Seyfert, '"His Majesty's Most Loyal Internees": The Internment and Deportation of German and Austrian Refugees as "Enemy Aliens". Historical, Cultural and Literary Aspects', in *Exile in Great Britain: Refugees from Hitler's Germany*, ed. by Gerhard Hirschfeld (Leamington Spa; Atlantic Highlands, 1984), pp. 163–93 (p. 175); Ronald Stent, *A Bespattered Page? The Internment of His Majesty's "Most Loyal Aliens"* (London, 1980), pp. 96–7.
143. Born to Simpson, 5 July 1940, MS S.P.S.L. 328/1, fol. 167^r.
144. Max F. Perutz, *Is Science Necessary? Essays on Science & Scientists* (London, 1989; repr. Oxford, 1992), pp. 102–3.
145. See Klaus Fuchs's internee record in TNA, KV 2/1253.
146. Robert Williams, *Klaus Fuchs*, p. 33. See also Dr. Glücksmann's paper cited in Beveridge, pp. 86–90.

147. Fröhlich, 'Moving On', p. 226.
148. 'Statement of Emil Julius Klaus Fuchs, of 17 Hillside, Harwell, Berkshire, who saith-', 27 January 1950, TNA, KV 2/1263, p. 6.
149. Sack to Tyndall, 5 July 1940. See also other correspondence between interned German-speaking, émigré scientists (and family members) and Arthur Tyndall: Fröhlich to Tyndall, 8 August 1940; Hoselitz to Tyndall, 20 August 1940; Maria Gross to Tyndall, 4 September 1940, H.H. Wills Physics Laboratory.
150. 'Interrogation by Mr. Serpell at Room 055, 23 March 1950', TNA, KV 2/1270, pp. 2–4. The description of a 'predominantly Jewish' atmosphere at Camps 'L' and 'N' contradicts Alan Moorehead's and Rudolf Peierls's statement that Fuchs was erroneously transferred to a camp for hardcore Nazi sympathizers in Canada; Moorehead, *The Traitors*, pp. 78–81; Peierls, *Bird of Passage*, p. 163.
151. 'Interrogation by Mr. Serpell at Room 055, 23 March 1950', TNA, KV 2/1270, p. 11; Perutz, pp. 102–3; *Prof. Dr. Klaus Fuchs: Kundschafter aus Überzeugung*, BstU, MfS ZAIG/Vi/227; Skardon, 'Emil Julius Klaus Fuchs', 22 December 1949, p. 2.
152. Louise Burletson, 'The State, Internment and Public Criticism in the Second World War', in *The Internment of Aliens in Twentieth Century Britain* (see Panayi, above), pp. 102–24 (pp. 106, 111–12, 115–16, 121).
153. 'Internee's Consent to Release and Return to the United Kingdom, RE: Fuchs, Klaus E. J.', n. d., but enclosed with letter, Brigadier-General E. de E. Panet, District Officer Commanding, Military District No. 4 to Director of Internment Operations, Department of the Secretary of State, Ottawa, Ontario, 17 December 1940, and Fuchs's personally signed medical waiver form (No. 581. Fuchs, K., 15 December 1940), TNA, KV 2/1253; The Under Secretary of State, Home Office (Aliens Department), London to the Chief Constable, City Police, Edinburgh, 17 October 1940; The Chief Constable, City Police Edinburgh to the Superintendent, Central Register of Aliens, Home Office, London, RE: Fuchs, Emil Julius Klaus, 24 January 1941; Macauley to Chief Constable, Birmingham City Police, RE: Control of Aliens. Emil Julius Klaus Fuchs – German, 16 August 1941, TNA, KV 2/1259.
154. Max Born [to Aliens Tribunal], 30 October 1939, TNA, KV 2/1259; Peierls, *Bird of Passage*, pp. 163–4.

2 Almost Accidental Beginnings: Professional Integration between Marginalization and British–American Nuclear Cooperation

1. Berghahn, p. 80; Hoch, 'Reception of Central European Refugee Physicists', pp. 234–9; Helge Kragh, *Quantum Generations: A History of Physics in the Twentieth Century* (Princeton, 1999), p. 249.
2. Alan D. Beyerchen, *Scientists under Hitler: Politics and the Physics Community in the Third Reich* (New Haven, 1977), pp. 6–9; Medawar and Pyke, p. 8.
3. Klaus Bärwinkel, 'Die Austreibung von Physikern unter der deutschen Regierung vor dem Zweiten Weltkrieg. Ausmaß und Auswirkung', in *Die Künste und die Wissenschaften im Exil 1933–1945* (see Horak, 'Filmkünstler im Exil'), pp. 569–99 (p. 581).

4. Peierls, *Bird of Passage*, pp. 16–32, 40–5, 90–8.
5. S. T. Keith and Paul Hoch, 'Formation of a Research School: Theoretical Solid State Physics at Bristol 1930–54', *British Journal for the History of Science*, 18. 3 (1986), 19–44 (pp. 19, 33–4); Bernard Lovell, 'Bristol and Manchester – The Years 1931–9', in *The Making of Physicists*, ed. by Rajkumari Williamson (Bristol, 1997), pp. 148–60 (p. 156); Tyndall to Secretary, AAC, 19 June 1934, MS S.P.S.L. 324/4, fol. 126^r.
6. Peierls, Interview by Weiner, p. 35.
7. Francis L. Carsten, 'German Refugees in Great Britain 1933–1945: A Survey', in *Exile in Great Britain* (see Seyfert, above), pp. 11–28 (p. 16); Hoch, 'Reception of Central European Refugee Physicists', pp. 231–4.
8. Hoch, 'Reception of Central European Refugee Physicists', pp. 237–8.
9. Stuart Hughes, *The Sea Change: The Migration of Social Thought, 1930–1965* (New York, 1975), p. 38; Helge Pröss, *Die deutsche akademische Emigration nach den Vereinigten Staaten, 1933–41* (Berlin, 1955), pp. 33–6.
10. Hoch, 'Reception of Central European Refugee Physicists', p. 219.
11. Gustav Born, 'The Effect of the Scientific Environment in Britain on Refugee Scientists from Germany and Their Effects on Science in Britain', *Berichte zur Wissenschaftsgeschichte* (hereafter *Ber. Wissenschaftsgesch.*), 7. 3 (1984), 129–43.
12. Paul Hoch and E. J. Yoxen, 'Schrödinger at Oxford: A Hypothetical National Cultural Synthesis which Failed', *Annals of Science*, 44. 6 (1987), 593–616.
13. Karl Popper, *Unended Quest: An Intellectual Biography* (La Salle, 1976), p. 108.
14. Gerard Hyland, 'Herbert Fröhlich, FRS: A Physicist Ahead of His Time', in *Herbert Fröhlich, FRS: A Physicist Ahead of His Time*, ed. by Gerard Hyland and Peter Rowlands (Liverpool, 2006), pp. 221–339 (p. 255).
15. Hoch, 'Reception of Central European Refugee Physicists', p. 231.
16. Martin J. Sherwin, *A World Destroyed: The Atomic Bomb and the Grand Alliance* (New York, 1975), p. 18.
17. Rudolf Peierls, 'Als Student bei Heisenberg', in *Werner Heisenberg in Leipzig 1927–1942*, ed. by Christian Kleint and Gerald Wiemers (=*Abhandlungen der Sächsischen Akademie der Wissenschaften zu Leipzig, Mathematisch-Naturwissenschaftliche Klasse*, 28. 2 (1993)), pp. 104–7; Peierls, *Bird of Passage*, pp. 34–5, 39–40; 'Verzeichnis der Studenten und Hörer bei Werner Heisenberg', ed. by Gerald Wiemers, in *Werner Heisenberg in Leipzig 1927–1942* (see Peierls, 'Als Student bei Heisenberg', above), pp. 144–72 (p. 151).
18. See 'Report on German Publications', n.d., attached to letter, Peierls to Chadwick, 23 September 1941; Peierls to Chadwick, 20 November 1941, CHAD I/19/6; 'Report on the 46th Meeting of the Deutschen Bunsen Gesellschaft in Zeitschrift für Electrochemie. Volume 47. December 1941. pp. 819–820'; 'Report on Current German Literature. February, 1942'; 'Report on current German literature. August 1942'; Peierls to Cherwell, 10 December 1942; Peierls to Simon, 8 March 1943, TNA, AB 3/94.
19. Peierls to Chadwick, 19 May 1944, TNA, AB 1/631.
20. Peierls, Interview by Weiner, p. 149.
21. Rudolf Peierls, Interview by Mark Walker, 7 April 1985, AIP, p. 21.
22. Klaus Fischer, 'Physik', in *Handbuch der deutschsprachigen Emigration 1933–1945* (see Krohn, above), pp. 824–36 (p. 834).
23. Max Born, *My Life*, p. 287.

24. *Professor Dr. Klaus Fuchs.*
25. Ibid.
26. Victor Weisskopf, *The Joy of Insight: Passions of a Physicist* (New York, 1991), pp. 118–19, 137; Hans Bethe, interview by Mario Balibrera, 9 November 1979, Los Alamos Historical Museum Archives, Los Alamos, New Mexico, United States (hereafter LAHM), p. 14; Lothar Nordheim, interview by Bruce Wheaton, 24 July 1977, AIP, p. 43.
27. Einstein to Roosevelt, 2 August 1939 (repr. in *The American Atom: A Documentary History of Nuclear Policies from the Discovery of Fission to the Present 1939–1984*, ed. by Robert Chadwell Williams and Philip Cantelon (Philadelphia, 1984), pp. 12–14); Richard Rhodes, *The Making of the Atomic Bomb* (New York, 1986), pp. 303–11.
28. Lawrence Badash, *Scientists and the Development of Nuclear Weapons: From Fission to the Limited Test Ban Treaty, 1939–1963* (Amherst, 1995), pp. 28–9; Jeremy Bernstein, *Oppenheimer: Portrait of an Enigma* (Chicago, 2004), pp. 66–9.
29. Gowing, 'James Chadwick and the Atomic Bomb', p. 86; Andrew Brown, 'A Tale of Two Documents', in *Remembering the Manhattan Project: Perspectives on the Making of the Atomic Bomb and Its Legacy*, ed. by Cynthia Kelly (London, 2004), pp. 41–6.
30. Gowing, *Britain and Atomic Energy*, pp. 33–4.
31. Robert Buderi, *The Invention that Changed the World: The Story of Radar from War to Peace* (New York, 1996; repr. London, 1998), pp. 77–97.
32. Ronald Clark, *The Birth of the Bomb: The Untold Story of Britain's Part in the Weapon that Changed the World* (London, 1961), p. 43.
33. Peierls, *Bird of Passage*, p. 153.
34. Frisch, *What Little I Remember*, pp. 126–7.
35. Leo Szilard, *Leo Szilard: His Version of the Facts: Selected Recollections and Correspondence*, ed. by Spencer Weart and Gertrud Weiss Szilard (Cambridge, MA, 1978), p. 17.
36. Sherwin, p. 20.
37. Frisch, Interview by Weiner, pp. 34–7; Otto Frisch and John Wheeler, 'The Discovery of Fission: How It All Began', *Physics Today*, 20. 11 (November 1967), 43–8 (pp. 47–8); Patricia Rife, *Lise Meitner and the Dawn of the Nuclear Age* (Boston: Birkhäuser, 1999), pp. 20–1, 26–7, 29–39, 178–96; Ruth Lewin Sime, *Lise Meitner: A Life in Physics* (Berkeley, 1996), pp. 231–58.
38. Lise Meitner and Otto Frisch, 'Disintegration of Uranium by Neutrons: A New Type of Nuclear Chain Reaction', *Nature*, 143. 3615 (11 February 1939), 239–40; Lise Meitner and Otto Frisch, 'Products of the Fission of the Uranium Nucleus', *Nature*, 143. 3620 (18 March 1939), 471–2.
39. Otto Frisch, Interview, 19 June 1973, LAHM, p. 8; Frisch and Wheeler, p. 48.
40. Lawrence Badash, Elizabeth Hodes and Adolph Tiddens, 'Nuclear Fission: Reaction to the Discovery in 1939', *Proceedings of the American Philosophical Society* (hereafter PAPS), 130. 2 (1986), 196–231 (p. 207).
41. Frisch, Interview by Weiner, p. 39.
42. C. A. Lyon, 'Scientists Make An Amazing Discovery', *Sunday Express*, 30 April 1939, p. 1. I am grateful to Peter Rowlands for pointing me towards this source. While the article does not mention it, it is possible that Lyon was inspired by an article in the *New York Times* published the day before; 'Vision

Earth Rocked by Isotope Blast: Scientists Say Bit of Uranium Could Wreck New York', *New York Times*, 29 April 1939, p. 1.

43. Lorna Arnold, 'The History of Nuclear Weapons: The Frisch-Peierls Memorandum on the Possible Construction of Atomic Bombs of February 1940', *Cold War History*, 3. 3 (April 2003), 111–26 (p. 113); Peierls, Interview by Weiner, pp. 90–1.

44. Peierls, interview by Weiner, p. 92.

45. Otto Frisch and Rudolf Peierls, 'Memorandum on the Properties of a Radioactive "Super-Bomb"', n.d.; Otto Frisch and Rudolf Peierls, 'On the Construction of a "Super-Bomb", based on a Nuclear Chain Reaction in Uranium', n.d., TNA, AB 1/210; Arnold, 'History of Nuclear Weapons', pp. 111–17. The first part was first reprinted in Ronald Clark, *Tizard* (London, 1965), pp. 215–17; and the second part in Gowing, *Britain and Atomic Energy 1939–1945*, pp. 389–93.

46. Peierls, Interview by Weiner, pp. 90–3.

47. Frisch and Peierls, 'Memorandum on the Properties of a Radioactive "Super-Bomb"', p. 1.

48. Einstein to Roosevelt, 2 August 1939, in *The American Atom: A Documentary History of Nuclear Policies from the Discovery of Fission to the Present, 1939–1984*, ed. by Robert Williams and Philip Cantelon (Philadelphia, 1984), pp. 12–14 (p. 13).

49. Chapman Pincher, 'Born in Britain – that Bomb', *Daily Express*, 24 September 1964, p. 12.

50. Arnold, 'History of Nuclear Weapons', p. 114; Gowing, *Britain and Atomic Energy*, p. 42; Sabine Lee, 'Birmingham – London – Los Alamos – Hiroshima: Britain and the Atomic Bomb', *Midland History*, 27 (2002), 146–64 (p. 151); Ferenc Szasz, *British Scientists and the Manhattan Project: The Los Alamos Years* (New York, 1992), p. 4.

51. Frisch and Peierls, 'Memorandum on the Properties of a Radioactive "Super-Bomb"', p. 2.

52. Arnold, 'History', p. 114.

53. Frisch and Peierls, 'Memorandum on the Properties of a Radioactive "Super-Bomb"', p. 2.

54. Arnold, 'History', p. 115.

55. Albert Camus, *Resistance, Rebellion, and Death*, tr. from the French and with an Introduction by Justin O' Brien (New York, 1960), p. 135.

56. See Samuel Walker, *Prompt and Utter Destruction: Truman and the Use of Atomic Bombs against Japan*, rev. edn (Chapel Hill, 2004).

57. Peter Rowlands, *120 Years of Excellence: The University of Liverpool Physics Department 1881 to 2001* (Liverpool, 2001), p. 20.

58. Ferenc Szasz, *The Day the Sun Rose Twice: The Story of the Trinity Site Explosion July 16, 1945* (Albuquerque, 1984), pp. 57, 77, 86.

59. Rudolf Peierls, 'Otto Robert Frisch, 1 October 1904–22 September 1979', *BMFRS*, 27 (November 1981), 283–306 (p. 291); H. M. Treasury, *Statements Relating to the Atomic Bomb* (London, 1945), p. 15; David Zimmerman, 'The Tizard Mission and the Development of the Atomic Bomb', *War in History*, 2. 3 (1995), 259–73 (p. 259).

60. Andrew Brown, *The Neutron and the Bomb: A Biography of Sir James Chadwick* (Oxford, 1997), pp. 195–215; Margaret Gowing, 'James Chadwick and the

Atomic Bomb', *Notes and Records of the Royal Society of London*, 47.1 (January 1993), 79–92 (p. 83).

61. Clark, *The Birth of the Bomb*, pp. 76–7; Frisch, *What Little I Remember*, p. 131; Peierls, *Bird of Passage*, p. 156.
62. Gowing, *Britain and Atomic Energy*, p. 45.
63. Frisch, *What Little I Remember*, p. 131; Gowing, *Britain and Atomic Energy*, pp. 46–54; 'Peierls's War Years', n.d., Peierls Papers, MS Eng. Misc. b. 197, A 1, fols 15^r–18^r (fol. 15^r); Thomson to Air Marshal R.H.M.S. Saundby, 3 May 1940, TNA, KV 2/1658.
64. Heitler to Simpson, 21 November 1940, MS S.P.S.L. 328/1, fol. 106^r.
65. Gowing, *Britain and Atomic Energy*, pp. 52, 53 note 1. London, for example, worked on the 'electrolytic method'; Simon to Chadwick, 27 September 1941, CHAD I/19/8.
66. Thomas Vincent Attwood, 'The 37 Inch Cyclotron and Nuclear Structure Research at Liverpool 1935 to 1960' (unpublished masters thesis, University of Liverpool, 1998), pp. 24–43; Frisch and Peierls, 'On the Construction of a "Super-Bomb"', p. 3; Gowing, *Britain and Atomic Energy*, pp. 27–8, 54; John R. Holt, 'James Chadwick in Liverpool', *Notes and Records of the Royal Society of London*, 48. 2 (July 1994), 299–308 (p. 304); Charles David King, 'Chadwick, Liverpool and the Bomb' (unpublished doctoral dissertation, University of Liverpool, 1997), pp. 67–137; Peierls, 'Otto Robert Frisch', p. 292; Rudolf Peierls, 'Outline of the Development of the British T.A. Project', n. d., attached to letter from Chadwick to Rickett, 31 July 1945, TNA CAB 126/1, p. 2; H.M. Treasury, p. 15. On the work conducted at the Cavendish see reports in CHAD I/12/6.
67. Gowing, *Britain and Atomic Energy*, pp. 60–1.
68. Rudolf Peierls, 'Recollections of James Chadwick', *Notes and Records of the Royal Society of London*, 48. 1 (January 1994), 135–41 (p. 137).
69. Bretscher to Chadwick, 20 February 1946, CHAD I/24/2.
70. Gowing, *Britain and Atomic Energy*, pp. 51, 59.
71. Peierls, 'Outline of the Development of the British T.A. Project', p. 2.
72. Clark, *The Birth of the Bomb*, p. 86.
73. 'Peierls's War Years', fol. 15^r; H.M. Treasury, p. 15.
74. Clark, *The Birth of the Bomb*, p. 118.
75. 'Peierls's War Years', fol. 15^r.
76. Gowing, *Britain and Atomic Energy*, pp. 56–7.
77. Ibid., p. 61.
78. Peierls, 'Recollections of James Chadwick', p. 137.
79. Clark, *The Birth of the Bomb*, p. 88; J. H. Sanders, 'Nicholas Kurti, C.B.E.: 14 May 1908–24 November 1998', *BMFRS*, 46 (November 2000), 301–15 (p. 306).
80. Nicholas Kurti, 'Franz Eugen Simon, 1893–1956', *BMFRS*, 4 (November 1958), 225–56 (pp. 225, 228–30); David Schoenberg, 'Heinz London, 1907–1970', *BMFRS*, 17 (1971), 441–61 (pp. 443, 445).
81. Frank Nabarro and A. S. Aragon, 'Egon Orowan, 2 August 1902–3 August 1989', *BMFRS*, 41 (November 1995), 316–40 (pp. 317–18); Sanders, p. 301.
82. Rider, pp. 148–9.
83. Brebis Bleaney, 'Heinrich Gerhard Kuhn, 10 March 1904–26 August 1994', *BMFRS*, 42 (November 1996), 221–32 (pp. 225–6).

84. Schoenberg, p. 444.
85. Farren and Thomson, p. 57; Mendelssohn, p. 1343.
86. Paul K. Hoch, 'Institutional Versus Intellectual Migrations in the Nucleation of New Scientific Specialities', *Studies in the History and Philosophy of Science*, 18. 4 (1987), 481–500; Strauss, 'Movement of People', pp. 57–8; Dieter Wuttke, 'Die Emigration der Kulturwissenschaftlichen Bibliothek Warburg und die Anfänge des Universitätsfaches Kunstgeschichte in Großbritannien', *Ber. Wissenschaftsgesch.*, 7. 3 (1984), 179–94.
87. Kurti, 'Franz Eugen Simon', p. 231.
88. Gowing, *Britain and Atomic Energy*, pp. 56–7.
89. H.M. Treasury, p. 16.
90. Rudolf Peierls, 'MS-12: Efficiency of Isotope Separation', n.d., TNA, AB 4/837.
91. See, for example, Klaus Fuchs and Rudolf Peierls, 'MS-12A: Separation of Isotopes', n.d., TNA, AB 4/838; Klaus Fuchs and Rudolf Peierls, 'MS-47A: Equilibrium Time in a Separation plant by Fuchs and Peierls', May 1942, TNA, AB 4/882.
92. Thomas Vincent Attwood, 'Uranium Separation in the U.K. during World War II' (unpublished doctoral dissertation, University of Liverpool, 2004), pp. 264–71; Clark, *The Birth of the Bomb*, p. 89; H.M. Treasury, pp. 16–17.
93. Peierls, Interview by Walker, pp. 25–6.
94. Gowing, *Britain and Atomic Energy*, p. 53 note 1.
95. 'Chain-Reactions in Uranium, by H. Fröhlich and W. Heitler, n.d.', H.H. Wills Physics Laboratory. I am grateful to Gerard J. Hyland for pointing me towards this document and to Brian Pollard providing me with a copy. A fragment of the document is reproduced in Hyland, p. 311. While Hyland dates the document to the period between March 1939 and June 1940 (p. 248 note 61), Nevill Mott dates it to the end of 1939; 'Walter Heinrich Heitler, 2 January 1904–15 November 1981', *BMFRS*, 28 (November 1982), 141–51 (p. 144).
96. Badash, Hodes and Tiddens, p. 207.
97. Gowing, *Britain and Atomic Energy*, pp. 55–6.
98. Peierls, 'Outline of the Development of the British T.A. Project', p. 4.
99. Peierls to Fröhlich, 7 February 1941; Fröhlich to Peierls, 10 February 1941; Peierls to Frisch, 21 February 1941, TNA, AB 1/574; Fröhlich and Heitler, 'Fission Produced By Cosmic Rays', TNA, AB 4/96.
100. Peierls to Fröhlich, 26 April 1946, TNA, AB 1/574; Mott, 'Herbert Fröhlich', pp. 149–50.
101. Clark, *The Birth of the Bomb*, p. 84.
102. Gowing, *Britain and Atomic Energy*, pp. 54–8; 'Peierls's War Years', fol. 15^r.
103. See, for example, 'Progress Report, March, 1941'; 'Progress Report, March, 1941', TNA, AB 1/494. Original typescripts of Rudolf Peierls's so-called MS reports can be found in AB 1/494.
104. Gowing, *Britain and Atomic Energy*, pp. 76, 85, 394–436.
105. Winston Churchill, *The Second World War*, 5 vols (London, 1948–52), IV (1951), 340; H.M. Treasury, p. 18.
106. H.M. Treasury, p. 18; 'Tube Alloys Project. Minutes of 5th Meeting of Technical Committee, 16 Old Queen Street, 7th May, 1942'; 'Tube Alloys Project. Minutes of 8th Meeting of Technical Committee, 16 Old Queen Street, 14th August, 1942', CHAD I/30/3.

107. H.M. Treasury, p. 19.
108. Peierls, 'Outline of the Development of the British T.A. Project', p. 5; H.M. Treasury, pp. 19–20, 22; M. M. R. Williams, 'Development of Nuclear Reactor Theory', pp. 239–322.
109. 'Discussion on Experimental Programme of T.A. Montreal Group, 16 Old Queen Street, August 12th 1943', CHAD I/30/3.
110. Schoenberg, p. 448; Geoffrey Taylor, 'Obituary: Other Lives, Franz Mandl', *Guardian*, 25 May 2009, p. 31.
111. H.M. Treasury, p. 22; 'Tube Alloys Project. Minutes of 10th Meeting, 13th October 1942'; 'T.A. Project. Minutes of 2nd Meeting of Diffusion Project Committee held at 16 Old Queen Street on 20th July 1943'; 'Tube Alloys Project. Minutes of 1st Meeting of Chemical Panel at 16 Old Queen Street on Dec. 18th 1941', CHAD I/30/3.
112. Chadwick to Rickett, 31 July 1945, TNA, CAB 126/1.
113. Gowing, *Britain and Atomic Energy*, p. 165.
114. Lee, 'Birmingham – London – Los Alamos – Hiroshima', p. 153.
115. John Baylis, *Anglo-American Defence Relations 1939–1984: The Special Relationship*, 2nd edn (London: Macmillan, 1984), p. 5; Szasz, *British Scientists*, p. 7.
116. Zimmerman, 'Tizard Mission', pp. 259–73.
117. Gowing, *Britain and Atomic Energy*, p. 123; Lee, 'Birmingham – London – Los Alamos – Hiroshima', p. 155.
118. Peierls, 'Outline of the Development of the British T.A. Project', p. 6; Thewlis to Chadwick, 8 October 1941, CHAD I/12/3.
119. Peierls, Interview by Weiner, p. 97.
120. Szasz, *British Scientists*, p. 8.
121. Conant to Akers, 15 December 1942; NWP to Gorell Barnes, RE: Prof. Peierls, 2 November 1942, TNA, AB 1/48.
122. Oppenheimer to Manley, 6 November 1942, (repr. in *Robert Oppenheimer: Letters and Correspondence*, ed. by Alice Kimball Smith and Charles Weiner (Cambridge, MA, 1980), pp. 236–7 (p. 237)).
123. Septimus Paul, p. 28.
124. Minute 128, 'Tube Alloys Project. Minutes of 18th Meeting of Technical Committee, 16 Old Queen Street, 26th July, 1943', CHAD I/30/3.
125. Perrin to Gorell Barnes, 'T.A. Staff for U.S.A.', 31 January 1944, p. 1; 'Staff Requirements for T.A. Project in U.S.A.', 7 February 1944, TNA, CAB 126/331, pp. 1–2.
126. J. S. M. Washington to War Cabinet Offices, London, 3 February 1944, TNA, CAB 126/331, p. 3.
127. Akers to the Chancellor of the Exchequer, 'Tube Alloys Project: Access to American Full-Scale Plant Information', 6 April 1944, TNA, CAB 126/331.
128. Gowing, *Britain and Atomic Energy*, pp. 176–7.
129. David Reynolds, *The Creation of the Anglo-American Alliance 1937–41: A Study in Competitive Co-Operation* (London, 1981), pp. 121, 145–68.
130. See Nicholas Cull, *Selling War: The British Propaganda Campaign against American "Neutrality" in World War II* (New York, 1995).
131. Septimus Paul, pp. 9–30.
132. Chadwick to Simon, 27 October 1941, CHAD I/19/8.
133. Akers to Conant, 15 December 1942, CHAD I/28/2.

134. Baylis, *Anglo-American Defence Relations*, p. 3; Septimus Paul, pp. 31–54.
135. Churchill, *The Second World War*, IV, 341.
136. Gowing, *Britain and Atomic Energy*, pp. 115–78, 439–40; Szasz, *British Scientists*, p. 11.
137. Jonathan Rosenberg, 'Before the Bomb and After: Winston Churchill and the Use of Force', in *Cold War Statesmen Confront the Bomb: Nuclear Diplomacy Since 1945*, ed. by John Lewis Gaddis and others (Oxford, 1999), pp. 171–93 (p. 180).
138. Perrin to Gorell Barnes, 'T.A. Staff for U.S.A.', 31 January 1944, TNA, CAB 126/331, p. 1.
139. Gowing, *Britain and Atomic Energy*, p. 174.
140. Gowing, pp. 126, 135–8, 142–3, 201–15; Septimus Paul, pp. 19, 62–4; Rhodes, *Making of the Atomic Bomb*, pp. 504–8.
141. Simone Turchetti, '"For Slow Neutrons, Slow Pay": Enrico Fermi's Patent and the U.S. Atomic Energy Program, 1938–1953', *Isis*, 97. 1 (2006), 1–27.
142. Baylis, *Anglo-American Defence Relations*, p. 15.

3   American Interlude: The Manhattan Project, the Atom Bomb and the Emergence of a New Approach to Nuclear Research

1. Fletcher to Ladd, p. 6; Moorehead, *The Traitors*, pp. 94–5.
2. Rhodes, *Making of the Atomic Bomb*, pp. 365–7.
3. Leslie Groves, *Now It Can Be Told: The Story of the Manhattan Project* (New York, 1962; repr. New York, 1983), pp. 3–18; Leland Johnson and Daniel Schaffer, *Oak Ridge National Laboratory: The First Fifty Years* (Knoxville, 1994), pp. 1–27; Michele Stenehjem Gerber, *On the Home Front: The Cold War Legacy of the Hanford Nuclear Site*, 3rd edn (Lincoln, 2007), pp. 31–6.
4. Gowing, *Britain and Atomic Energy*, p. 240; Peierls, *Bird of Passage*, p. 184; Szasz, *British Scientists*, p. 16.
5. Sanders, p. 307.
6. Robert Williams, *Klaus Fuchs*, pp. 67–70.
7. W. J. Skardon, 'Emil Julius Klaus Fuchs', 22 December 1949, p. 4.
8. Peierls, *Bird of Passage*, pp. 185–6.
9. Calder, pp. 231, 239–40, 276, 404–6.
10. Peierls, Interview by Weiner, p. 97.
11. Peierls, *Bird of Passage*, pp. 207–8.
12. Robert Williams, *Klaus Fuchs*, p. 16.
13. John Wirth and Linda Harvey Aldrich, *Los Alamos: The Ranch School Years, 1917–1943* (Albuquerque, 2003), pp. 157–8.
14. Szasz, *Day the Sun Rose Twice*, pp. 15, 179.
15. Charles Johnson and Charles Jackson, *City Behind a Fence: Oak Ridge, Tennessee 1942–1946* (Knoxville, 1981), p. xix.
16. Ruth Marshak, 'Secret City', in *Standing By and Making Do: Women of Wartime Los Alamos*, ed. by Jane S. Wilson and Charlotte Serber (Los Alamos, 1988), pp. 1–19 (p. 11).
17. Tarter, Donald, 'Peenemünde and Los Alamos: Two Studies', *History of Technology*, 14 (1994), 150–70.

18. Peierls, *Bird of Passage*, pp. 191–2.
19. Edward Teller, with Judith Shoolery, *Memoirs: A Twentieth-Century Journey in Science and Politics* (Cambridge, MA, 2001), p. 184.
20. Nancy Thorndike Greenspan, *The End of the Certain World: The Life and Science of Max Born* (Chichester, 2005), p. 262; Victor F. Weisskopf, 'Meine Göttinger Studienjahre mit Born und Franck', in *Max Born, James Franck, der Luxus des Gewissens: Physiker in ihrer Zeit*, ed. by Jost Lemmerich (Berlin (West), 1982), pp. 80–3.
21. Peierls to Placzek, 10 September 1944, TNA, AB 1/576.
22. Bethe to Peierls, 22 May 1944, TNA, AB 1/635.
23. Nordheim, Interview by Wheaton, pp. 39–44; Johnson and Schaffer, pp. 3, 17–21, 55–6.
24. King, p. 221; M. M. R. Williams, *Development of Nuclear Reactor Theory*, p. 312.
25. Perrin to Gorrell Barnes, 'T.A. Staff for U.S.A.', p. 3; Chadwick to Bünemann, 1 June 1945; Chadwick to Massey, 25 July 1945, CHAD IV/3/7.
26. Frisch, Interview, pp. 4–5; Jon Hunner, *Inventing Los Alamos: The Growth of an Atomic Community* (Norman, 2004), p. 36.
27. Hal Rothman, *On Rims & Ridges: The Los Alamos Area since 1880* (1992; repr. Lincoln, 1997), pp. 5–19.
28. Bernice Brode, *Tales of Los Alamos: Life on the Mesa 1943–1945* (Los Alamos, 1997), pp. 72–5, 124; Frisch, Interview, pp. 4, 18–19.
29. Françoise Ulam, Interview by Theresa Strottman, 1992, LAHM, p. 5.
30. Hunner, *Inventing Los Alamos*, p. 38.
31. Richard Melzer, *Breakdown: How the Secret of the Atomic Bomb Was Stolen during World War II* (Santa Fe, 2000), pp. 14–15; Edith C. Truslow, *Manhattan District History: Nonscientific Aspects of Los Alamos Project Y 1942 through 1946* (Los Alamos, 1997), pp. 2–3.
32. Bird and Sherwin, *American Prometheus*, pp. 223–35; Rhodes, *Making of the Atomic Bomb*, pp. 464, 502–3; and Teller, *Memoirs*, p. 180.
33. Fuchs to Esther Simpson, 29 September 1941, MS S.P.S.L. 328/1, fol. 181^r.
34. Fuchs to Skemp, 12 December 1945, MS S.P.S.L. 328/1, fol. 190^r.
35. Skemp to Fuchs, 14 January 1946, MS S.P.S.L., 328/1, fol. 191^r.
36. Frisch to the Secretary, Society for the Protection of Science and Learning, 27 November 1945, MS S.P.S.L. 327/10, fol. 519^r.
37. Akers to Heads of U.K. T.A. Teams, 11 April 1945, RTBT, B. 65A.
38. Richard Feynman, *'Surely You're Joking, Mr. Feynman!' Adventures of a Curious Character* (New York, 1985), pp. 107–55.
39. Hedy [*sic*] Bretscher, interview by John Bennett and Anne Shepherd, 21 July 1984, AIP, 35. Note that Bretscher's first name was Hanni and not Hedy, as erroneously stated in the interview. Mark Bretscher, email to author, 30 March 2007. Laura Fermi, *Illustrious Immigrants: The Intellectual Migration from Europe, 1930/41*, 2nd edn (Chicago, 1971), p. 201; Jane Wilson, 'Not Quite Eden', in *Standing By and Making Do*, p. 46; 'Alien Registration Form: Fuchs, Emil Julius Klaus', n.d., KV 2/1259, TNA. For Fuchs's Canadian internee record see KV 2/1253, TNA.
40. Cited in Brode, *Tales of Los Alamos*, p. 19.
41. Jennet Conant, *109 East Palace: Robert Oppenheimer and the Secret City of Los Alamos* (New York, 2005), p. 121; Hunner, *Inventing Los Alamos*, pp. 42–3;

Melzer, *Breakdown*, p. 112; and Alden Stevens, 'Cradle of the Bomb', *New Republic*, 17 March 1947, pp. 14–15.

42. Elsaesser, p. 113.
43. Szasz, *British Scientists*, p. 18.
44. Brown, *Neutron and the Bomb*, p. 286; Gowing, *Britain and Atomic Energy*, pp. 265–6; Szasz, *British Scientists*, p. 40.
45. Hawkins, p. 76.
46. Bethe, Interview by Balibrera, p. 10.
47. Gowing, *Britain and Atomic Energy*, pp. 250–6.
48. Hawkins, p. 197.
49. Otto R. Frisch, 'The First Nuclear Explosion', *New Scientist*, 6 August 1970, p. 274; Hoddeson and others, pp. 346–8.
50. Peierls, *Bird of Passage*, p. 39.
51. Klaus Fischer, *Changing Landscapes of Nuclear Physics: A Scientometric Study on the Social and Cognitive Position of German-Speaking Emigrants Within the Nuclear Physics Community, 1921–1947* (Berlin, 1993), pp. 25–7; Peierls, *Bird of Passage*, p. 188; M. M. R. Williams, 'Development of Nuclear Reactor Theory', pp. 255–6.
52. Szasz, *British Scientists*, p. 25.
53. Hoddeson and others, pp. 246–7, 399.
54. Szasz, *British Scientists*, p. 18.
55. Hawkins, p. 184.
56. Gowing, *Britain and Atomic Energy*, p. 262; Peierls, *Bird of Passage*, p. 200.
57. Szasz, *British Scientists*, p. 89.
58. Bethe, Interview by Balibrera, p. 8.
59. *Prof. Dr. Klaus Fuchs.*
60. Bethe, Interview by Balibrera, pp. 4–5.
61. Ben Dobbin, '"Last giant" of LANL dies at 98', *Santa Fe New Mexican*, 8 March 2005, pp. A1, A5 (p. A1).
62. Hoddeson and others, pp. 77, 408.
63. Bethe, Interview by Balibrera, p. 11; Szasz, *Day the Sun Rose Twice*, pp. 20–1, 57.
64. Hoddeson and others, pp. 312, 316.
65. Peierls, *Bird of Passage*, p. 191.
66. Bethe, Interview by Balibrera, p. 5; Hoddeson and others, p. 246.
67. J. David Jackson and Kurt Gottfried, *Victor Frederick Weisskopf, 1908–2002: A Biographical Memoir*, Biographical Memoirs, 84 (Washington, 2003), p. 14.
68. Hawkins, p. 7.
69. István Hargittai, *The Martians of Science: Five Physicists Who Changed the Twentieth Century* (New York, 2006), pp. 3–87.
70. Peierls, *Bird of Passage*, p. 192; Teller, *Memoirs*, p. 170; Eugene P. Wigner, *The Recollections of Eugene P. Wigner: as Told to Andrew Szanton* (New York, 1992), pp. 211, 222–5.
71. Teller, *Memoirs*, p. 178.
72. Peierls, *Bird of Peierls*, p. 200.
73. Gregg Herken, *Brotherhood of the Bomb: The Tangled Lives and Loyalties of Robert Oppenheimer, Ernest Lawrence, and Edward Teller* (New York, 2002), p. 539.

74. Kai Bird and Martin J. Sherwin, *American Prometheus: The Triumph and Tragedy of J. Robert Oppenheimer* (New York, 2005), pp. 282–3; Peter Goodchild, *J. Robert Oppenheimer: 'Shatterer of Worlds'* (London, 1980), p. 83.
75. Hoddeson and others, p. 204.
76. Frisch, *What Little I Remember*, p. 191; Hawkins, pp. 7, 90.
77. Hoddeson and others, pp. 148–56, 268–71.
78. Hawkins, p. 310.
79. Ruth Howes and Caroline Herzenberg, *Their Day in the Sun: Women of the Manhattan Project* (Philadelphia, 1999), p. 37; Peierls, *Bird of Passage*, p. 192.
80. Bethe, Interview by Balibrera, p. 16.
81. Hoddeson and others, p. 131.
82. Howes and Herzenberg, pp. 40, 47; Teller, *Memoirs*, p. 187.
83. Hoddeson and others, pp. 49, 134, 141; Szasz, *Day the Sun Rose Twice*, p. 9.
84. Teller, *Memoirs*, p. 180.
85. Teller, *Memoirs*, p. 190.
86. Hawkins, pp. 305–6.
87. 'Martin Deutsch, MIT Physicist Who Discovered Positronium, Dies at 85', *MIT Physics Annual* (2003), 12–15.
88. Hawkins, p. 349.
89. Hoddeson and others, pp. 4–5, 9–10; Teller, *Memoirs*, p. 171; Victor F. Weisskopf, Interview by Charles Weiner and Gloria Lubkin, 22 September and 5 December 1966, AIP, p. 31.
90. Bethe, Interview by Hoddeson, pp. 12–14.
91. Paul Hoch, 'Migration and the Generation of New Scientific Ideas', *Minerva*, 25. 3 (1987), 209–37 (pp. 212–13).
92. Richard Hewlett and Oscar Anderson, Jr, *A History of the United States Atomic Energy Commission*, vol. 1: The New World, 1939/1946 (University Park, 1962), p. 233.
93. Cornwell, p. 38.
94. Klaus Hoffmann, *J. Robert Oppenheimer: Schöpfer der ersten Atombombe* (Berlin, 1995), pp. 26–34; Elizabeth Noble Shor, 'Kistiakowsky, George Bogdan (18 Nov. 1900–7 Dec. 1982)', in *American National Biography*, ed. by John Garraty and Mark Carnes, 24 vols (Oxford, 1999), XII, 776–8 (p. 776).
95. Hoch, 'Reception of Central European Refugee Physicists', p. 238.
96. Paul Hoch, 'Some Contributions to Physics by German-Jewish Émigrés in Britain and Elsewhere', in *Second Chance* (see Stent, 'Jewish Refugee Organisations', above), pp. 229–41 (pp. 232–3).
97. Michael Eckert, *Die Atomphysiker: Eine Geschichte der theoretischen Physik am Beispiel der Sommerfeldschule* (Braunschweig, 1993), pp. 105–6, 199; Kragh, p. 249; Sharon Traweek, 'Big Science and Colonialist Discourse: Building High-Energy Physics in Japan', in *Big Science: The Growth of Large-Scale Research*, ed. by Peter Galison and Bruce Hevly (Stanford, 1992), pp. 100–28.
98. Hoddeson and others, pp. 9–10.
99. Hoch, 'Some Contributions to Physics', pp. 232–3.
100. Bethe, Interview by Balibrera, p. 4.
101. Hoch, 'Some Contributions to Physics', pp. 232–3.
102. Margaret Connell Szasz, 'Introduction', in *Between White and Indian Worlds: The Cultural Broker*, ed. by Szasz (Norman, 1994), pp. 3–20.

103. Peter Galison, 'Trading Zone: Coordinating Action and Belief', in *The Science Studies Reader*, ed. by Mario Biagioli (New York, 1999), pp. 137–60 (p. 146).
104. Roger Stuewer, 'Nuclear Physics in a New World: The Émigrés of the 1930s in America', *Ber. Wissenschaftsgesch.*, 7. 1 (1984), 23–40 (p. 33).
105. Paul Hoch and Jennifer Platt, 'Migration and the Denationalization of Science', in *Denationalizing Science: The Contexts of International Scientific Practice*, ed. by Elisabeth Crawford, Terry Shinn and Sverker Sörlin (Dordrecht, 1992), pp. 133–52 (pp. 135–9).
106. Peter Bacon Hales, *Atomic Spaces: Living on the Manhattan Project* (Urbana, 1997), pp. 244–5.
107. John Krige explores this phenomenon in *American Hegemony and the Postwar Reconstruction of Science in Europe* (Cambridge, 2006).
108. 'Peierls' War Years', fol. 18^r.
109. Groves, *Now It Can Be Told*, p. 38.
110. Edwin McMillan, 'Early Days at Los Alamos', in *Reminiscences of Los Alamos, 1943–1945*, ed. by Lawrence Badash, Joseph Hirschfelder and Herbert Broida (Dordrecht, 1980), pp. 13–19 (p. 16).
111. Hoddeson and others, pp. 86–90.
112. Hoddeson and others, pp. 206, 228; 'Martin Deutsch', p. 14.
113. Bethe, Interview by Balibrera, p. 6.
114. Frisch, *What Little I Remember*, p. 159; Lillian Hoddeson, 'Mission Change in the Large Laboratory: The Los Alamos Implosion Program, 1943–1945', in *Big Science* (see Traweek, 'Big Science', above), pp. 265–89 (pp. 272–5).
115. Lillian Hoddeson, 'Mission Change in the Large Laboratory', pp. 280–8.
116. Charles Thorpe, *Oppenheimer: The Tragic Intellect* (Chicago, 2006), pp. 134–8.
117. Bethe, Interview by Balibrera, pp. 7–8; Hoddeson and others, pp. 130–6; Robert Seidel, *Los Alamos and the Development of the Atomic Bomb* (Los Alamos, 1995), pp. 82–6.
118. Hoddeson, 'Mission Change', pp. 265–7, 271–2.
119. Bethe, Interview by Balibrera, p. 8; Peierls, *Bird of Passage*, p. 187.
120. Oppenheimer to Groves, 14 February 1944 (repr. in *Robert Oppenheimer: Letters and Correspondence* (see Oppenheimer to Manley, above), pp. 271–2 (p. 272)).
121. Bethe, Interview by Balibrera, p. 11.
122. Andrew Brown, *The Neutron and the Bomb: A Biography of Sir James Chadwick* (Oxford, 1997), p. 286; Szasz, *British Scientists and the Manhattan Project*, p. 40.
123. Hoddeson and others, pp. 163–77.
124. Ibid., pp. 294–301; McAllister Hull, with Amy Bianco, *Rider of the Pale Horse: A Memoir of Los Alamos and Beyond* (Albuquerque, 2005), pp. 29–30.
125. Bethe, Interview by Balibrera, p. 8.
126. Hoddeson and others, p. 307.
127. Szasz, *British Scientists*, p. 25.
128. Hoddeson and others, pp. 317, 331.
129. Ibid., pp. 156, 275, 278, 293.
130. See Szasz, *Day the Sun Rose Twice*.
131. James Chadwick, 'Initial Report by Sir James Chadwick', 16 July 1945, TNA, CAB 126/250.

132. Otto Frisch, 'Eye-Witness Report of Nuclear Explosion, July 16, 1945', TNA, CAB 126/250, p. 1.

133. James Tuck to James Chadwick, 17 July 1945, TNA, CAB 126/250, pp. 1, 3–5.

134. Chadwick to Bradbury, 23 January 1946; Bradbury to Chadwick, 5 February 1946; Chadwick to Fuchs, 24 January 1946, CHAD IV/3/6.

135. Placzek to Chadwick, 4 February 1946, CHAD IV/3/6.

136. Peierls to Fuchs, 29 May 1946, TNA, AB 1/574.

137. Bradbury to Chadwick, 5 February 1946, CHAD IV/3/6, p. 1.

138. Chadwick to Groves, 9 January 1946, CHAD IV/3/6.

139. Jeff Hughes, *The Manhattan Project: Big Science and the Atom Bomb* (Cambridge, 2002), p. 13. On the concept of 'Big Science', see James Chapshew and Karen Rader, 'Big Science: Price to the Present', in *Science after '40*, ed. by Arnold Thackray (=*Osiris*, 2nd series, 7 (1992)), pp. 2–25; *Big Science* (see Traweek, 'Big Science', above); Derek de Solla Price, *Little Science, Big Science* (New York, 1963).

140. Gowing, *Britain and Atomic Energy*, p. 240; Peierls, *Bird of Passage*, p. 184; Szasz, *British Scientists*, p. 16.

141. Alvin M. Weinberg, 'Impact of Large-Scale Science on the United States', *Science*, 134. 3473, 21 July 1961, pp. 161–4 (p. 161).

142. 'Text of Eisenhower's Farewell Address', *New York Times*, 18 January 1961, p. 22.

143. Pap Ndiaye, transl. by Elborg Forster, *Nylon and Bombs: DuPont and the March of Modern America* (Baltimore, 2007), pp. 141–78.

144. David Edgerton, *Warfare State: Britain, 1920–1970* (Cambridge, 2006); Johnson and Jackson, p. 8.

145. See Mark Walker, *German National Socialism and the Quest for Nuclear Power, 1939–49* (Cambridge, 1989).

146. Edward Teller, 'The Work of Many People', *Science*, 121 (25 February 1955), 267–75 (p. 267).

147. Hugh Gusterson, *People of the Bomb: Portraits of America's Nuclear Complex* (Minneapolis, 2004), pp. 188–9.

148. Gerald Nash, *The American West Transformed: The Impact of the Second World War* (Bloomington, 1985), p. 164.

149. Hales, *Atomic Spaces*, p. 3.

150. See *Cold War, Hot Science: Applied Research in Britain's Defence Laboratories, 1945–1990*, ed. by Robert Bud and Philip Gummett (Amsterdam; Abingdon, 1999; repr. London, 2002); Peter Galison, 'Physics Between War and Peace', in *Science, Technology and the Military*, ed. by Everett Mendelsohn, Merritt Roe Smith and Peter Weingart (=*Sociology of the Sciences*, 12. 1 (1988)), pp. 47–86; Silvan Schweber, 'The Mutual Embrace of Science and the Military: ONR and the Growth of Physics in the United States after World War II', in *Science, Technology and the Military* (see Galison, 'Physics Between War and Peace', above), pp. 1–45 (pp. 4–6).

151. Ferenc Szasz, *Larger Than Life: New Mexico in the Twentieth Century* (Albuquerque, 2006), pp. 125–50.

152. See Jack Holl, *Argonne National Laboratory, 1946–96* (Urbana, 1997); *History and Reflections of Engineering at Lawrence Livermore National Laboratory 1952–2002*, ed. by Camille Minichino (Livermore, 2002); Necah Stewart Furman, *Sandia National Laboratories: The Postwar Decade* (Albuquerque, 1990).

153. Robert Bud and Philip Gummett, 'Introduction: Don't You Know There's a War On', in *Cold War, Hot Science*, pp. 1–28 (p. 17).
154. 'A.E.R.E. Programme No. 3, September 1947', CHAD I/8/1.
155. Chadwick to Fuchs, 24 January 1946, TNA, AB 1/444.
156. Michael Goodman, 'The Grandfather of the Hydrogen Bomb?: Anglo-American Intelligence and Klaus Fuchs', *HSPS*, 34. 1 (2003), 1–22 (p. 15).
157. Joseph Rotblat, 'Memorandum on Nuclear Research Work for T.A. and Programme of Future Work at Liverpool', 6 February 1945. The DSIR was quite impressed by Rotblat's recommendations; A. K. Longair to Rotblat, 5 March 1945, Rotblat Papers (RTBT), B. 65A.
158. Lorna Arnold, with Katherine Pyne, *Britain and the H-Bomb* (Basingstoke, 2001), pp. 7–9; Szasz, *British Scientists*, p. 26; Herken, *Brotherhood of the Bomb*, p. 374 note 92.
159. Arnold, *Britain and the H-Bomb*, p. 38.
160. Arnold, *Britain and the H-Bomb*, pp. 40–1, 74–5.
161. Goodman, 'Grandfather of the Hydrogen Bomb', p. 16.
162. Michael Goodman, *Spying on the Nuclear Bear: Anglo-American Intelligence and the Soviet Bomb* (Stanford, 2007), pp. 58, 62–3.
163. 'The Case of Dr. Klaus Fuchs', 2 March 1950, TNA, KV 2/1253, p. 1.
164. Hoddeson and others, p. 345.
165. See Chadwick to Bretscher, 22 December 1945; Chadwick to Bragg, 15 January 1946, CHAD IV/3/6.
166. Bärwinkel, pp. 594–6.
167. Peierls to Chadwick, 26 February 1946, CHAD I/24, p. 1.
168. Peierls, *Bird of Passage*, pp. 322–4; Vick to Peierls, 21 February 1964; Peierls to Vick, 12 February 1964; Peierls to Sandford, 12 February 1964; Marshall to Oates, 11 February 1964; Marshall to Bretscher, 27 May 1964; Marshall to Peierls, 28 May 1964, Peierls Papers, MS Eng. Misc b. 223, F 3.
169. Bretscher to Chadwick, 2 July 1945, CHAD IV/3/6; Chadwick to Bretscher, 22 December 1945; Chadwick to Bragg, 15 January 1946; Bretscher to Chadwick, 20 February 1946; Bretscher to Chadwick, 28 February 1946; Bragg to Chadwick, 12 March 1946; Bretscher to Chadwick, 18 March 1946; Bretscher to Chadwick, 16 March 1946; Bretscher to Chadwick, 17 May 1946; Chadwick to Bretscher, 27 April 1946; Chadwick to Bragg, 2 April 1946, CHAD I/24/2.
170. Bretscher to Chadwick, 17 May 1946, CHAD I/24/2.
171. Chadwick to Nichols, 12 March 1946, CHAD IV/3/15.

4   A Nation Betrayed? The Klaus Fuchs Atomic Espionage Case Reconsidered

1. Gowing, *Independence and Deterrence*, II, 3, 144–5.
2. Jessica Wang, *American Science in an Age of Anxiety: Scientists, Anticommunism & the Cold War* (Chapel Hill, 1999), p. 297 note 3.
3. Garry Wills, *Bomb Power: The Modern Presidency and the National Security State* (New York, 2010), p. 1.
4. This is the title of Peter Hennessy's book *The Secret State: Preparing for the Worst 1945–2010*, 2nd edn (2002; London, 2010).
5. Gowing, *Independence and Deterrence*, II, 116–17.

6. 'The Case of Dr. Klaus Fuchs', 2 March 1950, TNA, KV 2/1253, p. 3.
7. United States. Cong. House. Joint Committee on Atomic Energy, *Soviet Atomic Espionage*, 82nd Cong., 1st sess. (Washington, DC: GPO, 1951), unpaginated.
8. Metropolitan Police Special Branch, 'Hearing, before The Lord Chief Justice, Lord Goddard, at the Old Bailey on 1st March 1950, of the case against Klaus Emil Julius FUCHS, arraigned on indictment (four counts) under the Official Secrets Act, 1911, Section 1', TNA, KV2/1264, p. 23.
9. Moorehead, *The Traitors*, p. 58.
10. Downing to Harbo, RE: Foocase, espionage, 9 March 1950, Klaus Fuchs FBI File, 65-58805, vol. 15, serials 720–80; Detlev Peukert, *Die Weimarer Republik: Krisenjahre der Klassischen Moderne* (Frankfurt a.M., 1987), pp. 243–65.
11. W. J. Skardon, 'Emil Julius Klaus Fuchs. Fourth, Fifth, Sixth and Seventh Interviews', 31 January 1950, TNA, KV 2/1250, p. 1; 'Statement of Emil Julius Klaus Fuchs', pp. 1–4.
12. Emil Fuchs, *Mein Leben*, II, 189, 201–02. The records of the local government office for registration of residents in the City of Kiel indicate that Emil Fuchs and his wife, two sons and one daughter moved from Eisenach to Kiel on 12 May 1931, 'Auskunft des Archivs des Einwohnermeldeamtes Kiel', Stadtarchiv Kiel, Kiel, Germany, Klaus Fuchs folder.
13. 'Statement of Emil Julius Klaus Fuchs', p. 1. Fuchs later reiterated this point; *Prof. Dr. Klaus Fuchs*.
14. Bernhardt Schell, 'Introduction', in *Einstein, Anschütz and the Kiel Gyro Compass, the Correspondence between Albert Einstein and Herrmann Anschütz-Kaempfe as well as other Documents*, ed. by Dieter Lohmeier and Bernhardt Schell, tr. by Anita Cervenák, 2nd rev. edn (Kiel, 2005), pp. 13–87 (pp. 57–66).
15. Einstein to Anschütz, 12 July 1922, in *Einstein, Anschütz and the Kiel Gyro Compass* (see Schell, above), pp. 168–9 (p. 169).
16. Michael Legband, 'Von der Provinz zum Bundesland – Schleswig-Holstein im 20. Jahrhundert', in *Schleswig-Holstein von den Ursprüngen bis zur Gegenwart: Eine Landesgeschichte*, ed. by Jann Markus Witt and Heiko Vosgerau (Hamburg, 2002), pp. 327–83 (p. 330).
17. Peter Wulf, 'Zustimmung, Mitmachen, Verfolgung und Widerstand – Schleswig-Holstein in der Zeit des Nationalsozialismus', in *Geschichte Schleswig-Holsteins: Von den Anfängen bis zur Gegenwart*, ed. by Ulrich Lange, 2nd edn (Neumünster, 2003), pp. 585–621 (p. 588).
18. Christoph Cornelißen, 'Die Universität Kiel im "Dritten Reich"', in *Wissenschaft an der Grenze: Die Universität Kiel im Nationalsozialismus*, ed. by Christoph Cornelißen and Carsten Mish (Essen, 2009), pp. 11–29; Peter Wulf, 'Die Stadt auf der Suche', pp. 357–8.
19. Dittrich, pp. 175–82. See also the Hedwig Gerth's testimony; 'Report, Made by Joseph C. Walsh, RE: Emil Julius Klaus Fuchs', 9 November 1950, Klaus Fuchs FBI File, 65-58805, vol. 41, serials 1457–500; 'Statement of Emil Julius Klaus Fuchs', pp. 1–3.
20. Reuven Golan, 'Aus der Erlebniswelt eines jüdischen Jugendlichen in Kiel Anfang der dreißiger Jahre', in *'Wir bauen das Reich': Aufstieg und erste Herrschaftsjahre des Nationalsozialismus in Schleswig-Holstein*, ed. by Erich Hoffmann and Peter Wulf (Neumünster, 1983), pp. 361–8 (p. 366).
21. Dittrich, pp. 180–2. See Eve Rosenhaft, *Beating the Fascists?: The German Communists and Political Violence, 1929–1933* (Cambridge, 1983).

22. Emil Fuchs, *Mein Leben*, II, 221; 'Statement of Emil Julius Klaus Fuchs', p. 3.
23. Emil Fuchs, *Mein Leben*, II, 200–1; *Prof. Dr. Klaus Fuchs*; 'Statement of Emil Julius Klaus Fuchs', p. 4.
24. SAC, NY, to Director, FBI, RE: Foocase, espionage, 26 July 1950, the Los Alamos National Laboratory Archives, Los Alamos, New Mexico, United States (hereafter LANL), VFA 529; Downing to Harbo.
25. 'Statement of Emil Julius Klaus Fuchs', p. 5.
26. SAC, NY, to Director, FBI, p. 26; Skardon, 'Emil Julius Klaus Fuchs. Fourth, Fifth, Sixth and Seventh Interviews', p. 2.
27. 'Secret Report, Soviet Embassy, GDR', 9 April 1986, Klaus Fuchs personal file ('Lichnoe delo Fuks Klaus'), the Comintern Archives, the Russian States Archive of Socio-Political History, Moscow, Russian Federation (hereafter RGASPI), Komintern, F. 495, op. 205, d. 6612, p. 4. I am grateful to Serge Simonov for translating the RGASPI files from Russian into English.
28. *Prof. Dr. Klaus Fuchs*; 'Statement by Klaus Fuchs to Hugh Clegg and Robert J. Lamphere, Wormwood Scrubs Prison, London, England, May 26, 1950', LANL, VFA 529, p. 1.
29. SAC, New York, to Director, FBI, pp. 6–7; Ruth Werner, *Sonjas Rapport*, new expanded edn (Berlin, 2006), p. 292.
30. 'The Case of Dr. Klaus Fuchs', p. 2. On Fuchs's espionage activities see also 'The Case of Dr. Klaus Fuchs – Appendix B: Summary of Information Obtained from Dr. Fuchs Regarding His Espionage Contacts', 2 March 1950, TNA, KV 2/1253.
31. Werner, *Sonjas Rapport*, new expanded edn, pp. 289–95. Ursula Kuczynski went by several names: after her marriage she became Ursula Hamburger but was also known as Ruth Werner and Sonja Ludwig; Ursula Kuczynski (Ursula Hamburger; Sonja Ludwig) personal file ('Lichnoe delo Kuchinski Ursula [Gamburger Ursula, Liudvig Sonia]'), RGASPI, Komintern, F. 495, op. 205, d. 1721.
32. Klaus Fuchs operated, at first, under the code name 'Rest' which was changed to 'Charles' in May 1944; Alexander Feklisov and Sergei Kostin, *The Man Behind the Rosenbergs: Memoirs of the KGB Spymaster Who also Controlled Klaus Fuchs and Helped Resolve the Cuban Missile Crisis*, tr. by Catherine Drop (New York, 2001), p. 416 note 54.
33. W. J. Skardon, 'Emil Julius Klaus Fuchs', 9 March 1950, TNA, KV 2/1879.
34. 'Jürgen Kuczynski', 11 March 1950, TNA, KV 2/1879, p. 3.
35. SAC, New York, to Director, FBI, pp. 1–3; W. J. Skardon, 'Interrogation of Dr. Fuchs by Officers of the F.B.I.', 9 June 1950, TNA, KV 2/1255, p. 1; Joseph C. Walsh, 'Emil Julius Klaus Fuchs, was', 10 October 1950, Klaus Fuchs FBI File, 65-58805, vol. 39, serials 1432–54.
36. 'The Case of Emil Julius Klaus Fuchs', 24 November 1950, TNA, KV 2/1256, pp. 5, 12.
37. Hyde, pp. 1–48.
38. Robert Lamphere and Tom Shachtmann, *The FBI-KGB War: A Special Agent's Story* (New York, 1986), pp. 85–6, 133. On the Venona files, see also John Earl Haynes and Harvey Klehr, *Venona: Decoding Soviet Espionage in America* (New Haven, 1999).
39. Hoover to McMahon, 6 April 1950, Klaus Fuchs FBI File, 65-588805, vol. 28, serials 1039–105.

40. Szasz, *British Scientists*, p. 83.
41. Gowing, *Independence and Deterrence*, II, 144–53.
42. J. C. Robertson, 'Emil Julius Klaus Fuchs', 7 September 1949; J. C. Robertson, 'Meeting with W/Cdr. Arnold on 9th September, 1949', 9 September 1949; J. C. Robertson, 'Klaus Fuchs. Further Investigation Plan', 12 September 1949, TNA, KV 2/1246.
43. Szasz, *British Scientists*, pp. 83–4.
44. Hyde, p. 99; J. C. Robertson, 'Proposed Interrogation of Fuchs', 16 December 1949, TNA, KV 2/1249.
45. *Prof. Dr. Klaus Fuchs.*
46. Skardon, 'Emil Julius Klaus Fuchs', 22 December 1949, pp. 3, 6.
47. J. C. Robertson, 'Action following Fuchs Interrogation', 21 December 1949, TNA, KV 2/1249.
48. W. J. Skardon, 'Emil Julius Klaus Fuchs. Second Interview', 2 January 1950, TNA, KV 2/1249; Skardon, 'Emil Julius Klaus Fuchs. Fourth, Fifth, Sixth and Seventh Interviews', p. 1.
49. Skardon, 'Emil Julius Klaus Fuchs. Fourth, Fifth, Sixth and Seventh Interviews', p. 1.
50. Rudolf Peierls, 'The Lessons of the Fuchs Case', n.d., Peierls Papers, b. 197, A 16, p. 3.
51. 'Statement of Emil Julius Klaus Fuchs'. A typescript of the statement is available in TNA, KV 2/1250.
52. 'Statement of Emil Julius Klaus Fuchs', pp. 6–7. Fuchs himself continued to apply the term 'controlled schizophrenia' to describe his actions. See, for example, Fuchs to Genia Peierls, 6 February 1950, Peierls Papers, sup. cat., D.52. In the MfS interview Fuchs provided a detailed overview of his tactics, including the befriending of his colleagues; *Prof. Dr. Klaus Fuchs.*
53. Robertson, 'Emil Julius Klaus Fuchs', 23 November 1949, pp. 11, 18.
54. Katrina Mason, *Children of Los Alamos: An Oral History of the Town Where the Atomic Age Began* (New York, 1995), p. 136.
55. Stephen Toulmin, 'The Conscientious Spy', *New York Review of Books*, 19 November 1987, pp. 54–60 (p. 56).
56. *Prof. Dr. Klaus Fuchs.*
57. Hedy [*sic*] Bretscher, interview by John Bennett and Anne Shepherd, 21 June 1984, AIP, p. 42.
58. *Prof. Dr. Klaus Fuchs.*
59. Fuchs to Rotblat, 2 October 1946; Rotblat to Fuchs, 10 October 1946; Fuchs to Rotblat, 24 December 1946; Rotblat to Fuchs, 6 January 1947; Fuchs to Rotblat, 11 February 1947, RTBT, D. 191; Rotblat to Fuchs, 18 February 1947, RTBT, D. 82A.
60. Fuchs to Bretscher, 13 May 1948, the Papers and Correspondence of Egon Bretscher, Churchill Archives Centre, Cambridge, United Kingdom (hereafter BRER), BRER, H. 29.
61. Hanni Bretscher, note, 1 April 1987, attached to letter, Fuchs to Bretscher, 13 May 1948, BRER, H. 29. Emil Fuchs visited England from autumn 1947 until spring 1948; *Mein Leben*, II, 287–9.
62. *Prof. Dr. Klaus Fuchs.* Herbert Skinner even offered Fuchs financial help to pay for a solicitor, J. C. Robertson, 'Subject: Klaus Emil Julius Fuchs', 6 February 1950, TNA, KV 2/1661, p. 1. Support was not restricted to Harwell

colleagues. F. C. Champion, for example, also offered help to find a solicitor for Fuchs; Peierls to Champion, 6 February 1950, Peierls Papers, MS Eng. Misc. b. 223, F6.

63. Metropolitan Police Special Branch, p. 24.
64. 'Record of Interview with Dr. K. Fuchs on 30th January, 1950 by M.W. Perrin'. The description given by Fuchs here is almost identical with that provided by him in *Prof. Dr. Klaus Fuchs*.
65. 'Statement of Emil Julius Klaus Fuchs', p. 10.
66. *Prof. Dr. Klaus Fuchs*; Henry DeWolf Smyth, *Atomic Energy for Military Purposes: The Official Report on the Development of the Atomic Bomb Under the Auspices of the United States Government, 1940–1945* (Princeton, 1945; repr. Stanford, 1989).
67. Goodman, 'Grandfather of the Hydrogen Bomb', p. 6; Goodman, *Spying on the Nuclear Bear*, pp. 20–1; 'Record of Interview with Dr. K. Fuchs on 30th January, 1950 by M.W. Perrin', p. 4; Jeffrey Richelson, *Spying on the Bomb: American Nuclear Intelligence from Nazi Germany to Iran and North Korea* (New York, 2006), p. 65; 'Statement of Emil Julius Klaus Fuchs', p. 10;
68. Hoover to Neal, Subject: Emil Julius Klaus Fuchs Espionage – R, 18 June 1950, Klaus Fuchs FBI File, 65-58805, vol. 36, serials 1346–66, pp. 1–2.
69. Alexander Feklisov, 'Podvig Klaus Fuksa', *Voenno-Istoricheskii Zhurnal*, 12 (1990), 22–9; Alexander Feklisov, 'Podvig Klaus Fuksa', *Voenno-Istoricheskii Zhurnal*, 1 (1991), 34–43.
70. David Holloway, *Stalin and the Bomb: The Soviet Union and Atomic Energy, 1939–1956* (New Haven, 1994), p. 222. Montgomery Hyde, for example, estimated that Fuchs helped save the Soviets about 18 months in their plutonium bomb project (p. 222).
71. Yuli Khariton and Yuri Smirnov, 'The Soviet Bomb: The Khariton Version', *Bulletin of the Atomic Scientists* (hereafter *BAS*), 49. 4 (May 1993), 20–31. Fuchs himself made a similar point; *Prof. Dr. Klaus Fuchs*.
72. Goodman, *Spying on the Nuclear Bear*, pp. 7–56.
73. 'Record of Interview with Dr. K. Fuchs on 30th January, 1950 by M.W. Perrin', p. 5.
74. Ladd to Hoover, Re: Emil Julius Klaus Fuchs, 1 February 1950, Klaus Fuchs FBI File, 65-58805, vol. 3, serials 83–171.
75. Holloway, pp. 174–8, 199, 222–3; Rainer Karlsch, *Uran für Moskau: Die Wismut – Eine populäre Geschichte* (Berlin, 2007).
76. This is discussed in Goodman, 'Grandfather of the Hydrogen Bomb', pp. 1–22.
77. Herken, *Brotherhood of the Bomb*, p. 219; Priscilla McMillan, *The Ruin of J. Robert Oppenheimer: And the Birth of the Modern Arms Race* (New York, 2005), pp. 65–6.
78. Arnold, *Britain and the H-Bomb*, pp. 24, 131–50; Richard Rhodes, *Dark Sun: The Making of the Hydrogen Bomb* (New York, 1995), pp. 523–5.
79. 'The Case of Dr. Klaus Fuchs. – Appendix A', 2 March 1950, TNA, KV 2/1253, pp. 1–2.
80. 'The Case of Dr. Klaus Fuchs. – Appendix A', p. 1.
81. Gowing, *Independence and Deterrence*, II, 147–8.
82. Goodman, 'Grandfather of the Hydrogen Bomb', pp. 15, 22.
83. 'The Case of Dr. Klaus Fuchs. – Appendix A', 2 March 1950, p. 2.
84. Robertson, 'Emil Julius Klaus Fuchs', 23 November 1949, pp. 6, 9, 19.

85. Max Born, *My Life*, p. 284. See also Nevill Mott's reply to Born's allegations in a note on the same page.
86. Nevill Mott, *A Life in Science* (London, 1986), p. 50.
87. Peierls, interview by Weiner, p. 153.
88. Mott, *A Life in Science*, p. 51.
89. SAC, New York, to Director, FBI, pp. 4–5; *Prof. Dr. Klaus Fuchs*; Skardon, 'Emil Julius Klaus Fuchs', 22 December 1949, p. 2.
90. Max Born, *My Life*, p. 284.
91. David Vincent, *The Culture of Secrecy: Britain, 1832–1998* (Oxford, 1998), p. 206.
92. Joan Mahoney, 'Civil Liberties in Britain During the Cold War: The Role of the Central Government', *American Journal of Legal History*, 33. 1 (1989), 53–100 (pp. 82–92).
93. Akers to Peierls, 2 May 1949, Peierls Papers, sup. cat., A. 13.
94. Peter Hennessy and Gail Brownfeld, 'Britain's Cold War Security Purge: The Origins of Positive Vetting', *Historical Journal*, 25. 4 (1982), 965–74 (pp. 968–70).
95. Vincent, pp. 1–25, 194–210.
96. Metropolitan Police Special Branch, pp. 3, 24, 25.
97. Robert Williams, *Klaus Fuchs*, p. 3.
98. Metropolitan Police Special Branch, pp. 11, 17–18.
99. 'Atomic Secrets Betrayed: Fuchs Sentenced to 14 Years', *The Times*, 2 March 1950, p. 6. This sentiment of betrayal lasted; see, for instance, Justin Atholl, 'How the Man to Whom Britain gave Shelter became Our Deadliest Traitor', *Sunday Express*, 3 March 1952, TNA, KV 2/1258.
100. Chapman Pincher, 'Fuchs Gave Bomb to Russia', *Daily Express*, 2 March 1950, p. 1. Fuchs's obituary, for example, also drew on this myth; John Ellison, 'Spy Who Gave Russia Atom Bomb Dies at 76', *Daily Express*, 29 January 1988, p. 15.
101. Weart, p. 121.
102. Chapman Pincher, 'World's No. 1 Spy – as He Is Today', *Daily Express*, 5 November 1965, p. 5. See also Chapman Pincher, 'The Strange New Twist in the Story of Fuchs the Traitor ... and Its Fascinating Question: Would You *Ever* Help a Spy?', *Daily Express*, 22 November 1955, p. 3; and Chapman Pincher, 'The Great Spy Purge Shambles: From Fuchs to Bettaney ... Moles Who Slipped the Nest', *Daily Express*, 19 April 1984, p. 10.
103. J. C. Robertson, 'Press Comments on DR. Fuchs' Espionage Methods & Comments.', 7 March 1950, TNA, KV 2/1254.
104. Christopher Andrew, *The Defence of the Realm: The Authorized History of MI5* (London, 2009), p. 388.
105. See, for example, an advertisement for a series of articles by Alan Moorehead, titled 'Atom-Bomb Traitors' and published in the *Sunday Times*, 20 June 1952, p. 5; Rebecca West, 'The Terrifying Impact of the Fuchs Case', *New York Times*, 4 March 1951, pp. 10, 29, 31.
106. Alan Moorehead, *The Traitors: The Double Life of Fuchs, Pontecorvo and Nunn May* (London, 1952; repr. New York, 1963).
107. Oliver Pilat, *The Atom Spies* (New York, 1952).
108. Justin Atholl, 'How Fuchs Was Found Out: Clue by Clue the World's Deadliest Spy Is Trapped in Britain', *Sunday Express*, 10 February 1952, TNA, KV 2/1258.

109. John Squire, 'Three Atomic Scientists and Their Betrayals', *Illustrated London News*, 26 July 1952, p. 2.
110. Rebecca West, 'The Traitors', *Evening Standard*, 4 June 1951, TNA, KV 2/1257.
111. Rebecca West, *The Meaning of Treason* (London, 1945); Rebecca West, *The Meaning of Treason*, 2nd enlarged and rev. edn (London, 1952); Rebecca West, *The New Meaning of Treason* (New York, 1964).
112. Rebecca West, 'The Hearts of Traitors', *Picture Post*, 17 January 1953, pp. 25, 27–9, 31, 33.
113. Rebecca West, 'The Terrifying Impact of the Klaus Fuchs Case', *New York Times Magazine*, 4 March 1951, pp. 6, 29–34 (pp. 31–2). An MI5 report drafted in November 1950 put forth a similar idea declaring that 'Fuchs was an ideological Communist and became a spy for that reason'; 'The Case of Emil Julius Klaus Fuchs', p. 5.
114. 'The Case of Emil Julius Klaus Fuchs', p. 5.
115. H. Wynn Parry, 'Deprivation of Citizenship Committee, British Nationality Act, 1948, re Klaus Emil Julius Fuchs', 30 January 1951, TNA, KV 2/1257.
116. Executive Producer to the Under Secretary of State, Home Office, Whitehall, 2 January 1951; Williams to the Under Secretary of State, Home Office, Whitehall, 3 February 1951, TNA, KV 2/1257.
117. Samuel Epstein and Beryl Williams, *The Real Book of Spies* (Garden City, 1953; repr. London, 1959).
118. E. D. O'Brien, 'A Christmas Hamper of Books for Children', *Illustrated London News*, 5 December 1959, p. 820.
119. David Reynolds, 'Great Britain', in *The Origins of the Cold War in Europe: International Perspectives*, ed. by David Reynolds (New Haven, 1994), 77–95 (pp. 80–1).
120. The development of this anti-Communist consensus is still debated, see Richard Thurlow, *The Secret State: British Internal Security in the Twentieth Century* (Oxford, 1994), pp. 285–6; Peter Weiler, *British Labour and the Cold War* (Stanford, 1988), pp. 189–229.
121. Douglas Hyde, *I Believed: The Autobiography of a Former British Communist* (London, 1950); *The God That Failed: Six Studies in Communism by Arthur Koestler, Ignazio Silone, André Gide Presented by Enid Starkie, Richard Wright, Louis Fischer, Stephen Spender*, ed. by Richard H. S. Crossman (London, 1950).
122. 'Has Professor J. B. S. Haldane Left the Communist Party? Report Last Night Follows Week of Mystery', *Sunday Express*, 12 November 1950, p. 1.
123. Skardon, 'Emil Julius Klaus Fuchs', 22 December 1949, p. 1.
124. Rebecca West, *New Meaning of Treason*, p. 361.
125. Moorehead, *The Traitors*, p. 58. Moorehead even called *The Traitors* in the new preface to the 1963 edition 'a book about conscience' (p. ix).
126. Herbert Skinner, 'The Atomic Bomb Conspiracy', n.d., Papers of Professor Herbert W. B. Skinner, FRS, Special Collections and Archives, Sidney Jones Library, University of Liverpool, Liverpool, United Kingdom (hereafter SJL), D.982/3/5, p. 6.
127. Stanley Goldberg, 'The Secret about Secrets', in *Secret Agents: The Rosenberg Case, McCarthyism & Fifties America*, ed. by Marjorie Garber and Rebecca L. Walkowitz (New York, 1995), pp. 47–58 (p. 51); Chapman Pincher,

Traitors: The Anatomy of Treason (New York, 1987), pp. 276–80; Chapman Pincher, *Too Secret Too Long* (London, 1984), p. 87; Richard Trahair, 'A Psychohistorical Approach to Espionage: Klaus Fuchs (1911–1988)', *Mentalities*, 9. 2 (1994), 28–49 (p. 29).

128. Skardon, 'Emil Julius Klaus Fuchs. Fourth, Fifth, Sixth and Seventh Interviews', p. 1; 'Record of Interview with Dr. K. Fuchs on 30th January, 1950 by M. W. Perrin', TNA, KV 2/1253, p. 5; 'Statement of Emil Julius Klaus Fuchs', p. 9.

129. 'Record of Interview with Dr. K. Fuchs on 30th January, 1950 by M. W. Perrin', p. 5.

130. Hyde, p. 103.

131. Lorna Arnold, *Britain and the H-Bomb*, pp. 25–6; Goodman, 'Grandfather of the Hydrogen Bomb', pp. 1–22. Both Cockroft and Peierls were also in favour of promoting Fuchs owing to his achievements; Cockroft to Peierls, 22 June 1948; Peierls to Cockroft, 24 June 1948, Peierls Papers, b. 205, C66.

132. Goodman, 'Grandfather of the Hydrogen Bomb', p. 22.

133. Michael Goodman, 'Who Is Trying to Keep What Secret from Whom and Why? MI5-FBI Relations and the Klaus Fuchs Case', *Journal of Cold War Studies*, 7. 3 (Summer 2005), 124–46 (pp. 132–4).

134. Goodman, 'Grandfather of the Hydrogen Bomb', p. 20. On the technical details Fuchs passed on to the Soviet Union, as he told the FBI, see, for example: Hoover to AEC, Re: Emil Julius Klaus Fuchs Espionage – R, 15 June 1950, Klaus Fuchs FBI file, 65-58805, vol. 36, serials 1346–66.

135. Hoover to Neal, p. 2. The FBI agents were accompanied by William Skardon; Skardon, 'Interrogation of Dr. Fuchs by Officers of the F.B.I.'

136. Vincent, p. 202.

137. Moorehead, *The Traitors*, p. 172.

138. Hennessy and Brownfeld, p. 970. On Britain and McCarthyism, see also John P. Rossi, 'The British Reaction to McCarthyism, 1950–54', *Mid-America*, 70. 1 (1988), 5–18.

139. Stephen Guy, '*High Treason* (1951): Britain's Cold War Fifth Column', *Historical Journal of Film, Radio and Television*, 13. 1 (1993), 35–47; Tony Shaw, *British Cinema and the Cold War: The State, Propaganda and Consensus* (London, 2001), pp. 40–5.

140. 'The Case of Dr. Klaus Fuchs. – Appendix A', pp. 1–2.

141. Peter Galison and Barton Bernstein, 'In Any Light: Scientists and the Decision to Build the Superbomb, 1952–1954', *HSPS*, 19. 2 (1989), 267–348 (pp. 310–12).

142. 'A Week of Shock and Decision: A Hint That We Have Lost the Arms race – the Order to Make the H-Bomb – a Plea to Wage Peace – the Espionage Case against an Atomic Spy', *Life*, 13 February 1950, pp. 36–7 (p. 36).

143. '14 Years For "Grossest Treachery": Fuchs Gets Limit under British Law – but It Does Not Conclude the Case', *Life*, 13 March 1950, pp. 42–3; 'A Look at the World's Week', *Life*, 6 July 1959, pp. 37–8, 40 (p. 40).

144. Groves, pp. 143–4.

145. Gowing, *Independence and Deterrence*, II, 145–50; Robert Williams, *Klaus Fuchs*, pp. 141–50.

146. John H. Manley, 'A New Laboratory Is Born', in *Reminiscences of Los Alamos* (see Edwin McMillan, above), pp. 21–40 (pp. 37–8).

147. *Prof. Dr. Klaus Fuchs.*

148. 'The Case of Dr. Klaus Fuchs. – Appendix A', p. 2.
149. David Reynolds, *Britannia Overruled: British Policy & World Power in the 20th Century* (London, 1991), p. 159.
150. Robert A. Goldberg, *Enemies Within: The Culture of Conspiracy in Modern America* (New Haven, 2001), p. 26.
151. Ellen Schrecker, 'Before the Rosenbergs: Espionage Scenarios in the Early Cold War', in *Secret Agents* (see Stanley Goldberg, above), pp. 127–41 (p. 136). On Communists as conspirators in American culture, see also Robert Goldberg, pp. 22–46.
152. Ellen Schrecker, *No Ivory Tower: McCarthyism and the Universities* (New York, 1986), pp. 8, 131, 142.
153. David Kaiser, 'The Atomic Secret in Red Hands? American Suspicions of Theoretical Physicists During the Early Cold War', *Representations*, 90 (Spring 2005), 28–60 (p. 28).
154. J. Edgar Hoover, 'The Crime of the Century: The Case of the A-Bomb Spies', *Reader's Digest*, 58 (May 1951), 149–68 (p. 155).
155. See, for example, 'Atomic Spy Trials: A Summary of the Trials', *BAS*, 7. 4 (April 1951), 125–6; 'Dr. Klaus Fuchs to Stand Trial', *BAS*, 6. 3 (March 1950), 68, 94; Eugene Rabinowitch, 'Atomic Spy Trials: Heretical Afterthoughts', *BAS*, 7. 5 (May 1951), 139–42, 157; 'Soviet Atomic Espionage', *BAS*, 7. 5 (May 1951), 143–8.
156. Lawrence Badash, 'Science and McCarthyism', *Minerva*, 38. 1 (2000), 53–80 (p. 60).
157. Joseph Albright and Marcia Kunstel, *Bombshell: The Secret Story of America's Unknown Atomic Spy Conspiracy* (New York, 1997), p. 112.
158. Richard Fried, *Nightmare in Red: The McCarthy Era in Perspective* (New York, 1990), p. 115.
159. See Cyndy Hendershot, *Anti-Communism and Popular Culture in Mid-Century America* (Jefferson, 2003).
160. Metropolitan Police Special Branch, pp. 24–5. For an assessment of the impact Klaus Fuchs's confession had on Anglo-American nuclear relations see also Richard J. Aldrich, *The Hidden Hand: Britain, America and Cold War Secret Intelligence* (London, 2001), pp. 380–4.
161. Makins to Colville, 24 May 1952; Colville to Makins, 26 May 1952, TNA, PREM 11/2799.
162. John Moser, *Twisting the Lion's Tail: American Anglophobia Between the World Wars* (New York, 1999), pp. 172–4.
163. Robert Norris, *Racing for the Bomb: General Leslie R. Groves, the Manhattan Project's Indispensable Man* (South Royalton, 2002), p. 327; Septimus Paul, p. 41. Whitehall was aware of the American attempts to exclude British scientists from sensitive parts of the project, see, for instance, Akers to Chancellor of the Exchequer, Re: Tube Alloy Project: Access to American Full-Scale Plant Information, 6 April 1944, TNA, CAB 126/331.
164. Rebecca West, *New Meaning of Treason*, p. 195.
165. Metropolitan Police Special Branch; a copy of the transcript was also forwarded to J. Edgar Hoover; Whitson to Director, FBI, Re: Foocase, 3 March 1950, Klaus Fuchs FBI File, 65-58805, vol. 21, serials 897–915.
166. On the Klaus Fuchs case and Anglo-American relations, see Goodman, 'Who Is Trying to Keep What Secret from Whom and Why', pp. 124–46.

167. 'Statement of Emil Julius Klaus Fuchs', p. 8.
168. Sillitoe to Rowlands, 19 January 1950, TNA, KV 2/1250.
169. Goodman, 'Who Is Trying to Keep What Secret from Whom and Why', pp. 130–1.
170. Metropolitan Police Special Branch, p. 22.
171. *Prof. Dr. Klaus Fuchs*.
172. Michael S. Goodman and Chapman Pincher, 'Research Note: Clement Attlee, Percy Sillitoe and the Security Aspects of the Fuchs Case', *Contemporary British History*, 19. 1 (2005), 67–77. The day following Fuchs's trial, the *Daily Mirror*, for example, had run the headline 'MI5 Duped for 6 Years – Why?' on its front page (2 March 1950). When a 1951 article by Rebecca West questioned MI5's practice of security checks in the Fuchs case again, Percy Sillitoe reaffirmed to the prime minister that it was not his service's fault that Fuchs was not exposed earlier; Sillitoe to Rickett, 5 June 1951, TNA, KV 2/1257. Whitehall and the Foreign Office also monitored American press and media coverage of Fuchs's trial; Sir O. Franks to Whitehall and Foreign Office, RE: Press and Radio Comment on the Trial and Sentence of Dr. Fuchs, 10 March 1950, TNA, KV 2/1254.
173. See the correspondence between MI5 and Alan Moorehead: 'Supplementary Notes on Fuchs' background given to Mr. Alan Moorehead on Monday, 24th September, 1951', TNA, KV 2/1257. On Fuchs's mother, see Emil Fuchs, *Mein Leben*, II, 204–6.
174. Belmont to Ladd, Re: 'The Traitors', Book by Alan Moorehead, British Author, 2 June 1952, Klaus Fuchs FBI File, 65-58805, vol. 42, serials 1501–66.
175. On the Pontecorvo case, see Simone Turchetti, 'Atomic Secrets and Governmental Lies: Nuclear Science, Politics and Security in the Pontecorvo Case', *British Journal for the History of Science*, 36. 4 (2003), 389–415.
176. Septimus Paul, pp. 166–87.
177. Turchetti, 'Atomic Secrets and Governmental Lies', pp. 408–9.
178. Central Office of Information, *Their Trade Is Treachery* (London, 1964), unpaginated picture inlet.
179. Ministry of Supply and Central Office of Information, *Harwell: The Atomic Energy Research Establishment* (London, 1952). The same year the New York-based Philosophical Society published the booklet under the same title in the United States.
180. 'Berlin Impressed by Allied Note', *The Times*, 15 May 1952, 6.
181. 'The Man to Whom Fuchs Confessed', *New Scientist*, 24 January 1957, 27–8 (p. 28).
182. Cited in 'Entente Cordiale Under Strain', *The Times*, 20 August 1959, p. 8.

5 Subject to Suspicion: Rudolf Peierls and the Klaus Fuchs Espionage Case

1. Simpson to Peierls, 27 March 1950, Peierls Papers, b. 226, F 47.
2. Rudolf Peierls, 'President's Report', *Atomic Scientists' News* (hereafter *ASN*), 4. 1 (August 1950), 6–8 (p. 7). Peierls later reiterated this point; see, for example, Rudolf Peierls, 'Britain in the Atomic Age', in *Alamogordo Plus Twenty-Five*

Years: The Impact of Atomic Energy on Science, Technology, and World Politics, ed. by Richard S. Lewis and Jane Wilson, with Eugene Rabinowitch (New York, 1970), pp. 91–105 (p. 102); Peierls, interview by Weiner, p. 153.

3. J. H. Marriott, 'Note', 6 February 1950, TNA, KV 2/1661.

4. Robertson, 'Subject: Klaus Emil Julius Fuchs', p. 1.

5. Ibid.

6. Peierls to Bethe, 15 February 1950, Peierls Papers, b. 202, C17.

7. Peierls to Burt, n.d. (received 6 February 1950), TNA, KV 2/1661; Peierls to the Governor, Brixton Prison, 27 February 1950, Peierls Papers, b. 207, C111. On Peierls's disillusionment with Fuchs, see also: Peierls to Taylor, 7 February 1950, Peierls Papers, suppl. cat., D. 52.

8. Genia Peierls to Fuchs, n.d. KV 2/1251. While the letter is not dated, the accompanying documents strongly suggest that it was written between 4 and 6 February 1950; J.C. Robertson, 'Note', 6 February 1950, TNA, KV 2/1251.

9. Fuchs to Genia Peierls, 6 February 1950, Peierls Papers, suppl. cat., D. 52.

10. Peierls to Brunt, 22 April 1950; Brunt to Peierls, 24 April 1950, Peierls Papers, b. 207, C111.

11. Peierls to Halsall, 13 February 1950; Halsall to Peierls, 14 February 1950; Peierls to Halsall, 21 February 1950, Peierls Papers, b. 207, C111.

12. Peierls to Fuchs, 15 June 1950, Peierls Papers, sup. cat., D. 52.

13. See, for example, Ward to Peierls, 10 March 1950, Peierls Papers, b. 207, C111.

14. Cited in Lansing Lamont, *Day of Trinity* (New York, 1965), p. 283.

15. Genia Peierls to Fuchs (c. 4–6 February 1950); Robertson, 'Note', 6 February 1950, Peierls Papers, suppl. cat., D. 52.

16. Sidney Rodin and Joseph Garrity, 'Perturbed Men: Foreign-born Atom Experts Disturbed by Pontecorvo Case', *Sunday Express*, 29 October 1950, pp. 1, 7.

17. Max Born, *My Life*, p. 288.

18. Peierls to the News Editor, *Sunday Express*, 6 November 1950; Cudlipp to Peierls, 7 November 1950; Peierls to Cudlipp, 8 November 1950. Peierls also distributed copies of his statement to Born, Frisch, Kurti, Rotblat and Simon; Peierls to Born, Frisch, Kurti, Rotblat, Simon, 9 November 1950, Peierls Papers, b. 197, A 16.

19. Rudolf Peierls, 'To a Just, Fair, and Steady Britain', *Sunday Express*, 12 November 1950, p. 4.

20. Arthur Bryant, 'Our Note Book', *Illustrated London News*, 25 November 1950, p. 852.

21. Peierls to the News Editor, *Daily Mail*, 2 March 1950. For the preceding exchange of letters, see Hallows to Peierls, 23 February 1950; Peierls to the News Editor, *Daily Mail*, 27 February 1950, Peierls Papers, b. 207, C111.

22. Temple to *Daily Mail*, 7 September 1951. Peierls also kept himself well informed about reports in the daily press; see, for instance, 'Extracts from "Intelligence Digest"', 1951, Peierls Papers, suppl. cat. A. 15.

23. Peierls, 'The Lesson of the Fuchs Case', p. 4.

24. Schlinck to Hoover, 20 November 1952, attached to letter, O'Brien to Marriott, 16 December 1952, TNA, KV 2/1662.

25. Hoover to Schlinck, 3 November 1952, Klaus Fuchs FBI File, 65-58805, vol. 15, serials 720–80.

26. 'Report', 20 January 1951, TNA, KV 2/1661. Rudolf Peierls cautiously observed anti-Semitism. See, for example, the exchange of letters with the Bishop of Oxford regarding the *Pamyat* movement in Russia; Peierls to the Rt. Rev. the Bishop of Oxford, 7 October 1989; The Bishop of Oxford to Peierls, 10 November 1989, Peierls Papers, suppl. cat. A. 19.

27. J. C. Robertson, 'Meeting with W/Cdr. Arnold on 9th September, 1949', 9 September 1949, TNA, KV 2/1246, p. 2; Robertson, 'Emil Julius Klaus Fuchs', 23 November 1949, p. 12; Arnold to Robertson, 7 October 1952, TNA, KV 2/1259; C. Grose-Hodge, 'Professor Rudolph [*sic*] Ernst Peierls, C.B.E. ('46), F.R.S. ('45), M.A. (Cantab), D.Phil. (Leipzig)', 6 December 1950, TNA, KV 2/1661, p. 4.

28. Robertson, 'Emil Julius Klaus Fuchs', 23 November 1949, pp. 12–14, 17.

29. J. C. Robertson, 'Press Comments on Dr. Fuchs' Espionage Methods & Contacts', 6 March 1950, TNA, KV 2/1254, p. 1.

30. Kaiser, p. 46.

31. Skardon, 'Emil Julius Klaus Fuchs', p. 4; SAC, New York, to Director, FBI, pp. 29–32, 34.

32. Skardon, 'Emil Julius Klaus Fuchs. Fourth, Fifth, Sixth and Seventh Interviews', p. 3; 'Record of Interview with Dr. K. Fuchs on 30th January, 1950 by M. W. Perrin', pp. 1, 4–5; SAC, New York, to Director, FBI, p. 37.

33. Carlton Lenz, 'Emil Julius Klaus Fuchs, was', 6 April 1950, Klaus Fuchs FBI File, 65-58805, vol. 27, serials 1031–78, p. 2.

34. 'Re: Rudolph Ernst Peierls', 24 January 1951, Klaus Fuchs FBI File, 65-58805, vol. 41, serials 1457–500.

35. Peter Maxson, 'Emil Julius Klaus Fuchs', 17 February 1950, Klaus Fuchs FBI File, 65-58805, vol. 6, serials 301–85; Ladd to the Director, 5 February 1950, Klaus Fuchs FBI File, 65-58805, vol. 8, serials 406–75; Nigel West, *Mortal Crimes: The Greatest Theft in History: The Soviet Penetration of the Manhattan Project* (New York, 2004), pp. 229–31.

36. Sillitoe to Morren, 13 March 1950, TNA, KV 2/1253.

37. Mott, *A Life in Science*, p. 50.

38. 'Re: Rudolph Ernst Peierls', 24 January 1951, attached to letter, Patterson to Director-General, 25 January 1951, TNA, KV 2/1661.

39. Patterson to Director-General of the Security Service, 16 November 1949, TNA, KV 2/1660. An FBI report states that Peierls's father Heinrich was alleged to have made pro-German remarks during the Second World War; 'Emil Julius Klaus Fuchs, was'; Rudolf Ernst Peierls', 31 October 1949, Klaus Fuchs FBI File, 65-58805, vol. 2, serials 27–80.

40. Metropolitan Police, Port of Harwich, to Special Branch, Metropolitan Police, Scotland House, 21 April 1938; Chief Constable, Birmingham Police to Major D. B. Dykes, 21 October 1943; D. B. Dykes, Major, to E. J. R. Corin, RE: Neils [*sic*] Henrik David Bohr and Aage Niels Bohr, 3 November 1943; Captain E. J. Corin to Major D. B. Dykes, 3 November 1943, TNA, KV 2/1658.

41. Although no evidence existed against either Peierls or Fuchs, the cases of both men were investigated by MI5; D.D.G. to D.C., C.2., D.B., B.1., 20 December 1946; Liddell to Maxwell, 23 January 1947, TNA, KV 2/1658.

42. 'Rudolph Ernest [*sic*] Peierls', 18 March 1954, TNA, KV 2/1663, p. 3.

43. R. H. Hollis to B.1.a, 14 January 1947; Mitchell to Allan, RE: Professor Rudolf Ernst Peierls, The University, Birmingham, 24 January 1947. For copies of

intercepted letters see, for example, Peierls to Fuchs, 22 March 1947; Peierls to Fuchs, 24 March 1947; Peierls to Fuchs, 10 May 1947; Fröhlich to Peierls, 22 September 1949, TNA, KV 2/1658. Note that the person inspecting the letter was unable to identify the sender as Herbert Fröhlich.

44. J. C. Robertson, 'Klaus Fuchs. Further Investigation Plan', 12 September 1949, TNA, KV 2/1246, p. 2; J. C. Robertson to Allan, 13 September 1949; J. C. Roberts to G. P. Saffery, 13 September 1949, TNA, KV 2/1658.
45. J. C. Robertson, 'Telephone Check Coverage following Fuchs' Arrest', 2 February 1950, TNA, KV 2/1250.
46. C.2.a. to B.2.a., 12 December 1949, TNA, KV 2/1660.
47. Ronnie Peierls to Rudolf Peierls, 1 January 1950, TNA, KV 2/1660.
48. See the correspondence between Sir Percy Sillitoe and E. J. Dodd, the Chief Constable of Birmingham, May to September 1950; Box 500, Parliamentary Street B.O., London, S.W.1 to the Chief Constable of Birmingham, 31 May 1950; Sir Percy Sillitoe to Dodd, 30 August 1950; Dodd to Sir Percy Sillitoe, 11 September 1950, TNA, KV 2/1661.
49. John H. Marriott to Lt. Colonel Collard, 10 September 1946, TNA, KV2/2589.
50. C. Grose-Hodge, 'Professor Rudolph [*sic*] Ernst Peierls, C.B.E. ('46), F.R.S. ('45), M.A. (Cantab), D.Phil. (Leipzig)', 6 December 1950, TNA, KV 2/1661, pp. 2–5.
51. J. C. Robertson, 'Note.', 22 November 1949, TNA, KV 2/1660.
52. 'Rudolph Ernest Peierls', p. 6. While Sabine Lee has reached the same conclusion, she did not examine the MI5 files ('The Spy That Never Was', *Intelligence and National Security*, 17. 4 (2002), 77–99).
53. R. H. Morton, 'Note', 3 January 1951, TNA, KV 2/1661.
54. Battersby to Patterson, 11 January 1951. Apart from Peierls, the American security services also investigated other British scientists such as P. M. S. Blackett, S. C. Curran, Otto Frisch and T. G. Pickavance with regard to their political beliefs; DeBardeleben to Badham, 28 December 1950, TNA, KV 2/1661.
55. Patterson to Director-General of the Security Service, Subject: Emil Fuchs, Rudolf Ernest [*sic*] Peierls, 2 December 1949, TNA, KV 2/1249.On the reconstruction of Rudolf Peierls's whereabouts see, for example, J. Jerome Maxwell, 'Emil Julius Klaus Fuchs, was', 20 October 1949, Klaus Fuchs FBI File, 65-58805, vol. 1, serials 1–26; J. Jerome Maxwell, 'Emil Julius Klaus Fuchs, was', 23 January 1950, Klaus Fuchs FBI File, 65-58805, vol. 2, serials 27–80, p. 6.
56. Badash, 'Science and McCarthyism', pp. 60–2.
57. Wang, p. 43.
58. Lawrence Badash, 'From Security Blanket to Security Risk: Scientists in the Decade after Hiroshima', *History and Technology*, 19. 3 (2003), 241–56 (pp. 247–8).
59. 'The Secrecy Clause in the Atomic Energy Bill', n.d. Peierls Papers, b. 223, F 8; Gowing, *Independence and Deterrence*, II, 118; Rudolf E. Peierls, 'The British Atomic Scientists' Association', *BAS*, 6. 2 (February 1950), 59; 'International Control of Atomic Energy (I)', *Nature*, 157. 3999 (22 June 1946), 817–20 (pp. 819–20); 'Comparison of British and American Atomic Energy Acts', *BAS*, 3. 2 (February 1947), 50–1.
60. W. J. Arrol and E. H. S. Burhop, 'Minutes of the First General Meeting of the Atomic Scientists' Association Held at the University, Edmund Street, Birmingham, on June 15th, 1946, at 2.15 p.m.', RTBT, K. 25, pp. 2–3.

61. Philip B. Moon and Rudolf E. Peierls, 'Atomic Energy: Second Reading of the Bill, Two Points of Criticism', *The Times*, 8 October 1946, p. 5. See also Philip B. Moon and Rudolf E. Peierls, 'Atomic Energy', *The Times*, 5 November 1946, p. 5.

62. James Chadwick, 'The Bomb: International Co-operation and Security', *ASN*, n.s. 1. 4 (March 1952), 125–7 (p. 127).

63. 'The Atomic Energy Bill: Statement Made by the Executive Committee of the A.Sc.W.', 17 September 1946, RTBT, K. 22.

64. Association of Scientific Workers, 'On the Case of Alan Nunn May', n.d., RTBT, K. 23, unpaginated; W.J. Arrol, 'Atomic Scientists' Association: Minutes of Council Meeting Held at Clarendon Laboratory, Oxford, on Saturday, July 6th, 1946 at 11.45 a.m.', n.d., RTBT, K. 27, p. 1.

65. Sir Henry Dale and R.E. Peierls, 'Freedom of Science', *BAS*, 5. 4 (April 1949), 106–9. For a digest of Peierls' and Dale's talks see also 'Freedom of Science', *ASN*, 2. 4 (7 January 1949), 88–93. Another British scientist, Michael Polanyi also published a more philosophical treatment of the issue; 'Freedom in Science', *BAS*, 6. 7 (July 1950), 195–8, 224.

66. Edward U. Condon, 'Un-American Activities', *ASN*, 1. 11 (4 June 1948), 178–80; 'American Scientists Involved in Security Investigations', *ASN*, 2. 3 (27 October 1948), 49–54.

67. 'Re: Rudolph Ernst Peierls', 24 January 1951, p. 7, attached to letter, Patterson to Director-General, 25 January 1951, TNA, KV 2/1661.

68. Lawrence Badash, 'From Security Blanket to Security Risk: Scientists in the Decade after Hiroshima', *History and Technology*, 19. 3 (2003), 241–56 (pp. 246–7).

69. Anonymous letter to Grose-Hodge, 'Professor Peierls', 15 January 1951, TNA, KV 2/1661. Note that the informant is referred to as a 'scientist' and the name 'Charwell' was added in handwriting. It could not be defined whether Lord Cherwell was in fact the anonymous informer.

70. Peierls, *Bird of Passage*, p. 321.

71. Atomic Scientists' Association, 'The Civil Service Purge', April 1950, RTBT, K. 51.; 'The Civil Service Purge', *ASN*, 3. 5 (May 1950), 108–9. The statement was even reprinted in the *BAS*; 'The Civil Service Purge in Britain: Statement by the British Atomic Scientists' Association', *BAS*, 6. 6 (June 1950), 185. See also Rudolf Peierls, 'Security and the Right of the Individual', n.d., RTBT, K. 49B.

72. Richard Beyler, Alexei Kojevnikov and Jessica Wang, 'Purges in Comparative Perspective: Rules for Exclusion and Inclusion in the Scientific Community under Political Pressure', *Osiris*, 2nd series, 20 (2005), 23–48 (p. 23).

73. F. C. Champion to Elizabeth A. Allen, 28 April 1949; Allen to Champion, 3 May 1949, RTBT, K. 63.

74. Rudolf Peierls, 'Bathwater and the Baby: Some Thoughts about the Cold War', *ASN*, n.s. 1. 1 (September 1951), 10–13 (p. 13).

75. 'British Physicists For Moscow', *The Times*, 11 May 1956, p. 8; Peierls, *Bird of Passage*, pp. 266–8.

76. Peierls to Penney, 6 March 1956, Peierls Papers, suppl. cat. A. 17, p. 3.

77. Greta Jones, *Science, Politics and the Cold War* (London, 1988), pp. 16–37.

78. Karl Popper, *The Open Society and Its Enemies*, 2 vols (1945; repr., London, 1962), I, p. viii.

79. Friedrich August Hayek, *The Road to Serfdom* (London, 1944), p. 178.

80. 'Dilemma of the Atomists', *Economist*, 24 July 1948, 140; 'Science and Its Social Relations', *Nature*, 162. 4111 (14 August 1948), 235–7 (p. 237). On the echo in the press see also: 'Some Press Reactions to Our memorandum on International Control', *ASN*, 2. 2 (3 September 1948), 31–4; '*Nature* and Our Memorandum on International Control', *ASN*, 2. 3 (27 October 1948), 64–6.

81. Rudolf Peierls, 'Dilemma of the Atomists', *Economist*, 4 September 1948, 375. See the accompanying correspondence in Peierls Papers, MS Eng. Misc. b. 223, F 6. See also Nevill Mott's letter to the editor of *Nature*; 'International Exchange of Scientific Information', *Nature*, 162. 4115 (11 September 1948), 417.

82. 'Some British Experiences: II. R. E. Peierls', in *American Visa Policy and Foreign Scientists*, ed. by Edward A. Shils (=*BAS*, 8.7 (October 1952)), pp. 229–30 (p. 229).

83. B. J. S. M. Washington to Cabinet Office, 23 August 1951, TNA, KV 2/1662. Both Peierls and Fuchs (before his confession) served on the Publication and Declassification Sub-Committee; J. H. Awbery and J. F. Jackson, 'Publication and Declassification Sub-Committee', 22 July 1948, CHAD I/19/8.

84. Cabinet Office to B. J. S. M. Washington, 27 August 1951, TNA, KV 2/1662.

85. Peierls, *Bird of Passage*, p. 321. Note that Peierls's account differs here from the aforementioned documents in the two preceding notes. Unlike the claims in his autobiography, he was not automatically granted a US visa to attend the declassification conference as a British government official.

86. Peierls, *Bird of Passage*, p. 322.

87. Badash, 'Science and McCarthyism', pp. 65–6.

88. The general problem is summarized in: Edward A. Shils, 'Editorial: America's Paper Curtain', in *American Visa Policy and Foreign Scientists* (see 'Some British Experiences', above), pp. 210–17.

89. 'Eminent American Scientists Give Their Views on American Visa Policy', in *American Visa Policy and Foreign Scientists* ('Some British Experiences', above), pp. 217–20.

90. John Krige, *American Hegemony and the Postwar Reconstruction of Science in Europe* (Cambridge, MA: MIT Press, 2006), p. 119.

91. John Krige, 'Building the Arsenal of Knowledge', *Centaurus*, 52. 4 (2010), 280–96 (p. 282).

92. 'Scientists and U.S. Visas', *Atomic Scientists' Journal* (hereafter *ASJ*), 4. 4 (March 1955), 202–3.

93. Badash, 'Science and McCarthyism', pp. 62–4.

94. Wang, pp. 277–8.

95. 'Dr. Burhop and the Press', ASN, n.s. 2. 6 (July 1953), 366–8 (p. 366).

96. Schrecker, *No Ivory Tower*, p. 139; Schrecker, 'Before the Rosenbergs', pp. 130–4.

97. While Gregg Herken has argued that Oppenheimer was a Communist (*Brotherhood of the Bomb*, pp. 43–62), Kai Bird and Martin J. Sherwin have refuted this claim (pp. 119–24).

98. Bird and Sherwin, p. 434; Priscilla McMillan, p. 66.

99. Rudolf Peierls, 'In the Matter of J. Robert Oppenheimer', *ASJ*, 4. 3 (January 1955), 145–54 (p. 154). Apart from Peierls's piece, this issue of the *ASJ* contained in its editorial two articles dealing with case, titled 'The Oppenheimer Case' (pp. 141–3) and 'The Hidden Struggle for the H-Bomb' (pp. 143–4), as well as a short bibliographical essay following Peierls's review titled 'Further Articles on the Oppenheimer Case', p. 154.

100. Peierls to Oppenheimer, 16 April 1954, Peierls Papers, suppl. cat. A. 16.
101. Peierls, *Bird of Passage*, p. 321.
102. Mattioli to Peierls, 31 May 1966; Mattioli to Peierls, 14 June 1966; Peierls to Mattioli, 16 June 1966, Peierls Papers, b. 219, D 22.
103. Young to Peierls, 18 June 1984; Peierls to Young, 19 June 1984; Peierls to Humphreys, 5 August 1984; Humphreys to Peierls, 13 August 1984; Peierls to Humphreys, 15 August 1984. The script of the documentary, 'The Brilliant Scientist', is also contained in the file; Peierls Papers, suppl. cat., D. 54.
104. Richard Deacon, *The British Connection: Russia's Manipulation of British Individuals and Institutions* (London, 1979); Peierls, *Bird of Passage*, pp. 324–5.
105. Sabine Lee, 'The Spy That Never Was', *Intelligence and National Security*, 17. 4 (Winter 2002), 77–99 (pp. 85–96).
106. Nigel West, 'Truth Behind the Atom Spies', *Daily Express*, 11 June 1999, pp. 32–3 (p. 33).
107. Hans Bethe, Kurt Gottfried and Roald Sagdeev, 'Did Bohr Share Nuclear Secrets?', *American Scientific*, (May 1995), pp. 84–90.
108. Andrew Brown, 'The Viennese Connection: Engelbert Broda, Alan Nunn May and Atomic Espionage', *Intelligence and National Security*, 24. 2 (2009), 173–93 (p. 178).

6 The Responsible Scientist: Rudolf Peierls and the Formation of the Atomic Scientists' Association

1. Robert Oppenheimer, 'The New Weapon: The Turn of the Screw', in *One World or None: A Report to the Public on the Full Meaning of the Atomic Bomb*, ed. by Dexter Masters and Katherine Way (New York, 1946), pp. 22–5 (p. 23).
2. Amy Staples, *The Birth of Development: How the World Bank, Food and Agriculture Organization, and World Health Organization Changed the World, 1945–1965* (Kent, 2006), pp. 1, 6, 9.
3. Peierls, *Bird of Passage*, p. 202.
4. Jon Hunner, 'Reinventing Los Alamos: Code Switching and Suburbia at America's Atomic City', in *Atomic Culture*, pp. 33–48 (p. 34).
5. *Prof. Dr. Klaus Fuchs.*
6. Samuel Goudsmit, *Alsos: The Failure of German Science* (London, 1947); Richelson, pp. 17–61; Mark Walker, *Nazi Science: Myth, Truth, and the German Atomic Bomb* (Cambridge, MA, 1995), pp. 183–241.
7. See Norris, pp. 373–94; Barton Bernstein, 'Reconsidering the "Atomic General": Leslie R. Groves', *Journal of Military History*, 67. 3 (2003), 883–920 (pp. 902–11).
8. Silvan Schweber, *In the Shadow of the Bomb: Oppenheimer, Bethe, and the Moral Responsibility of the Scientist* (Princeton, 2000), p. 153.
9. Joseph Rotblat, 'Leaving the Bomb Project', *BAS*, 41. 7 (August 1985), 16–19.
10. Thorpe, *Oppenheimer*, pp. 153–5.
11. Steven Shapin, *The Scientific Life: A Moral History of a Late Modern Vocation* (Chicago, 2008), p. 80.

12. Charles Thorpe and Steven Shapin, 'Who Was J. Robert Oppenheimer? Charisma and Complex Organization', *Social Studies of Science*, 30. 4 (August 2000), 545–90.
13. Rudolf Peierls, Interview by Charles Weiner, 11–13 August 1969, Oral History Collections, Niels Bohr Library and Archives, American Institute of Physics, College Park, Maryland, United States (hereafter AIP), pp. 149–50.
14. Leo Szilard, 'A Petition to the President of the United States', in *The Atomic Age: Scientists in National and World Affairs, Articles from the Bulletin of the Atomic Scientists 1945–1962*, ed. by Morton Grodzins and Eugene Rabinowitch (New York, 1963), pp. 28–9 (p. 28); Edward Teller, with Allen Brown, *The Legacy of Hiroshima* (Garden City, 1962), p. 13.
15. Peierls, interview by Weiner, p. 150.
16. 'A Report to the Secretary of War', *BAS*, 1. 5 (May 1946), 2–4, 16; Matt Price, 'Roots of Dissent: The Chicago Met Lab and the Origins of the Franck Report', *Isis*, 86. 2 (1995), 222–44 (pp. 222–3).
17. Samuel Walker, *Prompt and Utter Destruction*, pp. 60–1, 76–97.
18. Jonathan Rosenberg, 'Before the Bomb and After: Winston Churchill and the Use of Force', in *Cold War Statesmen Confront the Bomb: Nuclear Diplomacy Since 1945*, ed. by John Lewis Gaddis and others (Oxford, 1999), pp. 171–93 (pp. 180–2).
19. Peierls, *Bird of Passage*, p. 203. Hans Bethe and Victor Weisskopf shared these views; Hans Bethe, interview by Mario Balibrera, 9 November 1979, LAHM, p. 14; Weisskopf, *Joy of Insight*, pp. 127–8.
20. *Professor Dr. Klaus Fuchs.*
21. Peierls, *Bird of Passage*, pp. 204–5. Hans Bethe, for example, shared this view; interview by Balibrera, p. 14.
22. Paul Boyer's observation also applies to Britain, *By the Bomb's Early Light*, pp. 59–64. See 'Atomic Bomb in Use against Japs – Total Ruin Soon', *Daily Mirror*, 7 August 1945, p. 1; 'The Bomb That Has Changed the World', *Daily Express*, 7 August 1945, p. 1; 'First Atomic Bomb Hits Japan', *The Times*, 7 August 1945, p. 4; 'Scientists Whose Research Gave Britain and America the Secrets of Atomic Energy', *Picture Post*, 25 August 1945, pp. 12–13; 'Text of Statements by Truman, Stimson on Development of Atomic Bomb', *New York Times* 7 August 1945, p. 4; 'Atomic Bomb Used on Japan', *Manchester Guardian*, 7 August 1945, p. 5.
23. Peierls, interview by Weiner, p. 149.
24. Ibid., p. 151.
25. Lawrence Wittner, *The Struggle Against the Bomb*, 3 vols (Stanford: Stanford University Press, 1993–2003), I (1993), 20–2.
26. Hunner, *Inventing Los Alamos*, p. 112.
27. Cited in Robert Wilson, 'Hiroshima: The Scientists' Social and Political Reaction', *PAPS*, 140. 3 (1996), 350–7 (p. 352).
28. Hunner, *Inventing Los Alamos*, p. 112.
29. Johnson and Jackson, p. 177.
30. Lothar Nordheim, interview by Bruce Wheaton, 24 July 1977, AIP, pp. 42–3.
31. Sean Labat, 'Chicago Atomic Scientists and United States Foreign Policy, 1945–1950', *Journal of Illinois History*, 3 (Summer 2000), 121–40 (p. 126);

Alice Kimball Smith, 'Behind the Decision to Use the Atomic Bomb: Chicago 1944–45', *BAS*, 14. 8 (October 1958), 288–312 (p. 300).

32. '400 Experts Decry Lone Atom Policy; See "Unending" War', *New York Times*, 14 October 1945, pp. 1, 4; Robert Wilson, 'Hiroshima', p. 353.

33. R. Christy and others, 'The Committee proposes to submit this document first to the President's Interim Committee for Atomic power and subsequently to the public', n.d., attached to letter, E. Titterton and others to J. Chadwick, 6 September 1945, CHAD IV/12/2.

34. 'Memorandum from British Scientists at the Los Alamos Laboratory, New Mexico', n.d., pp. 1, 2, 4, 5, attached to letter, W. G. Marley to D. Rickett, 23 October 1945, CHAD IV/12/2.

35. Ibid., pp. 1, 2, 4, 5. Paul Boyer has examined fear as a tool used by the early American scientists' movement in *By the Bomb's Early Light*, pp. 65–75.

36. Wittner, I, 59–61.

37. Federation of American Scientists, 'Policy Statement by the Administrative Committee', 9 December 1946, RTBT, K. 29, p. 2.

38. Daniel Lang, *Early Tales of the Atomic Age* (Garden City, 1948), p. 78.

39. Rudolf Peierls, 'Britain in the Atomic Age', in *Alamogordo Plus Twenty-Five Years: The Impact of Atomic Energy on Science, Technology, and World Politics*, ed. by Richard Lewis and Jane Wilson, with Eugene Rabinowitch (New York, 1970), pp. 91–105 (p. 95).

40. Arrol to Rotblat, 27 March 1946, attached: 'Draft Pamphlet to Prosopective Members of the Association of Atomic Scientists', RTBT, K. 21; 'Committee of Atomic Scientists: Suggested Organization', 28 February 1946; 'Articles of Association of the Atomic Scientists' Association', 1946, attached to letter, Arrol to Rotblat, 5 June 1946, RTBT, K. 22.

41. R. Innes to Rotblat, 13 December 1945, RTBT, K. 21; 'Present Members of the Committee of Atomic Scientists', 1 January 1946, RTBT, K. 22.

42. Association of Scientific Workers, 'Press Statement: Committee of Atomic Scientists', 21 January 1946, attached to: Roy Innes, 'Minutes of the Second Meeting of the Committee Held on Saturday, 19th January, 1946 at 2.30 p.m. at Gas Industry House, London, S.W.1, RTBT, K. 23.

43. Roy Innes, 'Minutes of the Third Meeting of the Committee Held on Saturday, 23rd February, 1946 at 5. 0 p.m. at 15 Half Moon Street, RTBT, K. 23, p. 2.

44. G. O. Jones and N. Kurti to Rotblat, 9 February 1946; Rotblat to Kurti, 20 February 1946, RTBT, K. 24.

45. 'Atomic Scientists' Association', n.d., RTBT, K. 25.

46. G. O. Jones and N. Kurti to Rotblat, 27 February 1946; G. O. Jones to Rotblat, 2 March 1946, RTBT, K. 24.

47. Peierls to Chadwick, 26 February 1946, CHAD I 24/2.

48. 'Atomic Scientists' Association', *Nature*, 157. 3996 (1 June 1946), 725.

49. Chadwick to Peierls, 6 March 1946, CHAD I 24/2.

50. Rotblat to Kurti, 20 February 1946, RTBT, K. 24.

51. Peierls, *Bird of Passage*, p. 283.

52. G. N. Walton, 'Minutes of the 69th Council Meeting held in the Library, Physics Department, Imperial College of Science and Technology, London, S.W.7, on Saturday, December 18th, 1954, at 11 a.m.', 12 January 1954, RTBT, K. 113, p. 1; 'Proposed International Conference on the Social Implications

of Atomic Energy', 22 November 1954, attached to letter, E.H.S. Burhop to Rotblat, 8 December 1954, RTBT, K. 112.

53. Peierls to Chadwick, 12 March 1946, CHAD I 24/2.
54. Peierls, *Bird of Passage*, p. 284.
55. Talbot Hood to Peierls, 18 January 1950; Peierls to Talbot Hood, 26 January 1950, RTBT, K. 49B.
56. See correspondence in Peierls Papers, MS Eng. Misc. b 223, F 6; and RTBT, K. 55.
57. Atomic Scientists' Association, 'Report of Activities of Provisional Committee', n.d., p.1, attached to letter, Moon to ASA members, 15 July 1946, RTBT, K. 25.
58. E. H. S. Burhop, 'Minutes of Meeting of Atomic Scientists' Committee at Gas Industry House, Grosvenor Place, S.W. 1 on Friday, March 8th 1946', TNA, AB 16/52, p. 1; Joint Honorable Secretary to Rotblat, 30 April 1946, RTBT, K. 22.
59. Association of Atomic Scientists: Provisional Committee. Minutes of meeting held at University College, London on Saturday, March 16th, 1946 at 11 a.m.', TNA, AB 16/52, pp. 1–2.
60. Atomic Scientists' Association, *Atom Train*, unpaginated (last page).
61. 'List of Members', 14 May 1946; 'List of Members', 25 May 1946, RTBT, K. 24; 'A.S.A. members, 8th July 1946, RTBT, K. 26.
62. Rotblat to Burhop, 15 May 1946, attached: draft letter to ASA members, n.d., RTBT, K. 24; Arrol and Burhop to Rotblat , 10 April 1946; Rotblat to Members of the Association, n.d.; anonymous letter to ASA members, 6 June 1946.
63. W. J. Arrol and E. H. S. Burhop, 'Minutes of the First General Meeting of the Atomic Scientists' Association Held at the University, Edmund Street, Birmingham, on June 15th, 1946, at 2.15 p.m.', RTBT, K. 25, pp. 1–2.
64. Philip B. Moon to ASA members, 15 July 1946, RTBT, K. 25; Philip B. Moon and Eric H. S. Burhop, *Atomic Survey: A Short Guide to the Scientific and Political Problems of Atomic Energy* (n.p., 1946), p. 32.
65. 'Editorial', *ASN*, 2. 1 (15 July 1948), 1.
66. 'Annual General Meeting', *ASN*, 3. 1 (21 July 1949), 3–7 (p. 7); 'Annual General Meeting: Elections to Council', *ASN*, 4. 1 (August 1950), 6.
67. See Rudolf Peierls personal file 'Lichnoe delo Peierls Rudol'f', the Comintern Archives, the Russian States Archive of Socio-Political History, Moscow, Russian Federation (hereafter RGASPI), Komintern, F. 495, op. 198, d. 1811.
68. 'Memorandum and Articles of the Atomic Scientists' Association. Incorporated on the 15th day of April 1947', n.d., RTBT, K. 133.
69. Peierls, 'The British Atomic Scientists' Association', p. 59.
70. Peierls, *Bird of Passage*, p. 283.
71. 'Atomic Scientists' Association', *Nature*, 157. 3996 (1 June 1946), p. 725; Atomic Scientists' Association, *Atom Train: Guide to the Travelling Exhibition on Atomic Energy* (London: Atomic Scientists' Association, 1947), unpaginated (last page); Bryce Halliday, 'Professor Rotblat and the Atom Train', in *War and Peace: The Life and Work of Sir Joseph Rotblat*, ed. by Peter Rowlands and Vincent Attwood (Liverpool, 2006), pp. 139–44 (p. 139).
72. Peierls, 'The British Atomic Scientists' Association', p. 59.
73. Peierls, 'The British Atomic Scientists' Association', *BAS*, 6. 2 (February 1950), p. 59.
74. Jones to Rotblat, 27 January 1947, RTBT, K. 26; 'Editorial', *ASN*, 4. 1 (August 1950), 1–2 (p. 1).

75. Atomic Scientists' Association, 'Report of Activities of Provisional Committee', n.d., p. 1, attached to letter, Philip B. Moon to ASA members, 15 July 1946, RTBT, K. 25.

76. F. E. Simon to Rotblat, 25 June 1946, RTBT, K. 26. The conference had already been on the agenda of the ASA's first general meeting; W. J. Arrol and E. H. S. Burhop, 'Minutes of the First General Meeting of the Atomic Scientists' Association Held at the University, Edmund Street, Birmingham, on June 15th, 1946, at 2.15 p.m., RTBT, K. 25, p. 3.

77. N. Kurti, 'Notes on a Meeting Held in Oxford on May 22nd, 1946', 28 May 1946, RTBT, K. 25.

78. Atomic Scientists' Association, 'Oxford 29th–31st July 1946: Final Meeting and Reception', n.d., RTBT, K. 26. See also the transcripts of the discussions provided by W. J. Arrol to Council members: Atomic Scientists' Association, 'Oxford Conference, July 29th–31st, 1946: First Session, Monday, July 29th, 1946, 3 p.m., at Jesus College, Oxford. Professor R. E. Peierls, Executive Vice-President, in the Chair', n.d.; 'Tuesday, July 30th, 10 a.m. Chairman Professor Pryce. Subject: "Atomic Power."', n.d.; 'Open Meeting, Wednesday, July 31st, 1946. In the Chair, Professor Mott', n.d., RTBT, K. 27.

79. Hunner, *Inventing Los Alamos*, p. 114.

80. Rudolf E. Peierls, 'Preface', in *Atomic Survey*, unpaginated.

81. Stefan Collini, *English Pasts: Essays in History and Culture* (Oxford, 1999), pp. 305–25; Sophie Forgan, 'Atoms in Wonderland', *History and Technology*, 19. 3 (2003), 177–96 (p. 178).

82. Rudolf Peierls, 'Atomic Energy: Threat and Promise', *Endeavour*, 6 (April 1947), 51–7 (p. 51).

83. Paul Boyer's observation also applies to Britain; *Fallout: A Historian Reflects on America's Half-Century Encounter with Nuclear Weapons* (Columbus, 1998), p. 7; Wittner, I, 81.

84. John Hersey, *Hiroshima* (Harmondsworth, 1946); Wittner, I, 82–3.

85. *Gallup International Public Opinion Polls: Great Britain 1937–1975*, ed. by George H. Gallup, 2 vols (New York, 1976), I, 132, 183–4.

86. *The Effects of the Atomic Bombs at Hiroshima and Nagasaki: Report of the British Mission to Japan* (London, 1946); 'Hiroshima to Bikini', *Manchester Guardian*, 2 July 1946, 4.

87. Baylis, *Ambiguity and Deterrence*, p. 55.

88. Greta Jones, 'The Mushroom-Shaped Cloud: British Scientists' Opposition to Nuclear Weapons Policy, 1945–57', *Annals of Science*, 43. 1 (1986), 1–26 (p. 3).

89. Boyer, *By the Bomb's Early Light*, pp. 109–21.

90. David Dietz, *Atomic Energy in the Coming Era* (New York, 1945).

91. David Dietz, *Atomic Energy Now and in the Future* (London, 1946), pp. 14–15.

92. Peierls, 'Atomic Energy: Threat and Promise', pp. 51–7.

93. *Science News*, vol. 2, ed. by John Enogat and Rudolf Peierls (Harmondsworth, 1947).

94. Wittner, II, 80.

95. Kirk Willis, '"God and the Atom": British Churchmen and the Challenge of Nuclear Power 1945–1950', *Albion*, 29. 3 (1997), 422–57 (p. 424).

96. British Council of Churches, *The Era of Atomic Power: Report of a Commission Appointed by the British Council of Churches* (London, 1946). Two years later, the Anglican Church published its own report: Church of England, *The*

Church and the Atom: A Study of the Moral and Theological Aspects of Peace and War ([n.p.], 1948).

97. Willis, 'God and the Atom', pp. 445–7.
98. British Council of Churches, p. 7; International Control of Atomic Energy (II)', *Nature*, 157. 4000 (29 June 1946), 853–55 (p. 853).
99. Peierls, 'The Atomic Scientists' Association', p. 59.
100. Henry DeWolf Smyth, *Atomic Energy for Military Purposes*; H.M. Treasury.
101. Atomic Scientists' Association, 'Minutes of Council Meeting held at Clarendon Laboratory, Oxford, on Wednesday, 31st July, 1946', n.d., RTBT, K. 28.
102. R. R. Nimmo, 'The seventh meeting of the Council of the Atomic Scientists' Association was held in University College, London, on 1st March, 1947, at 11 a.m.', 3 March 1947, RTBT, K. 30, p. 1.
103. P. B. Moon, 'Minutes of the tenth Council meeting held at the University of Birmingham, Edmund Street, on Saturday, 21st June at 11.45 a.m.', 17 July 1947, RTBT, K. 31, p. 1.
104. N. F. Mott, 'The President's Report on the Year', *ASN*, 1.1 (11 July 1947), 1.
105. Peierls, 'The British Atomic Scientists' Association', p. 59.
106. Moyle to Peierls, 3 November 1950, RTBT, 49C.
107. See, for example, 'The International Control of Atomic Energy: Statement by the Council of the Atomic Scientists' Association, July, 1948', *ASN*, 2. 1 (15 July 1948), 13–14, was reprinted under the title 'British Atomic Scientists' Proposals for International Control of Atomic Energy', *BAS*, 3. 2 (February 1947), 42–3, 49; Peierls's article 'Bathwater and the Baby' was reprinted as guest editorial under the title 'Basic Science and the Cold War', *BAS*, 9. 3 (April 1953), 66–7.
108. See, for example, Joseph Hirschfelder, 'The Effects of Atomic Weapons', *BAS*, 6. 8–9 (August–September 1950), 236–40, 285–6, was reprinted under the same title in the *ASN*, 4. 2 (November 1950), 36–42; Leo Szilard, 'Calling for a Crusade ...', *BAS*, 3. 4–5 (April–May 1947), 102–6, 125, was published under the same title in the *ASN*, 1. 1 (11 July 1947), 6–7.
109. 'Editorial', *ASN*, n.s. 1. 1 (September 1951), 1.
110. 'Editorial', *ASJ*, 3. 1 (September 1953), 1–3.
111. 'Editorial: The Final Volume', *ASJ*, 5. 6 (July 1956), 355.
112. J. Howlett, 'Minutes of the 73rd Council Meeting held at the Society for Visiting Scientists, 5, Old Burlington St., London, W.1. on Saturday, 9th July at 11 a.m.', 18 September 1955, p. 1; Ford to Rosbaud, 25 September 1955; Maxwell to Ford, 4 October 1955; Ford to Members of the ASA Council, n.d.; G. W. K. Ford, 'Memorandum for special Council Meeting to be held on Friday, October 14th, at 7 p.m. at Professor Massey's Room, University College, Gower Street, London, W.C. 1', n.d., RTBT, K. 117, p. 1. The ASA ended its contract with Taylor and Francis in January 1956; Coffey to Ford, 6 January 1956; Hester to Rotblat, 20 January 1956, RTBT, K. 117.
113. Matterson to Fishenden, 20 September 1956, TNA, AB 27/6. H. R. Allen, 'Minutes of the 87th Council Meeting, held in the Physics Library, Imperial College, London, S.W.7. on Saturday February 2nd, 1957 at 10.45 a.m.', 3 February 1957, RTBT, K. 124, p. 1; Buckley to Rotblat, 4 January 1957, RTBT, K. 138.
114. Cudlipp to Rotblat, 2 November 1956; Cudlipp to Rotblat, 8 November 1956, RTBT, K. 117; H. R. Allen, 'Minutes of the 87th Council Meeting,

held in the Physics Library, Imperial College, London, S.W.7. on Saturday February 2nd, 1957 at 10.45 a.m.', 3 February 1957, RTBT, K. 124, p. 1; A. H. S. Matterson to R. M. Fishenden, 8 October 1956, TNA, AB 27/6; R.E. Peierls, 'A New Way to Fuse Atoms', *New Scientist*, 3 January 1957, pp. 36–7.

115. 'This Is Our Policy', *New Scientist*, 22 November 1956, p. 5.
116. 'Opening of the Atomic Energy Exhibition', *ASN*, 1. 5 (21 November 1947), 66–7.
117. 'Atomic Energy Exhibition: the Atom Train', *Nature*, 162. 4111 (14 August 1948), p. 267; Joseph Rotblat, 'The Atom Train: A Successful Experiment', *Atomic Scientists' News*, 2. 1 (15 July 1948), pp. 4–8 (p. 5); Rudolf Peierls, 'The British Atomic Scientists' Association', *Bulletin of the Atomic Scientists*, 6. 2 (February 1950), p. 59.
118. Rotblat, 'The Atom Train', pp. 4–5; Joseph Rotblat, 'The Atomic Energy Exhibition in Beirut', *ASN*, 2. 5 (9 March 1949), 120–5; Ayache to Rotblat, 24 January 1949; Rotblat to Shadid, 27 April 1949; Rotblat to Shadid, 9 June 1949, RTBT, K. 164.
119. P. B. Moon, 'The eighth meeting of the Council was held in the Clarendon Laboratory, Oxford, on Saturday, 3rd May, 1947', n.d., RTBT, K. 30, p. 2; 'Opening of the Atomic Energy Exhibition', *ASN*, 1. 5 (21 November 1947), 66–7; Joseph Rotblat, 'The Atomic Energy Exhibition Is Coming', *ASN*, 1. 3 (17 September 1947), 31–4 (p. 32). The Atom Train's importance for the ASA translated into considerable coverage in the *ASN*, see, for example, 'The Atomic Energy Exhibition', *ASN*, 1. 6 (19 December 1947), 80–1; 'Atomic Energy Exhibition', *ASN*, 1. 7 (30 January 1948), 104–5; 'The Atom Train', *ASN*, 1. 8 (6 March 1948), 109–10.
120. Peierls to Massey, 14 May 1946, Peierls Papers, MS Eng. Misc. b. 223, F 4.
121. Moon to Perrin, 13 January 1947, TNA, AB16/27.
122. Moon to ASA Council members, 31 March 1947; P.B. Moon, 'The eighth meeting of the Council was held in the Clarendon Laboratory, Oxford, on Saturday, 3rd May, 1947', n.d., RTBT, K.30, 1; J.E. Adamson to Godwin, 14 May 1947; J. E. Adamson, minute sheet, minute 19, 28 June 1947, TNA, AB16/27.
123. Atomic Scientists' Association, *Atom Train*, unpaginated acknowledgements section; 'Schedule of Equipment and Materials Loaned by A.E.R.E. Harwell for Use as Exhibits in the A.S.A. Train Exhibition', n.d., TNA, AB16/401; Thomas to Rotblat, 26 November 1947; Thomas to Rotblat, 3 December 1947, RTBT, K. 154.
124. Massey to Rotblat, 25 April 1947; Rotblat to Massey, 29 April 1947, RTBT, K. 153; P. B. Moon, 'The eighth meeting of the Council was held in the Clarendon Laboratory, Oxford, on Saturday, 3rd May, 1947', n.d., RTBT, K. 30, p. 2; G.O. Jones, 'Minutes of the thirteenth Council meeting held at the Clarendon Laboratory, Parks Road, Oxford, on Saturday, 4th October, 1947 at 11.45 a.m.', 7 October 1947, RTBT, K. 31., 1; D. E. H. Peirson to Eales, 17 November 1948, TNA, AB16/401; Atom Train Exhibition: Atomic Scientists' Association', n.d., attached to letter, Rotblat to Peirson, 30 September 1947; Peirson to Rotblat, 8 October 1947, TNA, AB16/27.
125. G. O. Jones, 'Minutes of the thirteenth Council meeting held at the Clarendon Laboratory, Parks Road, Oxford, on Saturday, 4th October, 1947 at 11.45 a.m.', 7 October 1947, RTBT, K. 31., p. 1.

126. Forgan, p. 177.
127. Rotblat, 'The Atom Train', p. 7; Joseph Rotblat, 'The Atomic Energy Exhibition Is Coming', *ASN*, 1. 3 (17 September 1947), 31–4 (pp. 33–4).
128. 'Atom Train', n.d., Peierls Papers, MS Eng. Misc. b. 223, F 5.
129. June Clayton, 'A Noble Man of Science, A Nobel Man of Peace: Professor Sir Joseph Rotblat, FRS, Nobel Peace Prize Laureate 1995' (unpublished master's thesis, University of Liverpool, 2003), p. 78.
130. 'Atom Train: A Travelling Exhibition on Atomic Energy Designed by Peter Moro and Robin Day', p. 435.
131. Atomic Scientists' Association, *Atom Train*, unpaginated introduction.
132. Atomic Scientists' Association, *Atom Train*, unpaginated.
133. Atomic Scientists' Association, *Atom Train*, pp. 18–20; 'Atom Train: A Travelling Exhibition on Atomic Energy Designed by Peter Moro and Robin Day', *The Architects' Journal*, 13 November 1947, pp. 434–5 (p. 435); 'Atom Train Exhibition: Atomic Scientists' Association', n.d., pp. 810, attached to letter, Rotblat to Peirson, 30 September 1947 TNA, AB16/27.
134. Atomic Scientists' Association, *Atom Train*, pp. 18–22.
135. Halliday, p. 142.
136. 'Professors on Atomic Power Stations in Britain', *Liverpool Echo*, 6 November 1947, University Press Cuttings, 5 November 1941–31 December 1947, SJL, S. 2523, fols 295–96^r (fol. 295^r). fol. 296^r.
137. 'Atom Train Exhibition: Atomic Scientists' Association', n.d., p. 13, attached to letter, Rotblat to Peirson, 30 September 1947 TNA, AB16/27.
138. 'The Atomic Bomb and Our Cities: From the Report of US Strategic Bombing Survey', *BAS*, 2. 3–4 (August 1946), 29–30; 'The City of Washington and an Atomic Attack', *BAS*, 6. 1 (January 1950), 29–30; Ralp Lapp, 'Atomic Bomb Explosions – Effects on an American City', *BAS*, 4. 2 (February 1948), 39–43, 48; 'Memorandum from British Scientists at the Los Alamos Laboratory, New Mexico', p. 1.
139. Atomic Scientists' Association, *Atom Train*, back sleeve.
140. Peierls, *Bird of Passage*, p. 283. It received, for example, considerable coverage in the first issue of the *ASN*, 1. 1 (11 July 1947).
141. 'International Control of Atomic Energy (I)', p. 820.
142. 'Memorandum from British Scientists at the Los Alamos Laboratory, New Mexico', pp. 2–3.
143. Peierls, untitled memorandum, 14 November 1945, attached to letter, Peierls to Anderson, 14 November 1945, TNA, AB I/572, p. 1.
144. Burhop to Rotblat, 30 April 1946; Joint Honorable Secretary to Rotblat, 30 April 1946, RTBT, K. 22.
145. Atomic Scientists' Association, 'Report of Activities of Provisional Committee', n.d., p. 3, attached to letter, Philip B. Moon to ASA members, 15 July 1946, RTBT, K. 25. Rotblat to Burhop, 11 April 1946; Burhop to Rotblat, 17 April 1946, attached: 'Proposed Alterations to Memorandum to U.N. Atomic Energy Commission', n.d., RTBT, K. 22.
146. Joint Honorable Secretary to Rotblat, 30 April 1945, RTBT, K. 22.
147. W. J. Arrol and Eric Burhop, 'Minutes of the First General Meeting of the Atomic Scientists' Association Held at the University, Edmund Street, Birmingham, on June 15th, 1946, at 2.15 p.m.', RTBT, K. 25, p. 3; 'International Control of Atomic Energy (I)', p. 817.

148. Boyer, *By the Bomb's Early Light*, pp. 53–7; James Chace, 'After Hiroshima: Sharing the Atom Bomb', *Foreign Affairs*, 75. 1 (1996), 129–44; Holloway, pp. 161–5.

149. 'Memorandum to United Nations Atomic Energy Commission', n.d., attached to letter, Joint Honorable Secretary to Rotblat, 30 April 1946, RTBT, K. 22.

150. 'Control of Atomic Energy: British Scientists' Scheme', *The Times*, 1 June 1946, p. 2; Ward to Rickett, 17 June 1946, TNA, CAB 126/209.

151. Rickett to Ward, 24 June 1946, TNA, CAB 126/209.

152. Lindsay to Moon, 10 January 1947. The ASA did not send the memorandum to UN representatives until RTBT, K. 30.

153. John Baylis, *Ambiguity and Deterrence*, pp. 37–45.

154. Peierls, *Bird of Passage*, p. 282.

155. 'Memorandum from the Cambridge Group', n.d., RTBT, K. 30; 'Minutes of the Second General Meeting (Extraordinary) held in the Clarendon Laboratory, Oxford, on Wednesday, 31st July, 1946', n.d., p. 2; Moon to Rotblat, 31 August 1946; Moon to ASA Council members, n.d.; 'Minutes of Council meeting held on 21st September, 1946, at the University, Birmingham, 3', n.d., RTBT, K. 28; Atomic Scientists' Association, 'Statement on International Control', n.d., attached to letter, Peierls to ASA Council members, RE: 'Third draft of statement on International Control', 21 November 1946, RTBT, K. 30.

156. 'Draft Statement on International Control', n.d., attached to letter, Peierls to Rotblat, 16 September 1946; 'Views Expressed at a Meeting of the Liverpool Branch of the A.S.A.', 29 August 1946; Rotblat to Jones, 30 September 1946, RTBT, K. 26; Atomic Scientists' Association, 'Statement on International Control', n.d., attached to letter, Moon and Peierls to ASA Council members, n.d., RTBT, K. 29; 'A.S.A. Council Meeting: Draft statement on International Control', n. d., attached to letter, Peierls to Chadwick, 20 September 1946; Atomic Scientists' Association, 'Statement on International Control', n.d., CHAD I 19/6.

157. Peierls to Members of ASA Council, n.d.; Rotblat to Peierls, 16 November 1946; Peierls to Rotblat, 19 November 1946; Herbert Skinner, 'Suggested Modification of Statement on International Control', n.d., attached to letter, Skinner to Rotblat, 20 November 1946, RTBT, K. 29. P.B. Moon, 'Minutes of the fifth Council meeting, held at the University, Edmund Street, Birmingham, on Saturday, November 9th, 1946', 18 November 1946, RTBT, K. 30, p. 2.

158. Peierls to Members of ASA Council, 3 December 1946, RTBT, K. 29.

159. 'Atomic Energy Control', *The Times*, 21 January 1947, p. 4; 'British Atomic Scientists' Proposals for International Control of Atomic Energy', *BAS*, 3. 2 (February 1947), 42–3, 49.

160. Moon, 'A.S.A. Council Memorandum on International Control', 18 March 1947; Moon to Members of ASA Council, 26 March 1947, RTBT, K. 30.

161. 'British Atomic Scientists' Proposals for International Control of Atomic Energy', pp. 42–3, 49.

162. Ibid., p. 49.

163. Hunner, *Inventing Los Alamos*, p. 113.

164. Boyer, *By the Bomb's Early Light*, pp. 33–45.

165. *Gallup International Public Opinion Polls*, I, 139.

166. Federation of American (Atomic) Scientists, 'Survival Is at Stake', in *One World or None* (see Oppenheimer, 'The New Weapons', above), pp. 78–9. The collection was published in Britain in 1947 by the London-based publisher Latimer House.

167. Peierls, Interview by Weiner, p. 152. Although Bohr provided the foreword to *One World or None*, he refrained from generating public fear as did many of the contributors; Boyer, *By the Bomb's Early Light*, p. 71.

168. 'Banning of Atomic Weapons: Request for International Authority', *The Times*, 8 March 1947, p. 4.

169. 'Statements on the Second Anniversary of Hiroshima', *BAS*, 3. 9 (September 1947), 235–6, 252 (p. 236).

170. P. B. Moon, 'Minutes of the tenth Council meeting held at the University of Birmingham, Edmund Street, on Saturday, 21st June at 11.45 a.m.', 17 July 1947, RTBT, K. 31, p. 2.

171. G. O. Jones, 'Minutes of the eleventh Council meeting held at the University of Birmingham, Edmund Street, on Wednesday, 30th July, 1947 at 11.45 a.m.', 2 August 1947, RTBT, K. 31, p. 2.

172. 'The International Control of Atomic Energy. Statement by the Council of the Atomic Scientists' Association, July 1948', RTBT, K. 48. The Minister of Supply also received a copy of the memorandum; Kurti to Strauss, 13 July 1948, TNA, AB 16/52.

173. Kurti to Cadogan, 12 July 1948; Kurti to Gromyko, 12 July 1948; Kurti to Innes, 10 July 1948; Kurti to Strauss, 13 July 1948; Kurti to Alexander, 13 July 1948; Kurti to Bevin, 12 July 1948; Kurti to Attlee, 12 July 1948; Kurti to Trygve Lie, 13 July 1948; Kurti to the Editor, *Nature*, 10 July 1948, RTBT, K. 48.

174. 'The International Control of Atomic Energy: Statement by the Council of the Atomic Scientists', pp. 13–14.

175. *Gallup International Public Opinion Polls*, I, 185.

176. 'Conference with the Federation of American Scientists', *ASN*, 4. 3 (January 1951), 51–5.

177. William A. Higinbotham, 'Scientists Discuss War and Peace', *BAS*, 6. 11 (November 1950), 350; Peierls to the Chairman of the Federation of American Scientists [Higinbotham], n.d., Peierls Papers, MS Eng. Misc. b. 223, F 7.

178. H. R. Allen, 'Minutes of the 87th Council Meeting, held in the Physics Library, Imperial College, London, S.W.7. on Saturday February 2nd, 1957 at 10.45 a.m.', 3 February 1957, RTBT, K. 124, p. 3.

179. L. F. Bates and others, 'Safeguards on Nuclear Production', *The Times*, 29 November 1957, p. 11.

180. Atomic Scientists' Association, *Atom Train: Guide to the Travelling Exhibition on Atomic Energy* (London: Atomic Scientists' Association, 1947), unpaginated (last page).

181. Peierls to Anderson, 14 November 1945, TNA, AB I/572.

182. 'Advisory Committee on Atomic Energy: Composition and Terms of Refernce. Note by the Secretary of the Cabinet', 20 August 1945, CHAD I 15/6.

183. Mary Jo Nye, p. 86.

184. Peierls to Massey, 14 May 1946, Peierls Papers, MS Eng. Misc. b. 223, F 4.

185. Peierls to Anderson, 29 May 1946, Peierls Papers, MS Eng. Misc. b. 223, F 4.

186. Anderson to Peierls, 30 May 1946, Peierls Papers, MS Eng. Misc. b. 223, F 4.
187. Fishenden to Schonland, 10 February 1958, TNA, AB 27/6.
188. Fishenden to Schonland, 4 April 1957, TNA, AB 27/6.
189. Cockcroft to Allan, 22 May 1959, TNA, AB 27/6.
190. Rudolf Peierls and F.C. Champion, 'Information on Atomic Energy', *The Times*, 23 January 1950, p. 5.
191. Frisch and Peierls, 'Memorandum on the properties of a radioactive "super-bomb"', p. 1.
192. Rudolf Peierls, 'Defence Against the Atomic Bomb', *Nature*, 158. 4011 (14 September 1946), 379. This article was a critique of D. G. Christopherson's article 'Defence Against the Atomic Bomb', *Nature*, 158. 4005 (3 August 1946), 151–3, which argued that the effects of nuclear arms were overestimated. Following the ASA's policy, the ASA Council had discussed the article prior to its publication; Peierls to ASA Council members, n.d., RTBT, K. 28.
193. Chapman Pincher, 'The Atomic Bomb: A Statement by the Man Who Is Back Here to Brief the Premier', *Daily Express*, 8 November 1945, University Press Cuttings, 5 November 1941–31 December 1947, SJL, S. 2523, fol. 154ʳ.
194. 'Atomic Weapons and Civil Defence', *ASN*, 3. 1 (21 July 1949), 10–16 (p. 10).
195. Home Office Civil Defence Department, *Civil Defence Manual of Basic Training. Vol. II: Atomic Warfare* (London: HMSO, 1950), p. 5.
196. D.F. Bracher, rev. of *Civil Defence Manual of Basic Training. Volume II: Atomic Warfare* by Home Office Civil Defence Department (London: HMSO, 1950), *ASN*, 4. 2 (November 1950), 42–6 (p. 46).
197. Jeff Hughes, 'The Strath Report: Britain Confronts the H-Bomb, 1954–1955', *History and Technology*, 19. 3 (2003), 257–75 (p. 263). See also Hennessy, *The Secret State*, pp. 167–78.
198. Matthew Grant, *After the H-Bomb: Civil Defence and Nuclear War in Britain, 1945–68* (Basingstoke, 2010), p. 7. Melissa Smith has also reached similar conclusions in her work on the Home Office Scientific Advisers' Branch; 'Architects of Armageddon: The Home Office Scientific Advisers' Branch and Civil Defence in Britain, 1945–68', *British Journal for the History of Science*, 43. 157 (2010), 149–80.

7 The 'Unpolitical' Scientist: Rudolf Peierls, the Concept of 'Objective' Science and the End of the Atomic Scientists' Association

1. Price, p. 244.
2. See Mark Walker, 'Legenden um die deutsche Atombombe', *Vierteljahrshefte für Zeitgeschichte*, 38. 1 (1990), 45–74 (p. 54).
3. Atomic Scientists' Association, 'The Atomic Energy Bill', n.d., p. 1, attached to letter, Peierls to ASA Council Members and Vice-Presidents, 16 October 1946, RTBT, K. 28.
4. Peierls, 'Bathwater and the Baby', pp. 10–13.
5. Ibid., p. 13.
6. Schweber, *In the Shadow of the Bomb*, pp. 16–17.
7. Robert Oppenheimer, 'Atomic Weapons', *PAPS*, 90. 1 (1946), 7–10 (p. 7).

8. Schweber, *In the Shadow of the Bomb*, p. 19.
9. Peierls, 'Bathwater and the Baby', pp. 12–13.
10. Peierls to Jackson, 26 November 1945; Jackson to Tuck, 30 November 1945; Joint Staff Mission, Washington, DC, to War Cabinet Offices, London, 6 December 1945, CHAD IV/12/2.
11. Ibid., p. 11.
12. 'British Atomic Scientists' Proposals for International Control of Atomic Energy', *BAS*, 3. 2 (February 1947), 42–3, 49 (p. 43).
13. Albert Einstein, 'On the Moral Obligation of the Scientist', *BAS*, 8. 2 (February 1952), 34–5 (p. 34).
14. Michael J. Neufeld, 'Wernher von Braun, the SS, and Concentration Camp Labor: Questions of Moral, Political, and Criminal Responsibility', *German Studies Review*, 25. 1 (2001), 57–78 (p. 73). These issues are also further explored in Rainer Eisfeld, *Mondsüchtig: Wernher von Braun und die Geburt der Raumfahrt aus dem Geist der Barbarei* (Reinbek, 1996).
15. Rudolf Peierls, 'Atom Bombs on Korea', *The Times*, 13 July 1950, p. 7. Peierls later reiterated this point, see, for example, L. F. Bates and others, 'Safeguards on Nuclear Production', *The Times*, 29 November 1957, p. 11.
16. 'International Control of Atomic Energy (I)', p. 819.
17. Rudolf Peierls, 'The Moral Question', *ASN*, 4. 1 (August 1950), 3. The issue also featured Peierls's letter to the editor of the *The Times*; Peierls, 'Atom Bombs on Korea', p. 7; and his article 'The Morality of Atomic Warfare', *ASJ*, 4. 1 (September 1954), 17–20 (p. 19).
18. Peierls, 'Atomic Energy: Threat and Promise', pp. 51–2.
19. Rudolf Peierls, 'Some Notes on the International Situation', *ASN*, 2. 5 (9 March 1949), 105–11 (p. 105). Peierls reiterated this point, for example, in his essay 'Bathwater and the Baby' (p. 10).
20. Peierls, 'The Morality of Atomic Warfare', p. 20; Rudolf Peierls, 'Agonising Misappraisal', *Spectator*, 28 April 1961, pp. 592–3.
21. See Ralf Dahrendorf, *Gesellschaft und Demokratie in Deutschland* (Munich, 1965); Fritz Stern, *The Failure of Illiberalism: Essays on the Political Culture of Modern Germany* (New York, 1972, especially pp. xi–xliv, 3–25); Jan Palmowski, 'The Politics of the "Unpolitical German": Liberalism in German Local Government, 1860–1880', *Historical Journal*, 42. 3 (1999), 675–704.
22. Thomas Mann, *Betrachtungen eines Unpolitischen*, 3rd edn (Frankfurt a.M., 2001).
23. Paul Forman, 'Scientific Internationalism and the Weimar Physicists: The Ideology and Its Manipulation in Germany after World War I', *Isis*, 64. 2 (1973), 150–80 (pp. 170–1).
24. See Klaus Hentschel, *The Mental Aftermath: The Mentality of German Physicists 1945–1949*, tr. by Ann M. Hentschel (Oxford, 2007).
25. Cathryn Carson, 'New Models for Science in Politics: Heisenberg in West Germany', *Historical Studies in the Physical and Biological Sciences*, 30. 1 (1999), 115–71; David Clay Large, *Germans to the Front: West German Rearmament in the Adenauer Era* (Chapel Hill, 1996), pp. 218–19, 224–5; Gabriele Metzler, *Internationale Wissenschaft und nationale Kultur: Deutsche Physiker in der internationalen Community 1900–1960* (Göttingen, 2000), pp. 220–1; Carl Friedrich von Weizsäcker, 'Should Germany Have Atomic Arms?', *BAS*, 13. 8

(October 1957), 283–91; Carl Friedrich von Weizsäcker, 'Do We Want to Save Ourselves?', *BAS*, 14. 5 (May 1958), 180–4.

26. 'Declaration of the German Nuclear Physicists', *BAS*, 13. 6 (June 1957), 228; Richard Beyler, 'The Demon of Technology, Mass Society, and Atomic Physics in West Germany, 1945–1957, *History and Technology*, 19. 3 (2003), 227–39 (pp. 232–7).

27. Max Born, 'Man and the Atom', *BAS*, 13. 6 (June 1957), 186–94.

28. Arne Schirrmacher, 'Physik und Politik in der frühen Bundesrepublik Deutschland: Max Born, Werner Heisenberg und Pascual Jordan als politische Grenzgänger', *Ber. Wissenschaftsgesch.*, 30. 1 (2007), 13–31 (pp. 15–19).

29. Metzler, pp. 196–8.

30. Mark Walker, 'Legenden um die deutsche Atombombe', pp. 54–5.

31. Lothar Burchhardt, 'Naturwissenschaftliche Universitätslehrer im Kaiserreich', in *Deutsche Hochschullehrer als Elite 1815–1945*, ed. by Klaus Schwabe (Boppard, 1988), pp. 151–214.

32. Georg Bollenbeck, *Bildung und Kultur: Glanz und Elend eines deutschen Deutungsmusters* (Frankfurt a.M., 1994); Fritz Ringer, '*Bildung*: The Social and Ideological Context of the German Historical Tradition', *History of European Ideas*, 10. 2 (1989), 193–202.

33. Fritz Ringer, *The Decline of the German Mandarins: The German Academic Community, 1890–1933* (Cambridge, MA, 1969), pp. 5–6.

34. Cathryn Carson, *Heisenberg in the Atomic Age: Science and the Public Sphere* (Washington; Cambridge, 2010), p. 175.

35. Mark Walker, *Nazi Science*, pp. 243–68.

36. Cathryn Carson, 'Objectivity and the Scientist: Heisenberg Rethinks', *Science in Context*, 16. 1–2 (2003), 243–69 (p. 244).

37. Holger Nehring, 'Cold War, Apocalypse and Peaceful Atoms: Interpretations of Nuclear Energy in the British and West German Anti-Nuclear Weapons Movements, 1955–1964', *Historical Social Research*, 29. 3 (2004), 150–70 (p. 161).

38. See Gregg Herken, *Cardinal Choices: Presidential Science Advising From the Atomic Bomb to SDI*, rev. and expanded edn (Stanford, 2000); Herbert York, *The Advisors: Oppenheimer, Teller & the Superbomb* (San Francisco, 1976).

39. See, for example, Hans Bethe, 'The Hydrogen Bomb', *BAS*, 6. 4 (April 1950), 99–104, 125; Albert Einstein, 'Arms Can Bring No Security', *BAS*, 6. 3 (March 1950), 71; Frederick Seitz and Hans A. Bethe, 'How Close Is the Danger?', in *One World or None* (see Oppenheimer, 'The New Weapon', above), pp. 42–6; Szilard, 'Calling for a Crusade ...', 102–6, 125; Leo Szilard, 'Letter to Stalin', *BAS*, 3. 12 (December 1947), 347–9, 376; Victor Weisskopf, 'On Avoiding Nuclear Holocaust', *Technology Review* (October 1980), 28–35; Victor Weisskopf, 'Science for Its Own Sake', *Scientific Monthly*, 78. 3 (March 1954), 133–5.

40. Edward Teller, 'Back to the Laboratories', *BAS*, 6. 3 (March 1950), 71–2.

41. 'Statements on the Second Anniversary of Hiroshima', *BAS*, 3. 9 (September 1947), 235–6, 252 (p. 252).

42. Kathleen Lonsdale, 'Use of Atomic Energy: The Choice Before Scientists', *ASN*, 1. 4 (17 October 1947), 47–9 (p. 47).

43. 'Atomic Energy and Society', *ASN*, 2. 4 (7 January 1949), 70–97 (p. 70); 'Atomic Energy and Society', *Nature*, 162. 4130 (25 December 1948), 1005–6 (p. 1005).

44. Peierls, 'Some Notes on the International Situation', p. 105. Peierls's later essay 'Bathwater and the Baby' was written in the same spirit.
45. 'Future of the Association', *ASN*, 2. 4 (7 January 1949), 93–7; P. F. D. Shaw, 'The Future of the Atomic Scientists' Association', *ASJ*, 3. 3 (January 1954), 149–51.
46. Marcus Oliphant, 'Control of Atomic Energy', *Nature*, 157. 3995 (25 May 1946), 679–80 (p. 679).
47. 'Editorial', *ASN*, 1. 9 (5 April 1948), 125; 'Association News: The Questionnaire', *ASN*, 1. 11 (4 June 1948), 169–70.
48. Champion to Peierls, 19 January 1950, RTBT, K. 49B.
49. Kathleen Londsdale and J. L. Michiels, 'Atomic Energy', *The Times*, 22 November 1950, 7; Herbert Skinner, 'The Policy of the Atomic Scientists' Association', *ASN*, 4. 4 (July 1951), 78–9.
50. 'Science and Its Social Relations', *Nature*, 162. 4111 (14 August 1948), 235–7 (p. 236).
51. 'Dilemma of the Atomists', *Economist*, 24 July 1948, 140.
52. Rudolf Peierls, 'The Answers to 11 Vital Questions on the Big, Big Bomb', *Sunday Express*, 5 February 1950, p. 8.
53. 'Election of Officers', *ASN*, 1. 1 (11 July 1947), 2; Greta Jones, 'The Mushroom-Shaped Cloud', p. 7; Mary Jo Nye, *Blackett: Physics, War, and Politics in the Twentieth Century* (Cambridge, MA, 2004), p. 161.
54. Greta Jones, *Science, Politics and the Cold War*, p. 39.
55. Patrick Blackett, *Military and Political Consequences of Atomic Energy* (London, 1948).
56. Association of Scientific Workers, *Atomic Attack: Can Britain Be Defended?* (London, 1950); Home Office Civil Defence Department, *Civil Defence Manual of Basic Training. Vol. II: Atomic Warfare* (London, 1950).
57. John Anderson, 'Need This Happen to Us?', *Evening Standard*, 19 June 1950, p. 8.
58. Eric Burhop, 'The Scientist and Dangerous Thoughts', *ASN*, 2. 6 (28 April 1949), 137–40. On reactions, see Nevill F. Mott's reply 'The Scientist and Dangerous Thoughts', *ASN*, 2. 6 (28 April 1949), 171–2; M. A. Short, letter to the editor, *ASN*, 2. 6 (28 April 1949), 172–3.
59. Peierls, *Bird of Passage*, p. 283.
60. Thomson to Peierls, 24 April 1950, RTBT, K. 49C.
61. Cherwell to Hon. General Secretary (F. C. Champion), 11 February 1950; Peierls defended the ASA's policy on reaching decisions on statements; Peierls to Cherwell, 23 March 1950, Peierls Papers, MS Eng. Misc. b. 223, F 6.
62. Peierls to Champion, 13 February 1950. See also Nabarro to Champion, 13 February 1950, Peierls Papers, MS Eng. Misc. b. 223, F 6.
63. Champion to Peierls, 14 February 1950, Peierls Papers, MS Eng. Misc. b. 223, F 6.
64. E. H. S. Burhop et al., 'Statement', n.d., attached to letter, Champion to The Editor, The Press Association, 20 February 1950, RTBT, K. 57.
65. Champion to Peierls, 14 February 1950, Peierls Papers, MS Eng. Misc. b. 223, F 6.
66. ASA to Council Members and Vice-Presidents, RE: 'Draft Statement Submitted by Sir George Thomson to Assist Council in Its Deliberations', 25 May 1950, RTBT, K. 51; 'Text of Statement', *New York Times*, 5.2, 1950, p. 3.

67. See, for example, 'The Soul-Searchers Find No Answer', *Life*, 27 February 1950, 37–8, 40; 'U.S. Physicists' Plea: Renunciation of First Use of Bomb', *The Times*, 6 February 1950, p. 6; William Laurence, '12 Physicists Ask U.S. Not to Be First to Use Super Bomb', *New York Times*, 5.2, 1950, p. 1.

68. 'British Scientists Answer the Question All World Is Asking: What Should Be Done About the Hydrogen Bomb?', *Picture Post*, 18 February 1950, pp. 34–5.

69. 'Atom Weapons "Are Contrary to God": Congregationalists' Appeal', *The Times*, 19 May 1955, p. 6. On British churches and nuclear weapons, see David Omrod, 'The Churches and the Nuclear Arms Race, 1945–1985', in *Campaigns for Peace: British Peace Movements in the Twentieth Century*, ed. by Richard Taylor and Nigel Young (Manchester, 1987), pp. 189–220.

70. 'Hydrogen Bomb Danger: Dr. Garbett on the Alternative', *The Times*, 6 February 1950, p. 6; 'The Hydrogen Bomb: Archdeacon on Crucial Election Issue', *The Times*, 20 February 1950, p. 3. But the Archbishop of York also flip-flopped in his position over the years, as Dianne Kirby has shown; 'The Church of England and the Cold War Nuclear Debate', *Twentieth Century British History*, 4. 3 (1993), 250–83 (pp. 273–7).

71. 'Control of Atomic Weapons: Government Urged to Initiate Talks', *The Times*, 21 April 1950, p. 3.

72. 'Danger of World Suicide: Churches' Appeal for New Peace Effort', *The Times*, 25 February 1950, p. 5.

73. The contributions included: 'American Scientists' Statements', *ASN*, 3. 4 (March 1950), 98; Max Born, 'The Position of the Scientist', *ASN*, 3. 4 (March 1950), 93; Otto Frisch, 'The Physics of the Hydrogen Bomb', *ASN*, 3. 4 (March 1950), 78–81.

74. 'Scientists Jib at H-Bomb Jobs', *Daily Mirror*, 25 March 1950, p. 1.

75. Rudolf Peierls, 'The Hydrogen Bomb and World Security', *ASN*, 3. 4 (March 1950), 85–7 (pp. 85–7).

76. Peter Galison and Barton Bernstein, 'In Any Light: Scientists and the Decision to Build the Superbomb, 1952–1954', *Historical Studies in the Physical and Biological Sciences*, 19. 2 (1989), 267–348 (pp. 267–347); Schweber, *In the Shadow of the Bomb*, pp. 156–68.

77. 'Editorial', *ASN*, 3. 5 (May 1950), 104–6 (p. 104).

78. 'The Hydrogen Bomb – American Reactions', *ASN*, 3. 5 (May 1950), 110–20. The H-bomb remained of interest so that the subsequent issue of the *ASN* featured another article from the *BAS*: Robert F. Bacher, 'The Hydrogen Bomb', *ASN*, 3. 6 (June 1950), 134–45, previously published under the same title in the *BAS*, 6. 5 (May 1950), 133–8.

79. Arnold, *Britain and the H-Bomb*, p. 115.

80. Bertrand Russell, 'The Tests Should Be Stopped', *New Scientist*, 28 March 1957, pp. 24–5 (p. 25).

81. Angus Maude, 'The Tests Must Go On', *New Scientist*, 28 March 1957, pp. 26–7 (p. 27).

82. Robert Divine, *Blowing on the Wind: The Nuclear Test Ban Debate 1954–1960* (New York, 1978), pp. 186–7.

83. Bulletin of the Atomic Scientists, 'Memo: International Congress of Scientists', 29 May 1954, RTBT, K. 112.

84. Peierls to Rotblat, 27 July 1954, RTBT, K. 112.

85. E. Rabinowitch, 'Memorandum from Dr Rabinowitch on Proposed Conference on Science and Society', n.d., RTBT, K. 114.
86. 'Notes on Meeting of Sub-Committee on International Conference on 28th October, 1954. (Mrs. Lonsdale, Walton, Hodgson and Rotblat)', n.d., RTBT, K. 114.
87. Rotblat to Rabinowitch, 29 October 1954, RTBT, K. 114.
88. G. N. Walton, 'Minutes of the 71st Council Meeting held in the Clarendon Laboratory Parks Road, Oxford on Saturday, April 2nd, 1955, at 11.30 a.m.', 6 May 1955, RTBT, p. 1.
89. H. R. Allen, 'Minutes of the 87th Council Meeting, held in the Physics Library, Imperial College, London, S.W.7. on Saturday February 2nd, 1957 at 10.45 a.m.', 3 February 1957, RTBT, K. 124, pp. 2–3.
90. H. R. Allen, 'Minutes of the 87th Council Meeting, held in the Physics Library, Imperial College, London, S.W.7. on Saturday February 2nd, 1957 at 10.45 a.m.', 3 February 1957, RTBT, K. 124, p. 3.
91. Thomson to Allen, 15 February 1957, RTBT, K. 124.
92. Allan to Vice-Presidents and Members of the Council, 4 March 1957, TNA, AB 27/6.
93. Cockcroft to Allan, 11 March 1957, TNA, AB 27/6.
94. Allan to Vice-Presidents and Members of the Council, 1 April 1957, TNA, AB 27/6.
95. Allan to ASA Council members, 5 April 1957; H. R. Allan, 'Minutes of the 89th Council meeting, held in the Physics Library, Imperial College, London, S.W. 7. on Saturday 13th April, 1957 at 10.45 a.m.', 10 May 1957, RTBT, K. 138.
96. 'Statement on Strontium Hazards', n.d., pp. 1–2, attached to letter, Fishenden to Schonland, 4 April 1957, TNA, AB 27/6.
97. 'Statement on Strontium Hazards', n.d., p. 3, attached to letter, Fishenden to Schonland, 4 April 1957, TNA, AB 27/6.
98. Peierls to Allan, 8 April 1957, RTBT, K. 124.
99. Cockcroft to Allan, 5 April 1957, TNA, AB 27/6.
100. Skinner to Allan, 10 April 1957, RTBT, K. 124.
101. Bates to Allan, 8 April 1957; Pryce to Allan, 5 April 1957; Mott to Allan, 5 April 1957, RTBT, K. 117.
102. Matterson to Schonland, 16 April 1957, TNA, AB 27/6.
103. Allan to Press Office, Foreign Office, 15 April 1957, TNA, FO 371/129239; Allen to Press Office, Ministry of Health, 15 April 1957; Allen to Office of the Lord President, 15 April 1957; Allen to Private Secretary to the Prime Minister, 15 April 1957, RTBT, K. 117.
104. Cockcroft to Plowden, 16 April 1957, TNA, AB 27/6.
105. Brown to Frost, 17 April 1957, TNA, FO 371/129239.
106. Note, 25 April 1957, TNA, AB 26/7.
107. Cockcroft to Massey, 2 May 1957, TNA, AB 27/6; 'Bone and Bone Cancer', *Manchester Guardian*, 18 April 1957, p. 8.
108. Massey to Cockcroft, 6 May 1957, TNA, AB 27/6.
109. See, for example, 'Strontium Risks From Bombs: Atomic Scientists' Findings', *The Times*, 17 April 1957, p. 7.
110. Aurand to Rotblat, 2 May 1957; Campbell to Rotblat, 10 May 1957; Cuissart de Grelle to Rotblat, 25 April 1957; Elkind to Rotblat, 20 April 1957; Errera

to Rotblat, 10 May 1957; Jay to Rotblat, 13 May 1957; Murata to Rotblat, 25 April 1957; Strange to Rotblat, 7 May 1957; Zimmermann to ASA, 27 May 1957, RTBT, K. 127.

111. Lonsdale and Rotblat to ASA Vice-Presidents, 14 November 1957, TNA, AB 27/6.
112. Cockcroft to Lonsdale, 19 November 1957, TNA, AB 27/6.
113. Skinner to Lonsdale, 22 November 1957, TNA, AB 27/6.
114. Skinner to Cockcroft, 17 November 1957, TNA, AB 27/6.
115. Cockcroft to Skinner, 21 November 1957, TNA, AB 27/6.
116. L. F. Bates and others, 'Safeguards on Nuclear Production', *The Times*, 29 November 1957, p. 11.
117. 'An Appeal by American Scientists to the Governments and the People of the World', *Bulletin of the Atomic Scientists*, 13. 7 (September 1957), pp. 264–6.
118. Peierls, *Bird of Passage*, p. 284.
119. Peierls, 'The British Atomic Scientists' Association', p. 59.
120. Peierls, 'Britain in the Atomic Age', pp. 94–5.
121. G. P. Thomson, 'The Russian Atomic Explosion', *ASN*, 3. 3 (21 December 1949), 59–60 (p. 59). The same issue contained a couple of articles on the topic: E. H. S. Burhop, 'International Control – The Present Position' (pp. 61–63); H. R. Allan, 'American Reactions' (pp. 63–7); K. Lonsdale, 'Russia's Atom Bomb' (pp. 68–72); 'Editorial', *ASN*, 4. 1 (August 1950), 1–2 (p. 1).
122. Boyer, *By the Bomb's Early Light*, pp. 96–8; Alice Kimball Smith, *A Peril and Hope: The Scientists' Movement in America, 1945–1947* (Chicago, 1965), p. 265.
123. Holger Nehring, 'National Internationalists: British and West German Protests against Nuclear Weapons, the Politics of Transnational Communications and the Social History of the Cold War, 1957–1964', *Contemporary European History*, 14. 4 (2005), 559–82.
124. See Jean-Jacques Salomon, 'The *Internationale* of Science', *Science Studies*, 1. 1 (1971), 23–43.
125. Metzler, p. 29. See also Elisabeth Crawford, *Nationalism and Internationalism in Science, 1880–1939: Four Studies of the Nobel Population* (Cambridge, 1992); Michael Desser, *Zwischen Skylla und Charybdis: Die 'scientific community' der Physiker 1919–1939* (Vienna, 1991).
126. 'Editorial', *ASN*, 4. 1 (August 1950), 1–2 (p. 1); Rudolf Peierls, 'President's Report', *ASN*, 4. 1 (August 1950), 6–8 (pp. 7–8).
127. C. I. Snow, 'The Future of the A.S.A.', *ASN*, 1. 9 (5 April 1948), 137–8 (p. 137).
128. Hawes to Massey, 17 January 1949, RTBT, K. 47.
129. Peierls to Chadwick, 12 March 1946, CHAD I 24/2.
130. Boyer's observation also applies within the British context; *By the Bomb's Early Light*, p. 99.
131. Rotblat to Cockcroft, 18 March 1953, RTBT, K. 80.
132. Rotblat to Dale, 14 October 1953; advertisement flier, n.d., RTBT, K. 84.
133. 'Editorial', *ASJ*, 3. 1 (September 1953), 1–3; 'Atomic Scientists' Association', *Nature*, 173. 4392 (2 January 1954), 18–19.
134. Ford, 'Memorandum to Members of the A.S.A. Council. Proposed Changes of Atomic Scientists' Association Title and Journal', 1 April 1955, RTBT, K. 116, p. 1. While the memorandum names no author, the minutes from the 71st ASA Council meeting identify Ford as the author; G. N. Walton, 'Minutes of the 71st Council Meeting held in the Clarendon Laboratory

Parks Road, Oxford on Saturday, April 2nd, 1955, at 11.30 a.m.', 6 May 1955, RTBT, K. 116, p. 2.

135. J. Howlett, 'Minutes of the 73rd Council Meeting held at the Society for Visiting Scientists, 5, Old Burlington St., London, W.1. on Saturday, 9th July at 11 a.m.', 18 September 1955, RTBT, K. 117, p. 1.
136. H. S. W. Massey, 'The Section's Purpose', *New Scientist*, 3 January 1957, p. 31.
137. Massey to ASA Vice-Presidents and officers, 20 June 1957, TNA, AB 27/6.
138. Cockcroft to Massey, 7 October 1957, TNA, AB 27/6.
139. Thomson to Cockcroft, 4 November 1957, TNA, AB 27/6.
140. Price to Schonland, 3 February 1958, TNA, AB 27/6.
141. Price to Fishenden, 24 February 1958, TNA, AB 27/6.
142. Allan to Full ASA Members, n.d., TNA, AB 27/6.
143. H. R. Allan, 'Minutes of the 103rd Council Meeting, held in the Physics Laboratory, Imperial College, London S.W. 7 on Saturday 14th March 1959 at 11.15 a.m.', 27 April 1959, RTBT, K. 138, p. 2; H.R. Allan, 'The Atomic Scientists' Association Limited: Resolution', 9 June 1959, TNA, 27/6.
144. J. B. Priestley, 'Britain and the Nuclear Bombs', *New Statesman*, 2 November 1957, pp. 554–6; Duncan Rees, 'Resisting the British Bomb: The Early Years', in *The British Nuclear Weapons Programme 1952–2002*, ed. by Douglas Holdstock and Frank Barnaby (London, 2003), pp. 56–63 (p. 57).
145. Mark Phythian, 'CND's Cold War', *Contemporary British History*, 15. 3 (2001), 133–56 (pp. 137–9); Richard Taylor, *Against the Bomb: The British Peace Movement, 1958–1965* (Oxford, 1988), pp. 5–112; Richard Taylor, 'The Labour Party and CND: 1957 to 1984', in *Campaigns for Peace* (see Omrod, above), pp. 100–30.
146. See Richard Toye, 'Developing Multilateralism: The Havana Charter and the Fight for the International Trade Organization, 1947–1948', *International History Review*, 25. 2 (2003), 282–305.
147. Peierls, 'Britain in the Atomic Age', pp. 95–6.
148. Powell, Bullard and Rotblat to Allan, 16 March 1959; Allan to Cockcroft, 26 April 1959, TNA, AB 26/7.
149. Peierls, interview by Weiner, p. 154.
150. William Glenn Gray, 'Floating the System: Germany, the United States, and the Breakdown of Bretton Woods, 1969–1973', *Diplomatic History*, 31. 2 (April 2007), 295–323.

Conclusions and Afterthoughts

1. Dalitz, 'Peierls, Sir Rudolf Ernst (1907–1995)'; Peierls, 'Britain in the Atomic Age', pp. 95–6; Joseph Rotblat, *Scientists in the Quest for Peace: A History of the Pugwash Conferences* (Cambridge, MA, 1972).
2. Peierls to Fröhlich, 20 May 1976, Papers of Herbert Fröhlich, FRS, SJL, D. 56, F. 222.
3. Dalitz, 'Peierls, Sir Rudolf Ernst (1907–1995)'; Peierls, *Bird of Passage*, p. 288; Wittner, III, 33.
4. Rudolf Peierls, 'Limited Nuclear War?', *BAS*, 38. 5 (May 1982), 2; Rudolf Peierls, 'Counting Weapons', *London Review of Books*, 5 March 1981, 16–18.

5. C. R. Hill and others, *Does Britain Need Nuclear Weapons?: A Report from the British Pugwash Group* (London, 1995), pp. 61–3. I am thankful to Professor Robert Hinde of the British Pugwash Group for providing me with a copy of the report.

6. Thorneycroft to Peierls, 15 February 1957; Peierls to Thorneycroft, 16 February 1957, Peierls Papers, Eng. Misc. b. 224, F 20; 'National Research Institute: Governing Body Named', *Times*, 13 March 1957, p. 4.

7. Chalfont to Peierls, 29 December 1964; Peierls to Chalfont, 31 December 1964, Peierls Papers, MS Eng. Misc. b. 223, F 1.

8. Crane to Peierls, 7 May 1965; Peierls to Crane, 14 May 1965, Peierls Papers, MS Eng. Misc. b. 223, F1.

9. Vick to Peierls, 21 February 1964; Peierls to Vick, 12 February 1964; Peierls to Sandford, 12 February 1964; Marshall to Oates, 11 February 1964; Marshall to Bretscher, 27 May 1964; Marshall to Peierls, 28 May 1964, Peierls Papers, MS Eng. Misc b. 223, F 3.

10. *A Breadth of Physics: The Proceedings of the Peierls 80th Birthday Symposium*, ed. by Richard Dalitz (Singapore, 1988); *Rudolf Peierls and Theoretical Physics: Proceedings of the Symposium Held in Oxford on July 11th, 1974 to Mark the Occasion of the Retirement of Professor Sir Rudolf E. Peierls, F.R.S., C.B.E.*, ed. by Ian J. R. Aitchison and J. E. Paton (Oxford, 1977); Peierls, *Bird of Passage*, p. 320.

11. Lange and Mörke, p. 40.

12. 'Prof. Dr. habil. Fuchs, Klaus', 24 September 1974, BStU, MfS, HA XVIII, Abt. 5-VSH; Günter Flach, 'Als Physiker aktiv im Kampf für den Frieden: Zum 75. Geburtstag von Prof. em. Dr. phil. Dr. rer. nat. habil. Klaus Fuchs', *Neues Deutschland*, 29 December 1986, p. 4.

13. 'Secret Report, Soviet Embassy, GDR', 9 April 1986, Klaus Fuchs personal file, RGASPI, Komintern, F. 495, op. 205, d. 6612, p. 4.

14. Markus Wolf, with Anne McElvoy, *Man without a Face: The Autobiography of Communism's Greatest Spymaster* (New York, 1997), p. 228.

15. 'Atomspion Klaus Fuchs behauptet: In zwei Jahren hat Bonn die A-Bombe', *Quick*, 21 November 1965, BStU, MfS, AP 4189/88, pp. 51–6; Klaus Fuchs, 'Wissenschaft und Produktion in der sozialistischen Revolution', *Sitzungsberichte der Akademie der Wissenschaften der DDR, Gesellschaftswissenschaften*, 2/G (1983), 3–21; Klaus Fuchs, 'Zur Bedeutung der theoretischen Physik für die Naturwissenschaften', *Sitzungsberichte der Akademie der Wissenschaften der DDR, Mathematik – Naturwissenschaften – Technik*, 5/N (1975), 5–16.

16. Ruth Werner, *Sonjas Rapport* (Berlin (East), 1977).

17. Ruth Werner, *Sonya's Report*, transl. by Renate Simpson (London, 1991); Werner, *Sonjas Rapport*, new expanded edn, pp. 289–95. The new German edition even includes interviews with Ruth Werner's children (pp. 345–71).

18. 'Central Committee Note', n.d., RGASPI, Komintern, F. 495, op. 205, d. 6612, p. 15.

19. Heinz Barwich and Elfi Barwich, *Das rote Atom: Als deutscher Wissenschaftler im Geheimkreis der russischen Kernphysik* (Munich, 1967), pp. 239–50; Paul Maddrell, 'The Scientist Who Came in from the Cold: Heinz Barwich's Flight from the GDR', *Intelligence and National Security*, 20. 4 (2005), 608–30.

20. Rudolf Peierls, 'Twentieth-Century Physics', in *Next Year in Jerusalem: Jews in the Twentieth Century*, ed. by Douglas Villiers (London, 1976), pp. 295–307 (p. 295).
21. See Andrew Nahum, '"I believe the Americans have not yet taken them all!"': The Exploitation of German Aeronautical Science in Postwar Britain', in *Tackling Transport*, ed. by Helmuth Trischler and Stefan Zeilinger (London, 2003), pp. 99–138.
22. *Gallup International Public Opinion Polls*, I, 492.

Bibliography

Archival Sources

Germany

Die Bundesbeauftragte für die Unterlagen des Staatessicherheitsdienstes der ehemaligen Deutschen Demokratischen Republik (BstU), Berlin
    Allgemeine Personenablage (AP)
    Hauptabteilung XVIII: Sicherung der Volkswirtschaft (HA XVIII)
    Zentrale Auswertungs- und Informationsgruppe (ZAIG)
Stadtarchiv Kiel, Kiel
    Klaus Fuchs folder
Landesarchiv Schleswig-Holstein, Schleswig
    Abteilung (Abt.) 455: Staatspolizeileitstelle für den Regierungsbezirk Schleswig

Russian Federation

The Russian State Archive of Socio-Political History (RGASPI), Moscow
    The Comintern Archives
        Personal Files by Countries

United Kingdom

Department of Western Manuscripts, Bodleian Library, University of Oxford, Oxford
    The Papers and Correspondence of Sir Rudolf Peierls, 1907–1995
    The Society for the Protection of Science and Learning Papers, 1933–1987
The Churchill Archives Centre, Churchill College, University of Cambridge, Cambridge
    The Papers of Sir James Chadwick, 1914–1974
    The Papers and Correspondence of Egon Bretscher
    The Papers of Professor Sir Joseph Rotblat
Departmental Archives of the H.H. Wills Physics Laboratory, University of Bristol
The National Archives, Kew, Richmond, Surrey
    The Records of the Atomic Weapons Establishment and Predecessors (AB)
    The Security Service: Personal (PF Series) Files (KV)
    The Records of the Cabinet Office (CAB)
    The Records of the Prime Minister's Office (PREM)
The Special Collections and Archives, Sydney Jones Library, University of Liverpool, Liverpool
    The Papers of Herbert Fröhlich, FRS
    The Papers of Professor Herbert W.B. Skinner, FRS
    University Press Cuttings

United States

The Center for Southwest Research, University of New Mexico, Albuquerque, New Mexico

The Ferenc Szasz Papers, 1894–2005
The Los Alamos Historical Museum Archives, Los Alamos, New Mexico
The Los Alamos National Laboratory Archives, Los Alamos, New Mexico
The Niels Bohr Library and Archives, American Institute of Physics, College Park, Maryland
Oral History Collections

Periodicals

American Scientific
Atomic Scientists' News
Atomic Scientists' Journal
Bulletin of the Atomic Scientists
Daily Express
Daily Mail
Daily Mirror
Economist
Foreign Affairs
Guardian
Illustrated London News
International Affairs
Life
London Review of Books
Manchester Guardian
MIT Physics Annual
Nature
Neues Deutschland
New Scientist
New Statesman
New York Review of Books
New York Times
Physics Today
Picture Post
Reichsgesetzblatt
Reader's Digest
Santa Fe New Mexican
Science
Scientific Monthly
Spectator
Sunday Express
Technology Review
Time
The Times
University of Liverpool Recorder

Films and Sound Recordings

Atomic Lunch. BBC Radio 4. 2005.
Atomic Physics. Dir. Derek Mayne. Gaumont-British Instructional. 1947.

Conspirator. Dir. Victor Saville. Metro-Goldwyn-Mayer. 1949.
Dr. Strangelove, or: How I Learned to Stop Worrying and Love the Bomb. Dir. Stanley Kubrick. Hawk Films. 1964.
High Treason. Dir. Roy Boulting. Paul Soskin Productions. 1951.
Ice Station Zebra. Dir. John Sturges. Filmways Pictures. 1968.
'I'm No Communist'. Comp. Scotty Wiseman. Perf. Carson Robinson and His Pleasant Valley Boys. Metro-Goldwyn-Mayer. 1952.
Nuclear Secrets. Dir. Chris Bould. BBC; Randy Murray Productions. 2007.
The Red Menace. Dir. R.G. Springsteen. Republic Pictures. 1949.
The Whip Hand. Dir. William Cameron Menzies. RKO Radio Pictures. 1951.

Printed Sources

Aitchison, Ian, and J. E. Paton (eds), *Rudolf Peierls and Theoretical Physics: Proceedings of the Symposium Held in Oxford on July 11th, 1974 to Mark the Occasion of the Retirement of Professor Sir Rudolf E. Peierls, F.R.S., C.B.E.* (Oxford, 1977).
ASA, *Atom Train: Guide to the Travelling Exhibition on Atomic Energy* (London, 1947).
Barwich, Heinz, and Elfi Barwich, *Das rote Atom: Als deutscher Wissenschaftler im Geheimkreis der russischen Kernphysik* (Munich, 1967).
Bethe, Hans, 'The Happy Thirties', in *Nuclear Physics in Retrospect: Proceedings of a Symposium on the 1930s*, ed. by Roger H. Stuewer (Minneapolis, 1977), pp. 11–31.
Beveridge, William, *A Defence of Free Learning* (London, 1959).
Bleaney, Brebis, 'Heinrich Gerhard Kuhn, 10 March 1904–26 August 1994', *Biographical Memoirs of Fellows of the Royal Society* (hereafter *BMFRS*), 42 (November 1996), 221–32.
Born, Max, *Mein Leben: Die Erinnerungen des Nobelpreisträgers* (Munich, 1975).
———, *My Life: Recollections of a Nobel Laureate* (London, 1978).
British Council of Churches, *The Era of Atomic Power: Report of a Commission Appointed by the British Council of Churches* (London, 1946).
Brode, Bernice, *Tales of Los Alamos: Life on the Mesa 1943–1945* (Los Alamos, 1997).
Camus, Albert, *Resistance, Rebellion, and Death*, transl. from the French and with an Introduction by Justin O' Brien (New York, 1960).
Central Office of Information, *Nuclear Energy in Britain*, 2nd edn (London, 1960).
Church of England, *The Church and the Atom: A Study of the Moral and Theological Aspects of Peace and War* (n.p., 1948).
Churchill, Winston, *The Second World War*, 5 vols (London, 1948–52).
Cmd. 9075, 'Statement on Defence 1954' (London, 1954).
Cooper, Ray M. (ed.), *Refugee Scholars: Conversations with Tess Simpson* (Leeds, 1992).
Critchfield, Charles L., 'The Robert Oppenheimer I Knew', in *Behind Tall Fences: Stories and Experiences About Los Alamos at Its Beginning*, ed. by Mary Mann and John C. Allred (Los Alamos, 1996), pp. 169–77.
Crossman, Richard (ed.), *The God That Failed: Six Studies in Communism by Arthur Koestler, Ignazio Silone, André Gide Presented by Enid Starkie, Richard Wright, Louis Fischer, Stephen Spender* (London, 1950).
Dalitz, Richard (ed.), *A Breadth of Physics: The Proceedings of the Peierls 80th Birthday Symposium* (Singapore, 1988).

Dietz, David, *Atomic Energy in the Coming Era* (New York, 1945).
———, *Atomic Energy Now and in the Future* (London, 1946).
The Effects of the Atomic Bombs at Hiroshima and Nagasaki: Report of the British Mission to Japan (London, 1946).
Everett, D. H., 'Eugen Glueckauf. 9 April 1906–12 September 1981', *BMFRS*, 30 (November 1984), 193–224.
Farren, William, and George Thomson, 'Frederick Alexander Lindemann, Viscount Cherwell. 1886–1957', *BMFRS*, 4 (November 1958), 45–71.
Feklisov, Alexander, 'Podvig Klaus Fuksa', *Voenno-Istoricheskii Zhurnal*, 12 (1990), 22–9.
———, 'Podvig Klaus Fuksa', *Voenno-Istoricheskii Zhurnal*, 1 (1991), 34–43.
Feklisov, Alexander, and Sergei Kostin, *The Man Behind the Rosenbergs: Memoirs of the KGB Spymaster Who also Controlled Klaus Fuchs and Helped Resolve the Cuban Missile Crisis*, transl. by Catherine Drop (New York, 2001).
Feynman, Richard, *'Surely You're Joking, Mr. Feynman!' Adventures of a Curious Character* (New York, 1985).
Flach, Günter, 'Klaus Fuchs – Sein Erbe bewahren', *Sitzungsberichte der Akademie der Wissenschaften der DDR, Mathematik – Naturwissenschaften – Technik*, 2/N (1990), 5–10.
Frisch, Otto, *What Little I Remember* (Cambridge, 1979).
Fröhlich, Fanchon, 'Biographical Notes', in *Cooperative Phenomena*, ed. by Hermann Haken and Max Wagner (Berlin, 1973), pp. 420–1.
Fuchs, Emil, *Christ in Catastrophe* (Wallingford, 1949).
———, *Mein Leben*, 2 vols (Leipzig, 1957–9).
Fuchs, Klaus, 'Zur Bedeutung der theoretischen Physik für die Naturwissenschaften', *Sitzungsberichte der Akademie der Wissenschaften der DDR, Mathematik – Naturwissenschaften – Technik*, 5/N (1975), 5–16.
———, 'Wissenschaft und Produktion in der sozialistischen Revolution', *Sitzungsberichte der Akademie der Wissenschaften der DDR, Gesellschaftswissenschaften*, 2/G (1983), 3–21.
Gallup, George (ed.), *The Gallup International Public Opinion Polls: Great Britain 1937–1975*, 2 vols (New York, 1976).
Golan, Reuven, 'Aus der Erlebniswelt eines jüdischen Jugendlichen in Kiel Anfang der dreißiger Jahre', in *'Wir bauen das Reich': Aufstieg und erste Herrschaftsjahre des Nationalsozialismus in Schleswig-Holstein*, ed. by Erich Hoffmann and Peter Wulf (Neumünster, 1983), pp. 361–8.
Gollancz, Victor, *Shall Our Children Live or Die?* (London, 1942).
Greene, Owen, Barry Rubin, Neil Turok, Philip Webber, and Graeme Wilkinson, *London After the Bomb: What a Nuclear Attack Really Means* (Oxford, 1982).
Groves, Leslie, *Now It Can Be Told The Story of the Manhattan Project* (New York, 1962; repr. New York, 1983).
Haddow, A. (ed.), *Biological Hazards of Atomic Energy: Being the Papers Read at the Conference Convened by the Institute of Biology and the Atomic Scientists' Association October 1950* (Oxford, 1952).
Halliday, Bryce, 'Professor Rotblat and the Atom Train', in *War and Peace: The Life and Work of Sir Joseph Rotblat*, ed. by Peter Rowlands and Vincent Attwood (Liverpool, 2006), pp. 139–44.
Hersey, John, *Hiroshima* (Harmondsworth, 1946).

Hill, C.R., R.S. Pease, Rudolf Peierls, and Joseph Rotblat, *Does Britain Need Nuclear Weapons?: A Report from the British Pugwash Group* (London, 1995).

H.M. Treasury, *Statements Relating to the Atomic Bomb* (London, 1945).

Home Office Civil Defence Department, *Civil Defence Manual of Basic Training. Vol. II: Atomic Warfare* (London, 1950).

Hull, McAllister, with Amy Bianco, *Rider of the Pale Horse: A Memoir of Los Alamos and Beyond* (Albuquerque, 2005).

Hyde, Douglas, *I Believed: The Autobiography of a Former British Communist* (London, 1950).

Jackson, David, and Kurt Gottfried, *Victor Frederick Weisskopf, 1908–2002: A Biographical Memoir*, Biographical Memoirs, 84 (Washington, 2003).

Kellermann, Bernhard, *The Tunnel* (London, 1915).

Kurti, Nicholas, 'Franz Eugen Simon, 1893–1956', *BMFRS*, 4 (November 1958), 225–56.

Lafitte, François, *The Internment of Aliens*, new edn (London, 1988).

Lamphere, Robert, and Tom Shachtmann, *The FBI-KGB War: A Special Agent's Story* (New York, 1986).

Lange, Gert and Joachim Mörke, *Wissenschaft im Interview: Gespräche mit Akademiemitgliedern über ihr Leben und Werk* (Leipzig, 1979).

Lee, Sabine (ed.), *The Bethe-Peierls Correspondence* (Singapore, 2007).

——— (ed.), *Sir Rudolf Peierls: Selected Private and Scientific Correspondence*, 2 vols (Singapore, 2007-9).

Lohmeier, Dieter, and Bernhardt Schell (eds), *Einstein, Anschütz and the Kiel Gyro Compass, the Correspondence between Albert Einstein and Herrmann Anschütz-Kaempfe as well as other Documents*, transl. by Anita Cervenák, 2nd rev. edn (Kiel, 2005).

Manley, John, 'A New Laboratory Is Born', in *Reminiscences of Los Alamos, 1943–1945*, ed. by Lawrence Badash, Joseph Hirschfelder and Herbert Broida (Dordrecht, 1980), pp. 21–40.

Mann, Thomas. *Betrachtungen eines Unpolitischen*, 3rd edn (Frankfurt a.M., 2001).

Marshak, Ruth, 'Secret City', in *Standing By and Making Do: Women of Wartime Los Alamos*, ed. by Jane Wilson and Charlotte Serber (Los Alamos, 1988), pp. 1–19.

Masters, Dexter, and Katherine Way (eds), *One World or None: A Report to the Public on the Full Meaning of the Atomic Bomb* (New York, 1946).

McMillan, Edwin, 'Early Days at Los Alamos', in *Reminiscences of Los Alamos* (see John Manley, above), pp. 13–19.

Moon, Philip, and Eric Burhop, *Atomic Survey: A Short Guide to the Scientific and Political Problems of Atomic Energy* (n.p., 1946).

Moorehead, Alan, *The Traitors: The Double Life of Fuchs, Pontecorvo and Nunn May* (London, 1952; repr. New York, 1963).

Mott, Nevill, 'Walter Heinrich Heitler, 2 January 1904–15 November 1981', *BMFRS*, 28 (November 1982), 141–51.

———, *A Life in Science* (London, 1986).

Mott, Nevill, and Rudolf Peierls, 'Werner Heisenberg: 5 December 1901–1 February 1976', *BMFRS*, 23 (November 1977), 212–57.

Nabarro, Frank, and A. S. Aragon, 'Egon Orowan, 2 August 1902–3 August 1989', *BMFRS*, 41 (November 1995), 316–40.

Oppenheimer, Robert, 'Atomic Weapons', *Proceedings of the American Philosophical Society* (hereafter *PAPS*), 90. 1 (1946), 7–10.

Peierls, Rudolf, 'The Scientific and Technical Backgrounds: I. The Scientific Background', in *Atomic Energy: Its International Implications. A Discussion by a Chatham Study Group*, ed. by Royal Institute of International Affairs (London, 1948), pp. 29–36.

——, 'Britain in the Atomic Age', in *Alamogordo Plus Twenty-Five Years: The Impact of Atomic Energy on Science, Technology, and World Politics*, ed. by Richard Lewis and Jane Wilson, with Eugene Rabinowitch (New York, 1970), pp. 91–105.

——, 'Twentieth-Century Physics', in *Next Year in Jerusalem: Jews in the Twentieth Century*, ed. by Douglas Villiers (London, 1976), pp. 295–307.

——, 'Otto Robert Frisch, 1 October 1904–22 September 1979', *BMFRS*, 27 (November 1981), 283–306.

——, *Bird of Passage: Recollections of a Physicist* (Princeton, 1985).

——, 'Als Student bei Heisenberg', in *Werner Heisenberg in Leipzig 1927–1942*, ed. by Christian Kleint and Gerald Wiemers (=*Abhandlungen der Sächsischen Akademie der Wissenschaften zu Leipzig, Mathematisch-naturwissenschaftliche Klasse*, 28. 2 (1993)), pp. 104–7.

——, 'Recollections of James Chadwick', *Notes and Records of the Royal Society of London*, 48. 1 (January 1994), 135–41.

——, *Atomic Histories* (Woodbury; New York, 1997).

Perutz, Max, *Is Science Necessary? Essays on Science & Scientists* (London, 1989; repr. Oxford, 1992).

Pilat, Oliver, *The Atom Spies* (New York, 1952).

Popper, Karl, *Unended Quest: An Intellectual Biography* (La Salle, 1976).

Rotherham, Leonard, 'Hans Kronberger, 1920–1970', *BMFRS*, 18 (November 1972), 412–26.

Sanders, J.H., 'Nicholas Kurti, C.B.E.: 14 May 1908–24 November 1998', *BMFRS*, 46 (November 2000), 301–15.

Schoenberg, David, 'Heinz London, 1907–1970', *BMFRS*, 17 (1971), 441–61.

Smith, Alice Kimball, and Charles Weiner (eds), *Robert Oppenheimer: Letters and Correspondence* (Cambridge, 1980).

Smyth, Henry DeWolf, *Atomic Energy for Military Purposes: The Official Report on the Development of the Atomic Bomb under the Auspices of the United States Government, 1940–1945* (Princeton, 1945; repr. Stanford, 1989).

Szilard, Leo, 'A Petition to the President of the United States', in *The Atomic Age: Scientists in National and World Affairs, Articles from the Bulletin of the Atomic Scientists 1945–1962*, ed. by Morton Grodzins and Eugene Rabinowitch (New York, 1963), pp. 28–9.

——, *Leo Szilard: His Version of the Facts: Selected Recollections and Correspondence*, ed. by Spencer R. Weart and Gertrud Weiss Szilard (Cambridge, 1978).

Teller, Edward, with Allen Brown, *The Legacy of Hiroshima* (Garden City, 1962).

Teller, Edward, with Judith Shoolery, *Memoirs: A Twentieth-Century Journey in Science and Politics* (Cambridge, 2001).

United States. Cong. House. Joint Committee on Atomic Energy, *Soviet Atomic Espionage*, 82nd Cong., 1st sess. (Washington, 1951).

Vansittart, Robert, *Black Record: Germans Past and Present* (London, 1941).

'Verzeichnis der Studenten und Hörer bei Werner Heisenberg', ed. by Gerald Wiemers, in *Werner Heisenberg in Leipzig 1927–1942* (see Peierls, 'Als Student bei Heisenberg', above), pp. 144–72.

Weisskopf, Victor, *The Joy of Insight: Passions of a Physicist* (New York, 1991).
———, 'Meine Göttinger Studienjahre mit Born und Franck', in *Max Born, James Franck, der Luxus des Gewissens: Physiker in ihrer Zeit*, ed. by Jost Lemmerich (Berlin [West], 1982), pp. 80–3.
Wells, H. G., *The World Set Free* (London, 1914).
Werner, Ruth, *Sonjas Rapport* (Berlin [East], 1977).
———, *Sonya's Report*, transl. by Renate Simpson (London, 1991).
———, *Sonjas Rapport*, new expanded edn (Berlin, 2006).
West, Rebecca, *The Meaning of Treason*, 2nd enlarged and rev. edn (London, 1952).
———, *The New Meaning of Treason* (New York, 1964).
Wigner, Eugene, *The Recollections of Eugene P. Wigner: As Told to Andrew Szanton* (New York, 1992).
Williams, Robert Chadwell, and Philip Cantelon (eds), *The American Atom: A Documentary History of Nuclear Policies from the Discovery of Fission to the Present 1939–1984* (Philadelphia, 1984).
Wilson, Robert, 'Hiroshima: The Scientists' Social and Political Reaction', *PAPS*, 140. 3 (1996), 350–7.
Wolf, Markus, with Anne McElvoy, *Man without a Face: The Autobiography of Communism's Greatest Spymaster* (New York, 1997).

Secondary Sources

Albright, Joseph, and Marcia Kunstel, *Bombshell: The Secret Story of America's Unknown Atomic Spy Conspiracy* (New York, 1997).
Aldrich, Richard, *The Hidden Hand: Britain, America and Cold War Secret Intelligence* (London, 2001).
Arnold, Lorna, with Katherine Pyne, *Britain and the H-Bomb* (Basingstoke, 2001).
Arnold, Lorna, *Windscale 1957: Anatomy of a Nuclear Accident*, 2nd edn (Basingstoke, 1995).
———, 'The History of Nuclear Weapons: The Frisch-Peierls Memorandum on the Possible Construction of Atomic Bombs of February 1940', *Cold War History*, 3. 3 (April 2003), 111–26.
Arnold, Lorna, and Mark Smith, *Britain, Australia and the Bomb: The Nuclear Tests and Their Aftermath*, rev. edn (Basingstoke, 2006).
Ash, Mitchell, and Alfons Söllner (eds), *Forced Migration and Scientific Change: Émigré German-Speaking Scientists and Scholars after 1933* (Washington; New York, 1996).
———, 'Introduction: Forced Migration and Scientific Change after 1933', in Ash and Söllner, pp. 1–19.
Asper, Helmut G., 'Film', in *Handbuch der deutschsprachigen Emigration Handbuch der deutschsprachigen Emigration 1933–1945*, ed. by Claus-Dieter Krohn, Patrik von zur Mühlen, Gerhard Paul and Lutz Winckler (Darmstadt, 1998), pp. 957–70.
Attwood, Thomas Vincent, 'The 37 Inch Cyclotron and Nuclear Structure Research at Liverpool 1935 to 1960' (unpublished master's thesis, University of Liverpool, 1998).
———, 'Uranium Separation in the U.K. during World War II' (unpublished doctoral dissertation, University of Liverpool, 2004).

Avery, Donald Howard, 'Atomic Scientific Co-Operation and Rivalry Among Allies: The Anglo-Canadian Montreal Laboratory and the Manhattan Project, 1943–1946', *War in History*, 2. 3 (1995), 274–305.

Badash, Lawrence, *Scientists and the Development of Nuclear Weapons: From Fission to the Limited Test Ban Treaty, 1939–1963* (Amherst, 1995).

——, 'Science and McCarthyism', *Minerva*, 38. 1 (2000), 53–80.

——, 'From Security Blanket to Security Risk: Scientists in the Decade after Hiroshima', *History and Technology*, 19. 3 (2003), 241–56.

Badash, Lawrence, Elizabeth Hodes and Adolph Tiddens, 'Nuclear Fission: Reaction to the Discovery in 1939', *PAPS*, 130. 2 (1986), 196–231.

Ball, S.J., 'Military Nuclear Relations between the United States and Great Britain under the Terms of the McMahon Act, 1946–1958', *Historical Journal*, 38. 2 (1995), 439–54.

Bärwinkel, Klaus, 'Die Austreibung von Physikern unter der deutschen Regierung vor dem Zweiten Weltkrieg. Ausmaß und Auswirkung', in *Die Künste und die Wissenschaften im Exil 1933–1945*, ed. by Edith Böhme and Wolfgang Motzkau-Valeton (Gerlingen, 1992), pp. 569–99.

Baylis, John, *Anglo-American Defence Relations 1939–1984: The Special Relationship*, 2nd edn (London, 1984).

——, *Ambiguity and Deterrence: British Nuclear Strategy 1945–1964* (Oxford, 1995).

Beck, Ulrich, *Risikogesellschaft: Auf dem Weg in eine andere Moderne* (Frankfurt a.M., 1986).

Berghahn, Marion, *Continental Britons: German-Jewish Refugees from Nazi Germany* (Oxford, 1988).

Bernstein, Barton, 'Reconsidering the "Atomic General": Leslie R. Groves', *Journal of Military History*, 67. 3 (2003), 883–920.

Bernstein, Jeremy, *Oppenheimer: Portrait of an Enigma* (Chicago, 2004).

Beyerchen, Alan, *Scientists under Hitler: Politics and the Physics Community in the Third Reich* (New Haven, 1977).

Beyler, Richard, 'The Demon of Technology, Mass Society, and Atomic Physics in West Germany, 1945–1957, *History and Technology*, 19. 3 (2003), 227–39.

Beyler, Richard, Alexei Kojevnikov and Jessica Wang, 'Purges in Comparative Perspective: Rules for Exclusion and Inclusion in the Scientific Community under Political Pressure', *Osiris*, 2nd series, 20 (2005), 23–48.

Bird, Kai, and Martin Sherwin, *American Prometheus: The Triumph and Tragedy of J. Robert Oppenheimer* (New York, 2005).

Bollenbeck, Georg, *Bildung und Kultur: Glanz und Elend eines deutschen Deutungsmusters* (Frankfurt a.M., 1994).

Born, Gustav, 'The Effect of the Scientific Environment in Britain on Refugee Scientists from Germany and Their Effects on Science in Britain', *Ber. Wissenschaftsgesch.*, 7. 3 (1984), 129–43.

Boyer, Paul, *By the Bomb's Early Light: American Thought and Culture at the Dawn of the Atomic Age* (New York, 1985; repr. Chapel Hill, 1994).

——, *Fallout: A Historian Reflects on America's Half-Century Encounter with Nuclear Weapons* (Columbus, 1998).

Brinson, Charmian, 'The Gestapo and the German Political Exiles in Britain During the 1930s: The Case of Hans Wesemann – and Others', *German Life and Letters*, 51. 1 (1998), 43–64.

Brown, Andrew, *The Neutron and the Bomb: A Biography of Sir James Chadwick* (Oxford, 1997).

———, 'A Tale of Two Documents', in *Remembering the Manhattan Project: Perspectives on the Making of the Atomic Bomb and Its Legacy*, ed. by Cynthia Kelly (London, 2004), pp. 41–6.

———, 'The Viennese Connection: Engelbert Broda, Alan Nunn May and Atomic Espionage', *Intelligence and National Security*, 24. 2 (2009), 173–93.

Bud, Robert, *Penicillin: Triumph and Tragedy* (Oxford, 2007).

Bud, Robert, and Philip Gummet (eds), *Cold War, Hot Science: Applied Research in Britain's Defence Laboratories, 1945–1990* (London, 2002).

———, 'Introduction: Don't You Know There's a War On', in Bud and Gummet, pp. 1–28.

Buderi, Robert, *The Invention that Changed the World: The Story of Radar from War to Peace* (New York, 1996; repr. London, 1998).

Burchhardt, Lothar, 'Naturwissenschaftliche Universitätslehrer im Kaiserreich', in *Deutsche Hochschullehrer als Elite 1815–1945*, ed. by Klaus Schwabe (Boppard, 1988), pp. 151–214.

Burletson, Louise, 'The State, Internment and Public Criticism in the Second World War', in *The Internment of Aliens in Twentieth Century Britain*, ed. by David Cesarani and Tony Kushner (London, 1993), pp. 102–24.

Calder, Angus, *The People's War: Britain 1939–1945* (London, 1969; repr. London, 1997).

Carson, Cathryn, 'New Models for Science in Politics: Heisenberg in West Germany', *Historical Studies in the Physical and Biological Sciences* (hereafter *HSPS*), 30. 1 (1999), 115–71.

———, 'Objectivity and the Scientist: Heisenberg Rethinks', *Science in Context*, 16. 1–2 (2003), 243–69.

———, *Heisenberg in the Atomic Age: Science and the Public Sphere* (Washington; Cambridge, 2010).

Carsten, Francis, 'German Refugees in Great Britain 1933–1945: A Survey', in *Exile in Great Britain: Refugees from Hitler's Germany*, ed. by Gerhard Hirschfeld (Leamington Spa; Atlantic Highlands, 1984), pp. 11–28.

Cassidy, David, 'Understanding the History of Special Relativity: Bibliographical Essay', *HSPS*, 16. 1 (1986), 177–95.

Chapshew, James, and Karen Rader, 'Big Science: Price to the Present', in *Science after '40*, ed. by Arnold Thackray (=*Osiris*, 2nd series, 7 (1992)), pp. 2–25.

Cesarani, David, 'An Alien Concept? The Continuity of Anti-Alienism in British Society Before 1940', in *The Internment of Aliens in Twentieth Century Britain* (see Burletson, above), pp. 25–52.

Clark, Ian, and Nicholas Wheeler, *The British Origins of Nuclear Strategy 1945–1955* (Oxford, 1989).

Clark, Ronald, *The Birth of the Bomb: The Untold Story of Britain's Part in the Weapon that Changed the World* (London, 1961).

———, *Tizard* (London, 1965).

Clayton, June, 'A Noble Man of Science, A Nobel Man of Peace: Professor Sir Joseph Rotblat, FRS, Nobel Peace Prize Laureate 1995' (unpublished master's thesis, University of Liverpool, 2003).

Colley, Linda, *Britons: Forging the Nation 1707–1837*, new edn (London, 2003).

Collini, Stefan, *English Pasts: Essays in History and Culture* (Oxford, 1999).

Conekin, Becky, Frank Mort and Chris Waters, 'Introduction', in *Moments of Modernity: Reconstructing Britain 1945–1964*, ed. by Conekin, Mort and Waters (London, 1999), pp. 1–21.

Cornwell, John, *Hitler's Scientists: Science, War, and the Devil's Pact* (New York, 2004).

Crawford, Elizabeth, *Nationalism and Internationalism in Science, 1880–1939: Four Studies of the Nobel Population* (Cambridge, 1992).

Crawford, John, '"A Political H-Bomb": New Zealand and the British Thermonuclear Weapon Test of 1957–58', *Journal of Imperial and Commonwealth History*, 26. 1 (1998), 127–50.

Cull, Nicholas, *Selling War: The British Propaganda Campaign against American "Neutrality" in World War II* (New York, 1995).

Dahrendorf, Ralf, *Gesellschaft und Demokratie in Deutschland* (Munich, 1965).

Dalitz, Richard, 'Peierls, Sir Rudolf Ernst (1907–1995)', in *Oxford Dictionary of National Biography*, Oxford University Press, Sept 2004; online edn, Jan 2008 <http://www.oxforddnb.com/view/article/60076> [accessed 10 August 2011].

Daunton, Martin, and Bernhard Rieger (eds), *Meanings of Modernity: Britain from the Late-Victorian Era to World War II* (Oxford, 2001).

Deacon, Richard, *The British Connection: Russia's Manipulation of British Individuals and Institutions* (London, 1979).

Desser, Michael, *Zwischen Skylla und Charybdis: Die 'scientific community' der Physiker 1919–1939* (Vienna, 1991).

Dittrich, Irene, 'Die "Revolutionäre Studentengruppe" an der Christian-Albrechts-Universität zu Kiel (1930–1933)', *Demokratische Geschichte*, 4 (1989), 175–84.

Divine, Robert, *Blowing on the Wind: The Nuclear Test Ban Debate, 1954–60* (New York, 1978).

Eckert, Michael, *Die Atomphysiker: Eine Geschichte der theoretischen Physik am Beispiel der Sommerfeldschule* (Braunschweig, 1993).

Edgerton, David, *Warfare State: Britain, 1920–1970* (Cambridge, 2006).

Eisfeld, Rainer, *Mondsüchtig: Wernher von Braun und die Geburt der Raumfahrt aus dem Geist der Barbarei* (Reinbek, 1996).

Elsaesser, Thomas, 'Ethnicity, Authenticity, and Exile: A Counterfeit Trade? German Filmmakers and Hollywood', in *Home, Exile, Homeland: Film, Media, and the Politics of Place*, ed. by Hamid Naficy (New York, 1999), pp. 97–123.

Fermi, Laura, *Illustrious Immigrants: The Intellectual Migration from Europe, 1930/41*, 2nd edn (Chicago, 1971).

Fischer, Klaus, *Changing Landscapes of Nuclear Physics: A Scientometric Study on the Social and Cognitive Position of German-Speaking Emigrants Within the Nuclear Physics Community, 1921–1947* (Berlin, 1993).

——, 'Physik', in *Handbuch der deutschsprachigen Emigration 1933–1945* (see Asper, above), pp. 824–36.

Flowers, Mary, 'Fuchs, (Emil Julius) Klaus (1911–1988)', in *Oxford Dictionary of National Biography*, Oxford University Press, 2004; online edn, May 2008 <http://www.oxforddnb.com/view/article/40698> [accessed 10 August 2011].

Forgan, Sophie, 'Atoms in Wonderland', *History and Technology*, 19. 3 (2003), 177–96.

Forman, Paul, 'Scientific Internationalism and the Weimar Physicists: The Ideology and Its Manipulation in Germany after World War I', *Isis*, 64. 2 (1973), 150–80.

———, 'Behind Quantum Electronics: National Security as Basis for Physical Research in the United States, 1940–1960', *HSPS*, 18. 1 (1987), 149–229.

Fort, Adrian, *Prof: The Life of Frederick Lindemann* (London, 2004).

Fried, Richard, *Nightmare in Red: The McCarthy Era in Perspective* (New York, 1990).

Friedmann, Ronald, *Der Mann, der kein Spion war: Das Leben des Kommunisten und Wissenschaftlers Klaus Fuchs* (Rostock, 2005).

'Fuchs, Klaus Emil Julius', in *Biographisches Handbuch der deutschsprachigen Emigration nach 1933*, 3 vols, ed. by Werner Röder and Herbert Strauss (Munich, 1980–83), I (1980), p. 206.

Furman, Necah Stewart, *Sandia National Laboratories: The Postwar Decade* (Albuquerque, 1990).

Galbreath, Ross, 'The Ruherford Connection: New Zealand Scientists and the Manhattan and Montreal Projects', *War in History*, 2. 3 (1995), 306–19.

Galison, Peter, 'Physics Between War and Peace', in *Science, Technology and the Military*, ed. by Everett Mendelsohn, Merritt Roe Smith and Peter Weingart (=*Sociology of the Sciences*, 12. 1 (1988)), pp. 47–86.

———, 'Trading Zone: Coordinating Action and Belief', in *The Science Studies Reader*, ed. by Mario Biagioli (New York, 1999), pp. 137–60.

Galison, Peter, and Barton Bernstein, 'In Any Light: Scientists and the Decision to Build the Superbomb, 1952–1954', *HSPS*, 19. 2 (1989), 267–348.

Geertz, Clifford, *The Interpretation of Cultures* (New York, 1973; repr. London, 1975).

Gerber, Michele Stenehjem, *On the Home Front: The Cold War Legacy of the Hanford Nuclear Site*, 3rd edn (Lincoln, 2007).

Gillman, Peter, and Leni Gillman, *'Collar the Lot!' How Britain Interned and Expelled Its Wartime Refugees* (London, 1980).

Goldberg, Robert, *Enemies Within: The Culture of Conspiracy in Modern America* (New Haven, 2001).

Goldberg, Stanley, 'The Secret about Secrets', in *Secret Agents: The Rosenberg Case, McCarthyism & Fifties America*, ed. by Marjorie Garber and Rebecca Walkowitz (New York, 1995), pp. 47–58.

Goodchild, Peter, *J. Robert Oppenheimer: 'Shatterer of Worlds'* (London, 1980).

Goodman, Michael, 'The Grandfather of the Hydrogen Bomb?: Anglo-American Intelligence and Klaus Fuchs', *HSPS*, 34. 1 (2003), 1–22.

———, 'Who Is Trying to Keep What Secret from Whom and Why? MI5-FBI Relations and the Klaus Fuchs Case', *Journal of Cold War Studies*, 7. 3 (Summer 2005), 124–46.

———, *Spying on the Nuclear Bear: Anglo-American Intelligence and the Soviet Bomb* (Stanford, 2007).

Goodman, Michael, and Chapman Pincher, 'Research Note: Clement Attlee, Percy Sillitoe and the Security Aspects of the Fuchs Case', *Contemporary British History*, 19. 1 (2005), 67–77.

Goudsmit, Samuel, *Alsos: The Failure of German Science* (London, 1947).

Gowing, Margaret, *Britain and Atomic Energy, 1939–1945* (London, 1964).

———, *Independence and Deterrence: Britain and Atomic Energy, 1945–1952*, 2 vols (London, 1974).

———, 'Britain and the Bomb: The Origins of Britain's Determination to Be a Nuclear Power', *Contemporary Record*, 2. 2 (Summer 1988), 36–40.

———, 'James Chadwick and the Atomic Bomb', *Notes and Records of the Royal Society of London*, 47. 1 (January 1993), 79–92.

Gray, William Glenn, 'Floating the System: Germany, the United States, and the Breakdown of Bretton Woods, 1969–1973', *Diplomatic History*, 31. 2 (April 2007), 295–323.

Greenspan, Nancy Thorndike, *The End of the Certain World: The Life and Science of Max Born* (Chichester, 2005).

Gusterson, Hugh, *People of the Bomb: Portraits of America's Nuclear Complex* (Minneapolis, 2004).

Guy, Stephen, *'High Treason* (1951): Britain's Cold War Fifth Column', *Historical Journal of Film, Radio and Television*, 13. 1 (1993), 35–47.

Hales, Peter Bacon, *Atomic Spaces: Living on the Manhattan Project* (Urbana, 1997).

Hargittai, István, *The Martians of Science: Five Physicists Who Changed the Twentieth Century* (New York, 2006).

Hawkins, David, *Project Y: The Los Alamos Story, Part I: Toward Trinity* (Los Angeles, 1983).

Haynes, John Earl, and Harvey Klehr, *Venona: Decoding Soviet Espionage in America* (New Haven, 1999).

Hecht, Gabrielle, *The Radiance of France: Nuclear Power and National Identity after World War II* (Cambridge, 1998).

Heilbron, John, and Robert Seidel, *Lawrence and His Laboratory: A History of the Lawrence Berkeley Laboratory* (Berkeley, 1989).

Hendershot, Cyndy, *Anti-Communism and Popular Culture in Mid-Century America* (Jefferson, 2003).

Henriksen, Margot, *Dr. Strangelove's America: Society and Culture in the Atomic Age* (Berkeley, 1997).

Hennessy, Peter, *The Secret State: Preparing for the Worst 1945–2010*, 2nd edn (2002; London, 2010).

Hennessy, Peter, and Gail Brownfeld, 'Britain's Cold War Security Purge: The Origins of Positive Vetting', *Historical Journal*, 25. 4 (1982), 965–74.

Hentschel, Klaus, *The Mental Aftermath: The Mentality of German Physicists 1945–1949*, transl. by Ann M. Hentschel (Oxford, 2007).

Herken, Gregg, *Brotherhood of the Bomb: The Tangled Lives and Loyalties of Robert Oppenheimer, Ernest Lawrence, and Edward Teller* (New York, 2002).

Herken, Gregg, *Cardinal Choices: Presidential Science Advising from the Atomic Bomb to SDI*, rev. and expanded edn (Stanford, 2000).

Hirschfeld, Gerhard, 'German Refugee Scholars in Great Britain, 1933–1945', in *Refugees in the Age of Total War*, ed. by Anna Bramwell (London, 1988), pp. 152–63.

———, '"A High Tradition of Eagerness ...": British Non-Jewish Organisations in Support of Refugees', in *Handbuch der deutschsprachigen Emigration 1933–1945*, ed. by Werner E. Mosse and Julius Carlebach (Tübingen, 1991), pp. 599–610.

Hoch, Paul, 'The Reception of Central European Refugee Physicists of the 1930s: U.S.S.R., U.K., U.S.A.', *Annals of Science*, 40. 3 (1983), 217–46.

———, 'Institutional Versus Intellectual Migrations in the Nucleation of New Scientific Specialities', *Studies in the History and Philosophy of Science*, 18. 4 (1987), 481–500.

———, 'Migration and the Generation of New Scientific Ideas', *Minerva*, 25. 3 (1987), 209–37.

———, 'Some Contributions to Physics by German-Jewish Émigrés in Britain and Elsewhere', in *Second Chance* (see Hirschfeld, '"A High Tradition of eagerness ..."',' above), pp. 229–41.

Hoch, Paul, and Jennifer Platt, 'Migration and the Denationalization of Science', in *Denationalizing Science: The Contexts of International Scientific Practice*, ed. by Elisabeth Crawford, Terry Shinn and Sverker Sörlin (Dordrecht, 1992), pp. 133–52.

Hoch, Paul, and E.J. Yoxen, 'Schrödinger at Oxford: A Hypothetical National Cultural Synthesis which Failed', *Annals of Science*, 44. 6 (1987), 593–616.

Hoddeson, Lillian, 'Mission Change in the Large Laboratory: The Los Alamos Implosion Program, 1943–1945', in *Big Science: The Growth of Large-Scale Research*, ed. by Peter Galison and Bruce Hevly (Stanford, 1992), pp. 265–89.

Hoddeson, Lillian, Paul W. Hendrikson, Roger A. Meade, and Catherine L. Westfall, *Critical Assembly: A Technical History of Los Alamos During the Oppenheimer Years, 1943–1945* (Cambridge, 1993).

Hoffmann, Klaus, *J. Robert Oppenheimer: Schöpfer der ersten Atombombe* (Berlin, 1995).

Holl, Jack, *Argonne National Laboratory, 1946–96* (Urbana, 1997).

Holloway, David, *Stalin and the Bomb: The Soviet Union and Atomic Energy, 1939–1956* (New Haven, 1994).

Holt, John, 'James Chadwick in Liverpool', *Notes and Records of the Royal Society of London*, 48. 2 (July 1994), 299–308.

Horak, Jan-Christopher, 'Filmkünstler im Exil: Ein Weg nach Hollywood', in *Die Künste und die Wissenschaften im Exil 1933–1945* (see Bärwinkel, 'Die Austreibung von Physikern', above), pp. 231–54.

———, 'On the Road to Hollywood: German-Speaking Filmmakers in Exile 1933–1950', in *Kulturelle Wechselbeziehungen im Exil – Exile Across Cultures*, ed. by Helmut F. Pfanner (Bonn, 1986), pp. 240–8.

Howes, Ruth, and Caroline Herzenberg, *Their Day in the Sun: Women of the Manhattan Project* (Philadelphia, 1999).

Hughes, Jeff, *The Manhattan Project: Big Science and the Atom Bomb* (Cambridge, 2002).

———, 'The Strath Report: Britain Confronts the H-Bomb, 1954–1955', *History and Technology*, 19. 3 (2003), 257–75.

Hughes, Stuart, *The Sea Change: The Migration of Social Thought, 1930–1965* (New York, 1975).

Hunner, Jon, *Inventing Los Alamos: The Growth of an Atomic Community* (Norman, 2004).

———, 'Reinventing Los Alamos: Code Switching and Suburbia at America's Atomic City', in *Atomic Culture: How We Learned to Stop Worrying and Love the Bomb*, ed. by Scott Zeman and Michael Amundson (Boulder, 2004), 33–48.

Hyde, Montgomery, *The Atom Bomb Spies* (London, 1980).

Hyland, Gerard, 'Herbert Fröhlich, FRS: A Physicist Ahead of His Time', in *Herbert Fröhlich, FRS: A Physicist Ahead of His Time*, ed. by Gerard Hyland and Peter Rowlands (Liverpool, 2006), pp. 221–339.

Johnson, Charles, and Charles Jackson, *City Behind a Fence: Oak Ridge, Tennessee 1942–1946* (Knoxville, 1981).

Johnson, Leland, and Daniel Schaffer, *Oak Ridge National Laboratory: The First Fifty Years* (Knoxville, 1994).

Jones, Greta, 'The Mushroom-Shaped Cloud: British Scientists' Opposition to Nuclear Weapons Policy, 1945–57', *Annals of Science*, 43. 1 (1986), 1–26.

Jones, Matthew, 'Anglo-American Relations after Suez: The Rise and Decline of the Working Group Experiment and the French Challenge to NATO, 1957–59', *Diplomacy & Statecraft*, 14. 1 (March 2003), 49–78.

Kaiser, David, 'The Atomic Secret in Red Hands? American Suspicions of Theoretical Physicists During the Early Cold War', *Representations*, 90 (Spring 2005), 28–60.

Karlsch, Rainer, *Uran für Moskau: Die Wismut – Eine populäre Geschichte* (Berlin, 2007).

Kavanagh, Dennis, and Peter Morris, *Consensus Politics: From Attlee to Major*, 2nd edn (Oxford, 1994).

Keith, S. T., and Paul Hoch, 'Formation of a Research School: Theoretical Solid State Physics at Bristol 1930–54', *British Journal for the History of Science*, 18. 3 (1986), 19–44.

Kevles, Daniel, *The Physicists: The History of a Scientific Community in Modern America*, new edn (Cambridge, MA, 1995).

King, Charles David, 'Chadwick, Liverpool and the Bomb' (unpublished doctoral dissertation, University of Liverpool, 1997).

Kirsch, Scott, *Proving Grounds: Project Plowshare and the Unrealized Dream of Nuclear Earthmoving* (New Brunswick, 2005).

Kragh, Helge, *Quantum Generations: A History of Physics in the Twentieth Century* (Princeton, 1999).

Krige, John, *American Hegemony and the Postwar Reconstruction of Science in Europe* (Cambridge, 2006).

Krohn, Claus-Dieter, 'Vereinigte Staaten von Amerika', in *Handbuch der deutschsprachigen Emigration 1933–1945* (see Asper, above), pp. 446–66.

'Kronberger, Hans', in *Biographisches Handbuch der deutschsprachigen Emigration nach 1933* (see 'Fuchs, Klaus Emil Julius', above), II (1983), p. 668.

Kushner, Tony, and David Cesarani, 'Alien Internment in Britain During the Twentieth Century: An Introduction', in *The Internment of Aliens in Twentieth Century Britain* (see Burletson, above), pp. 1–22.

——, 'Clubland, Cricket Tests and Alien Internment, 1939–40', in *The Internment of Aliens in Twentieth Century Britain* (see Burletson, above), pp. 79–101.

Labat, Sean, 'Chicago Atomic Scientists and United States Foreign Policy, 1945–1950', *Journal of Illinois History*, 3 (Summer 2000), 121–40.

Lamont, Lansing, *Day of Trinity* (New York, 1965).

Large, David Clay, *Germans to the Front: West German Rearmament in the Adenauer Era* (Chapel Hill, 1996).

Lee, Sabine, 'Birmingham – London – Los Alamos – Hiroshima: Britain and the Atomic Bomb', *Midland History*, 27 (2002), 146–64.

——, 'The Spy That Never Was', *Intelligence and National Security*, 17. 4 (2002), 77–99.

——, 'Rudolf Ernst Peierls, 5 June 1907–19 September 1995', *BMFRS*, 53 (December 2007), 265–84.

Legband, Michael, 'Von der Provinz zum Bundesland – Schleswig-Holstein im 20. Jahrhundert', in *Schleswig-Holstein von den Ursprüngen bis zur Gegenwart: Eine Landesgeschichte*, ed. by Jann Markus Witt and Heiko Vosgerau (Hamburg, 2002), pp. 327–83.

Lekan, Thomas, 'German Landscape: Local Promotion of the *Heimat* Abroad', in *The Heimat Abroad: The Boundaries of Germanness*, ed. by Krista O'Donnell, Renate Bridenthal and Nancy Reagin (Ann Arbor, 2005), pp. 141–66.

London, Louise, 'British Immigration Control Procedures and Jewish Refugees 1933–1939', in *Second Chance* (see Hirschfeld, '"A High Tradition of Eagerness ..."',' above), pp. 485–517.

———, *Whitehall and the Jews, 1933–1948: British Immigration Policy and the Holocaust* (Cambridge, 2000).

Lovell, Bernard, 'Bristol and Manchester – The Years 1931-9', in *The Making of Physicists*, ed. by Rajkumari Williamson (Bristol, 1997), pp. 148–60.

Maddrell, Paul, 'The Scientist Who Came in from the Cold: Heinz Barwich's Flight from the GDR', *Intelligence and National Security*, 20. 4 (2005), 608–30.

Mahoney, Joan, 'Civil Liberties in Britain During the Cold War: The Role of the Central Government', *American Journal of Legal History*, 33. 1 (1989), 53–100.

Malet, Marian, and Anthony Grenville (eds), *Changing Countries: The Experience and Achievement of German-Speaking Exiles from Hitler to Britain, from 1933 to Today* (London, 2002).

Maranta, Alessandro, Michael Guggenheim, Priska Gisler, and Christian Pohl, 'The Reality of Experts and the Imagined Lay Person', *Acta Sociologica*, 46. 2 (2003), 150–65.

Mariner, Rosemary, and Kurt Piehler (eds), *The Atomic Bomb and American Society: New Perspectives* (Knoxville, 2009).

Mason, Katrina, *Children of Los Alamos: An Oral History of the Town Where the Atomic Age Began* (New York, 1995).

McInerney, Claire, Nora Bird and Mary Nucci, 'The Flow of Scientific Knowledge from Lab to the Lay Public: The Case of Genetically Modified Food', *Science Communication*, 26. 1 (2004), 44–74.

McMillan, Priscilla, *The Ruin of J. Robert Oppenheimer: And the Birth of the Modern Arms Race* (New York, 2005).

Medawar, Jean, and David Pyke, *Hitler's Gift: Scientists Who Fled Nazi Germany* (London, 2000).

Metzler, Gabriele, *Internationale Wissenschaft und nationale Kultur: Deutsche Physiker in der internationalen Community 1900–1960* (Göttingen, 2000).

Minichino, Camille, ed., *History and Reflections of Engineering at Lawrence Livermore National Laboratory 1952–2002* (Livermore, 2002).

Moser, John, *Twisting the Lion's Tail: American Anglophobia between the World Wars* (New York, 1999).

Moss, Norman, *Klaus Fuchs: The Man Who Stole the Atom Bomb* (New York, 1987).

Nahum, Andrew, '"I believe the Americans have not yet taken them all!": The Exploitation of German Aeronautical Science in Postwar Britain', in *Tackling Transport*, ed. by Helmuth Trischler and Stefan Zeilinger (London, 2003), pp. 99–138.

Nash, Gerald, *The American West Transformed: The Impact of the Second World War* (Bloomington, 1985).

Ndiaye, Pap, transl. by Elborg Forster, *Nylon and Bombs: DuPont and the March of Modern America* (Baltimore, 2007).

Nehring, Holger, 'Cold War, Apocalypse and Peaceful Atoms: Interpretations of Nuclear Energy in the British and West German Anti-Nuclear Weapons Movements, 1955–1964', *Historical Social Research*, 29. 3 (2004), 150–70.

——, 'National Internationalists: British and West German Protests against Nuclear Weapons, the Politics of Transnational Communications and the Social History of the Cold War, 1957–1964', *Contemporary European History*, 14. 4 (2005), 559–82.

Neufeld, Michael, 'Wernher von Braun, the SS, and Concentration Camp Labor: Questions of Moral, Political, and Criminal Responsibility', *German Studies Review*, 25. 1 (2001), 57–78.

Nicosia, Francis, 'Nazi Persecution in Germany and Austria, 1933–1939', in *The Holocaust: Introductory Essays*, ed. by David Scrase and Wolfgang Mieder (Burlington, 1996), pp. 51–64.

Norris, Robert, *Racing for the Bomb: General Leslie R. Groves, the Manhattan Project's Indispensable Man* (South Royalton, 2002).

Nye, David, *American Technological Sublime* (Cambridge, MA, 1994).

——, *Narratives and Spaces: Technology and the Construction of American Culture* (New York, 1997).

Nye, Mary Jo, *Blackett: Physics, War, and Politics in the Twentieth Century* (Cambridge, MA, 2004).

Omrod, David, 'The Churches and the Nuclear Arms Race, 1945–1985', in *Campaigns for Peace: British Peace Movements in the Twentieth Century*, ed. by Richard Taylor and Nigel Young (Manchester, 1987), pp. 189–220.

O'Neill, Dan, *The Firecracker Boys* (New York, 1994).

Palmowski, Jan, 'The Politics of the "Unpolitical German": Liberalism in German Local Government, 1860–1880', *Historical Journal*, 42. 3 (1999), 675–704.

Panayi, Panikos, 'An Intolerant Act by an Intolerant Society: The Internment of Germans in Britain During the First World War', in *The Internment of Aliens in Twentieth Century Britain* (see Burletson, above), pp. 53–75.

Panitz, Eberhard, *Treffpunkt Banbury oder wie die Atombombe zu den Russen kam: Klaus Fuchs, Ruth Werner und der größte Spionagefall der Geschichte* (Berlin, 2003).

Paul, Gerhard, 'Die Gestapozentrale in der Düppelstraße 23: Die Zentrale des NS-Terrors in Schleswig-Holstein', in *Täter und Opfer unter dem Hakenkreuz: Eine Landespolizei stellt sich der Geschichte*, ed. by Förderverein 'Freundeskreis zur Unterstützung der Polizei Schleswig-Holstein e.V.' (Kiel, 2001), pp. 43–50.

Paul, Septimus H., *Nuclear Rivals: Anglo-American Atomic Relations, 1941–1952* (Columbus, 2000).

Peukert, Detlev, *Die Weimarer Republik: Krisenjahre der Klassischen Moderne* (Frankfurt a.M., 1987).

Phythian, Mark, 'CND's Cold War', *Contemporary British History*, 15. 3 (2001), 133–56.

Pickering, Andrew, *The Mangle of Practice: Time, Agency and Science* (Chicago, 1995).

Pincher, Chapman, *Too Secret Too Long* (London, 1984).

——, *Traitors: The Anatomy of Treason* (New York, 1987).

Porter, Dilwyn, '"Never-Never Land": Britain under the Conservatives 1951–1954', in *From Blitz to Blair: A New History of Britain Since 1939*, ed. by Nick Tiratsoo (London, 1998), pp. 102–31.

Price, Derek de Solla, *Little Science, Big Science* (New York, 1963).

Price, Matt, 'Roots of Dissent: The Chicago Met Lab and the Origins of the Franck Report', *Isis*, 86. 2 (1995), 222–44.

Pröss, Helge, *Die Deutsche Akademische Emigration nach den Vereinigten Staaten, 1933–41* (Berlin, 1955).

Rees, Duncan, 'Resisting the British Bomb: The Early Years', in *The British Nuclear Weapons Programme 1952–2002*, ed. by Douglas Holdstock and Frank Barnaby (London: Cass, 2003), pp. 56–63.

Reynolds, David, *The Creation of the Anglo-American Alliance 1937–41: A Study in Competitive Co-Operation* (London, 1981).

———, *Britannia Overruled: British Policy & World Power in the 20th Century* (London, 1991).

———, 'Great Britain', in *The Origins of the Cold War in Europe: International Perspectives*, ed. by David Reynolds (New Haven, 1994), pp. 77–95.

Rhodes, Richard, *Dark Sun: The Making of the Hydrogen Bomb* (New York, 1995).

———, *The Making of the Atomic Bomb* (New York, 1986).

Richelson, Jeffrey, *Spying on the Bomb: American Nuclear Intelligence from Nazi Germany to Iran and North Korea* (New York, 2006).

Rider, Robin, 'Alarm and Opportunity: Emigration of Mathematicians and Physicists to Britain and the United States, 1933–1945', *HSPS*, 15 (1985), 101–76.

Rieger, Bernhard, *Technology and the Culture of Modernity in Britain and Germany 1890–1945* (Cambridge, 2005).

Ringer, Fritz, *The Decline of the German Mandarins: The German Academic Community, 1890–1933* (Cambridge, MA, 1969).

———, '*Bildung*: The Social and Ideological Context of the German Historical Tradition', *History of European Ideas*, 10. 2 (1989), 193–202.

Rife, Patricia, *Lise Meitner and the Dawn of the Nuclear Age* (Boston, 1999).

Ritchie, James, *German Exiles: British Perspectives* (New York, 1997).

Rose, Sonya, *Which People's War? National Identity and Citizenship in Wartime Britain 1939–1945* (Oxford, 2003).

Rosenberg, Jonathan, 'Before the Bomb and After: Winston Churchill and the Use of Force', in *Cold War Statesmen Confront the Bomb: Nuclear Diplomacy since 1945*, ed. by John Lewis Gaddis, Philip Gordon, Ernest May, and Jonathan Rosenberg (Oxford, 1999), pp. 171–93.

Rosenhaft, Eve, *Beating the Fascists?: The German Communists and Political Violence, 1929–1933* (Cambridge, 1983).

Rossi, John, 'The British Reaction to McCarthyism, 1950–54', *Mid-America*, 70. 1 (1988), 5–18.

Rotblat, Joseph, *Scientists in the Quest for Peace: A History of the Pugwash Conferences* (Cambridge, MA, 1972).

Rothman, Hal, *On Rims & Ridges: The Los Alamos Area since 1880* (1992; repr. Lincoln, 1997).

Rowlands, Peter, *120 Years of Excellence: The University of Liverpool Physics Department 1881 to 2001* (Liverpool, 2001).

Schell, Bernhardt, 'Introduction', in Lohmeier and Schell, pp. 13–87.

Schirrmacher, Arne, 'Physik und Politik in der frühen Bundesrepublik Deutschland: Max Born, Werner Heisenberg und Pascual Jordan als politische Grenzgänger', *Ber. Wissenschaftsgesch.*, 30. 1 (2007), 13–31.

Schrecker, Ellen W., 'Before the Rosenbergs: Espionage Scenarios in the Early Cold War', in *Secret Agents* (see Stanley Goldberg, above), pp. 127–41.

Schrecker, Ellen, *No Ivory Tower: McCarthyism and the Universities* (New York, 1986).

Schweber, Silvan, *In the Shadow of the Bomb: Oppenheimer, Bethe, and the Moral Responsibility of the Scientist* (Princeton, 2000).

——, 'The Mutual Embrace of Science and the Military: ONR and the Growth of Physics in the United States after World War II', in *Science, Technology and the Military* (see Galison, 'Physics Between War and Peace', above), pp. 1–45.

Seidel, Robert, *Los Alamos and the Development of the Atomic Bomb* (Los Alamos, 1995).

Seyfert, Michael, '"His Majesty's Most Loyal Internees": The Internment and Deportation of German and Austrian Refugees as "Enemy Aliens". Historical, Cultural and Literary Aspects', in *Exile in Great Britain* (see Carsten, above), pp. 163–93.

Shaw, Tony, *British Cinema and the Cold War: The State, Propaganda and Consensus* (London, 2001).

Sherwin, Martin, *A World Destroyed: The Atomic Bomb and the Grand Alliance* (New York, 1975).

Shor, Elizabeth Noble, 'Kistiakowsky, George Bogdan (18 Nov. 1900–7 Dec. 1982)', in *American National Biography*, ed. by John A. Garraty and Mark C. Carnes, 24 vols (Oxford, 1999), XII, 776–8.

Sime, Ruth Lewin, *Lise Meitner: A Life in Physics* (Berkeley, 1996).

Smith, Alice Kimball, *A Peril and Hope: The Scientists' Movement in America, 1945–1947* (Chicago, 1965).

Snowman, Daniel, *The Hitler Émigrés: The Cultural Impact on Britain of Refugees from Nazism* (London, 2002).

Staples, Amy, *The Birth of Development: How the World Bank, Food and Agriculture Organization, and World Health Organization Changed the World, 1945–1965* (Kent, 2006).

Stent, Ronald, *A Bespattered Page? The Internment of His Majesty's "Most Loyal Aliens"* (London, 1980).

——, 'Jewish Refugee Organisations', in *Second Chance* (see Hirschfeld, '"A High Tradition of Eagerness ...,"' above), pp. 579–598.

Stern, Fritz, *The Failure of Illiberalism: Essays on the Political Culture of Modern Germany* (New York, 1972).

Strauss, Herbert, 'The Movement of People in a Time of Crisis', in *The Muses Flee Hitler: Cultural Transfer and Adaptation 1930–1945*, ed. by Jarrell C. Jackman and Carla M. Borden (Washington, 1983), pp. 45–59.

Stuewer, Roger, 'Nuclear Physics in a New World: The Émigrés of the 1930s in America', *Ber. Wissenschaftsgesch.*, 7. 1 (1984), 23–40.

Szasz, Ferenc, *The Day the Sun Rose Twice: The Story of the Trinity Site Explosion July 16, 1945* (Albuquerque, 1984).

——, *British Scientists and the Manhattan Project: The Los Alamos Years* (New York, 1992).

——, *Larger Than Life: New Mexico in the Twentieth Century* (Albuquerque, 2006).

Szasz, Margaret Connell, 'Introduction', in *Between White and Indian Worlds: The Cultural Broker*, ed. by Szasz (Norman, 1994), pp. 3–20.

Taylor, Richard, *Against the Bomb: The British Peace Movement, 1958–1965* (Oxford, 1988).

——, 'The Labour Party and CND: 1957 to 1984', in *Campaigns for Peace* (see Omrod, above), pp. 100–30.

Thorpe, Charles, *Oppenheimer: The Tragic Intellect* (Chicago, 2006).

Thorpe, Charles, and Steven Shapin, 'Who Was J. Robert Oppenheimer? Charisma and Complex Organization', *Social Studies of Science*, 30. 4 (August 2000), 545–90.

Thurlow, Richard, *The Secret State: British Internal Security in the Twentieth Century* (Oxford, 1994).

Tomlinson, Jim, 'Reconstructing Britain: Labour in Power 1945–1951', in *From Blitz to Blair: A New History of Britain Since 1939*, ed. by Nick Tiratsoo (London, 1998), pp. 77–101.

Trahair, Richard, 'A Psychohistorical Approach to Espionage: Klaus Fuchs (1911–1988)', *Mentalities*, 9. 2 (1994), 28–49.

Traweek, Sharon, 'Big Science and Colonialist Discourse: Building High-Energy Physics in Japan', in *Big Science* (see Hoddeson, above), pp. 100–28.

Turchetti, Simone, 'Atomic Secrets and Governmental Lies: Nuclear Science, Politics and Security in the Pontecorvo Case', *British Journal for the History of Science*, 36. 4 (2003), 389–415.

——, '"For Slow Neutrons, Slow Pay": Enrico Fermi's Patent and the U.S. Atomic Energy Program, 1938–1953', *Isis*, 97. 1 (2006), 1–27.

Uhlig, Ralph (ed.), *Vertriebene Wissenschaftler der Christian-Albrechts-Universität zu Kiel (CAU) nach 1933: Zur Geschichte der CAU im Nationalsozialismus – Eine Dokumentation bearbeitet von Uta Cornelia Schmatzler und Matthias Wieben* (Frankfurt a. M., 1991).

VanDeMark, Brian, *Pandora's Keepers: Nine Men and the Atomic Bomb* (New York, 2003).

Vincent, David, *The Culture of Secrecy: Britain, 1832–1998* (Oxford, 1998).

Vinzent, Jutta, *Identity and Image: Refugee Artists from Nazi Germany in Britain 1933–1945* (Weimar, 2006).

Walker, Mark, *German National Socialism and the Quest for Nuclear Power, 1939–49* (Cambridge, 1989).

——, 'Legenden um die deutsche Atombombe', *Vierteljahrshefte für Zeitgeschichte*, 38. 1 (1990), 45–74.

——, *Nazi Science: Myth, Truth, and the German Atomic Bomb* (Cambridge, MA, 1995).

Walker, Samuel, *Prompt and Utter Destruction: Truman and the Use of Atomic Bombs against Japan*, rev. edn (Chapel Hill, 2004).

Wallace, Ian (ed.), *German-Speaking Exiles in Great Britain* (Amsterdam, 1999).

Wang, Jessica, *American Science in an Age of Anxiety: Scientists, Anticommunism and the Cold War* (Chapel Hill, 1999).

Wasserstein, Bernard, 'Intellectual Émigrés in Britain, 1933–1939', in *The Muses Flee Hitler* (see Strauss, 'Movement of People', above), pp. 251–2.

Weart, Spencer, *Nuclear Fear: A History of Images* (Cambridge, MA, 1988).

Weiler, Peter, *British Labour and the Cold War* (Stanford, 1988).

Weiner, Charles, 'A New Site for the Seminar: The Refugees and American Physics in the 1930s', in *The Intellectual Migration: Europe and America, 1930–1960*, ed. by Donald H. Fleming and Bernard Baylin (Cambridge, MA, 1969), pp. 190–233.

Weldes, Jutta, Mark Laffey, Hugh Gusterson, and Raymond Duvall, 'Introduction: Constructing Insecurity', in *Cultures of Insecurity: States, Communities, and the Production of Danger*, ed. by Jutta Weldes, Mark Laffey, Hugh Gusterson, and Raymond Duvall (Minneapolis, 1999), pp. 1–33.

West, Nigel, *Mortal Crimes: The Greatest Theft in History: The Soviet Penetration of the Manhattan Project* (New York, 2004).

Williams, Bill, '"Displaced Scholars": Refugees at the University of Manchester', *Melilah: Manchester Journal of Jewish Studies*, 3 (2005) <http://www.mucjs.org/MELILAH/2005/3.pdf> [accessed 10 August 2011], 1–29.

Williams, M. M. R., 'The Development of Nuclear Reactor Theory in the Montreal Laboratory of the National Research Council of Canada (Division of Atomic Energy) 1943–1946', *Progress in Nuclear Energy*, 36. 3 (2000), 239–322.

Williams, Robert, *Klaus Fuchs, Atom Spy* (Cambridge, MA, 1987).

Willis, Kirk, 'The Origins of British Nuclear Culture, 1895–1939', *Journal of British Studies*, 34. 1 (1995), 59–89.

——, '"God and the Atom": British Churchmen and the Challenge of Nuclear Power 1945–1950', *Albion*, 29. 3 (1997), 422–57.

Winkler, Allan, *Life under a Cloud: American Anxiety about the Atom* (New York, 1993; repr. Urbana, 1999).

Wirth, John, and Linda Harvey Aldrich, *Los Alamos: The Ranch School Years, 1917–1943* (Albuquerque, 2003).

Wittner, Lawrence, *The Struggle Against the Bomb*, 3 vols (Stanford, 1993–2003).

Wolff, Stefan, 'Frederick Lindemanns Rolle bei der Emigration der aus Deutschland vertriebenen Physiker', in *German-Speaking Exiles in Great Britain*, ed. by Anthony Grenville (Amsterdam, 2000), pp. 25–58.

Wulf, Peter, 'Die Stadt auf der Suche nach ihrer neuen Bestimmung (1918 bis 1933)', in *Geschichte der Stadt Kiel*, ed. by Jürgen Jensen and Peter Wulf (Neumünster, 1991), pp. 303–58.

——, 'Zustimmung, Mitmachen, Verfolgung und Widerstand – Schleswig-Holstein in der Zeit des Nationalsozialismus', in *Geschichte Schleswig-Holsteins: Von den Anfängen bis zur Gegenwart*, ed. by Ulrich Lange, 2nd edn (Neumünster, 2003), pp. 585–621.

Wuttke, Dieter, 'Die Emigration der Kulturwissenschaftlichen Bibliothek Warburg und die Anfänge des Universitätsfaches Kunstgeschichte in Großbritannien', *Ber. Wissenschaftsgesch.*, 7. 3 (1984), 179–94.

York, Herbert, *The Advisors: Oppenheimer, Teller & the Superbomb* (San Francisco, 1976).

Zeman, Scott, and Michael Amundson (eds), *Atomic Culture: How We Learned to Stop Worrying and Love the Bomb* (Boulder, 2004).

Zimmerman, David, 'The Tizard Mission and the Development of the Atomic Bomb', *War in History*, 2. 3 (1995), 259–73.

——, 'The Society for the Protection of Science and Learning and the Politicization of British Science in the 1930s', *Minerva*, 44. 1 (2006), 25–45.

Index

Academic Assistance Council (AAC), 18, 20–1, 26, 110
Acheson, Dean, 142–4, 147
Acheson-Lilienthal Report, 143
  see also international control of atomic energy
Adams, Walter, 19
Adenauer, Konrad, 156
Advisory Committee on Atomic Energy, 147–8, 159–60
Air Ministry (AM), 41–2
Akers, Wallace, 48–9, 52–3, 95
  see also Tube Alloys (TA)
Aldermaston, see Atomic Weapons Research Establishment (AWRE) Aldermaston
Alison, Samuel, 65
Allen, Elizabeth, 119
Allen, H.R., 170
Allen, Jack, 17
ALSOS teams, 127
Anderson, John, 26, 142–3, 145, 147–9, 160
Anderson, Oscar Jr., 68
Anglican Church, 137
  see also British Council of Churches
Anschütz-Kaempfe, Herrmann, 84
  see also Einstein, Albert
  see also gyro compass
anti-Semitism, 3, 22, 65
Appleton, Edward, 53, 147
Argonne National Laboratories (Illinois; US), 78
Arlan-Baykov, Inna, 116
Arms, H.S., 49
Arnold, Lorna, 40
Arnold, Henry, 87–8, 90
Arnold, William, 38
Association of Atomic Scientists, 132
  see also Atomic Scientists' Association
Association of Los Alamos Scientists (ALAS; US), 129–30, 135–6, 145

Association of Scientific Workers (AScW), 118, 131–3, 135, 146, 154, 159–60
Atomic Attack: Can Britain Be Defended? (1950), 159
Atom Train, 102, 139–42, 148, 150, 169
  see also Atomic Scientists' Association (ASA)
Atomic Energy Act (1946), 117, 152
Atomic Energy Commission (AEC; US), 122, 140, 143
Atomic Energy Research Establishment (AERE) Harwell, 8, 10–11, 28, 46, 79–80, 90, 94, 139–40, 148, 166, 173, 175
Atomic Energy Weeks, 140
  see also Atom Train
Atomic Physics (1947), 140
Atomic Scientists' Association (ASA), 6, 8, 11, 117–18, 121, 125–71, 174
  advising political decision-makers, 147–50
  American origins of, 126–31
  Atom Train, 102, 139–42, 148, 150, 169
  Atomic Age, 138
  Atomic Scientists' Journal (ASJ), 121–2, 138, 169–70
  Atomic Scientists' News (ASN), 118, 121–2, 138, 149, 152–3, 157–60, 162, 169–70
  civil service purge (1948), 90, 115, 118, 160
  confronts hydrogen bomb, 159–64
  creation of, 131–5
  disbandment of, 151–71
  international control of atomic energy, 6, 11, 119–20, 129–30, 134, 141–7, 158–60, 164, 166–8, 174
  Nuclear Age, 170

257

Atomic Scientists' Association – *continued*
  Nuclear Age Society, 170
  nuclear testing, 148, 163–7
  public outreach and education,
    135–42
  see also names of individual
    scientists
  see also Peierls, Rudolf
Atomic Scientists' Committee (ASC),
    131–2
Atomic Weapons Establishment
    (AWE) Aldermaston, 6
  see also Atomic Weapons Research
    Establishment (AWRE)
    Aldermaston
Atomic Weapons Research
    Establishment (AWRE)
    Aldermaston, 6, 78, 82
  march (1958), 6
  see also Atomic Weapons
    Establishment (AWE)
    Aldermaston
Attlee, Clement, 7, 95, 99, 107, 136,
    144, 146
Australia, 9, 17, 25, 27, 135
Australian Association of Scientific
    Workers, 135
Austria, 4, 27–8, 66, 114, 119
Austro-Hungarian Empire, 4, 65
  see also Austria
  see also Czechoslovakia
  see also Hungary
Auxiliary Fire Service, 24

Bacher, Robert, 65
Bainbridge, Kenneth, 65
Baker Test, *see* Bikini Atoll
    (South Pacific)
Banbury, 86, 175
Baruch, Bernard, 143–5, 167
  see also international control of
    nuclear energy
Baruch Plan, *see* Baruch, Bernard
Barwich, Heinz, 175
  see also Fuchs, Klaus
  see also Rossendorf (GDR)
Bates, L.F., 147, 166–7
Baumgarten, Otto, 14
Baykov, Alexander, 116
Baylis, John, 54

Beck, Ulrich, 6
Beirut (Lebanon), 139
Belgium, 12, 26, 166
Bennett, Curtis, 83, 97
Berg, Wolfgang, 20
Berghahn, Marion, 16
Bergman, Ingrid, 103
Berlin, 20, 23, 32, 34, 67, 85, 105,
    109, 146
  Berlin blockade, 105
  Berlin University, 32, 34, 69, 85
  Physikalisch Technisches Institut,
    20
  West Berlin, 109
Bernal, John D., 100, 131
Bethe, Hans, 4, 12–13, 17–19, 21,
    32–3, 36, 40, 58–60, 63–6,
    67–71, 72–4, 91, 114–15, 121,
    152, 157, 161, 168
Bethe, Rose, 59
Beyler, Richard, 119
  see also Kojevnikov, Alexei
  see also Wang, Jessica
Bhaba, Homi, 135
Big Science, 3, 11, 56, 62, 68, 76–7,
    79, 81, 173
Bikini Atoll (South Pacific), 76, 129,
    136, 141, 165
  Baker Test, 141
  see also nuclear testing
Birkbeck College, 33
Birmingham, 11, 16, 18, 21, 24, 30,
    32, 33, 36–8, 42, 44, 49, 61,
    69, 80, 82, 91, 134, 138, 173
  Birmingham University, 11, 18,
    21, 30, 32–3, 36–8, 42, 44,
    49, 61, 69, 80, 82, 91, 134,
    137–8, 173
Blackett, Patrick, 41, 131, 134, 144,
    146–7, 159–60
  *Military and Political Consequences of
    Atomic Energy* (1948), 159
Blitz, 39, 113
Bloch, Felix, 4, 58, 60, 66–7
Boag, J.W., 165
Bohm, David, 121
Bohr, Aage, 59
Bohr, Niels, 17, 36, 38, 41, 43, 59, 67,
    123, 129, 145
  Institute in Copenhagen, 32, 38

Bondi, Hermann, 28
Born, Hedwig, 36
Born, Max, 3, 4, 15, 17–19, 21–2,
    25–7, 30, 33, 35–6, 59, 94–5,
    112, 116, 156
Boulting, Roy, 102
    High Treason (1951), 102
Brabazon, Lord, 135
Bradbury, Norris, 75, 76, 112
Bragg, Lawrence, 13, 17–18
Brandt, Willy, *see* Frahm, Herbert
Breslau, 45
    Technische Hochschule
        (Polytechnic Institute), 45
Bretscher, Egon, 4, 33, 42–4, 48, 54,
    59–60, 63–4, 66, 68, 79–80,
    90, 115, 123, 129–30, 173
Bretscher, Hanni, 89–90
Bristol, 20, 22, 27, 32, 33, 42, 46,
    48–9, 94–5, 115, 130
    Bristol University, 27, 32, 33, 42,
        46, 48–9, 95, 115
    see also Mott, Nevill
    see also Wills Laboratory
    German Consulate, 20, 22
Britain
    and anti-communism, 99
    and atomic weapons decision, 7,
        101, 136, 144, 163
    relations with United States, 2–3, 9,
        10, 49–81, 92, 101–08, 120,
        124, 160, 172–3
    Extradition Treaty, 107
    intelligence relations, *see*
        Military Intelligence, Section
        5 (MI5)
    nuclear cooperation with United
        States, 2–3, 9, 10, 49–81, 92,
        101, 106, 108, 120, 124, 160,
        172–3
    see also McMahon Act (1946)
    and nuclear testing, 164
    see also nuclear testing
    and thermonuclear programme, 49,
        79, 101
    as third atomic power, 7
    as third thermonuclear power, 7
    see also Atomic Scientists'
        Association (ASA)
    see also British nuclear culture

    see also Fuchs, Klaus
    see also international control of
        nuclear energy
    see also Peierls, Rudolf
British Association for the
    Advancement of Science
    (BA), 135, 164, 169–70
    cooperation with Atomic Scientists'
        Association, 164, 170–71
    Division for Social and
        International Relations,
        170–71
British Council of Churches,
    135, 137
    Atomic Energy Committee, 135
    The Era of Atomic Power (1946),
        137
    see also Anglican Church
British Museum, 140
British Empire, 7, 9, 45
British nuclear culture
    atomic imperialism, 9
    Britishness of, 6–9
    and Cold War context, 3, 6, 7, 10,
        82–171, 174–5, 177
    'cultures of insecurity' (Weldes and
        others), 6
    definition of, 5
    and impact on political and social
        cultures, 2, 5–6, 82–171
    origins of, 6–7
    pluralism of, 8
    and 'scientific culture' (Pickering),
        5, 7, 12–81
    and Second World War context, 2,
        5, 7, 9, 11, 12–81
    and inherent tension between
        modernity and tradition,
        7–9
    United States as reference culture
        for, 9–10
    Willis, Kirk and definition of, 5
    see also Britain
    see also Fuchs, Klaus
    see also Peierls, Rudolf
British Thomson-Houston Co. Ltd.,
    140
Broda, Engelbert, 123
Bryant, Arthur, 113
Budapest (Hungary), 45, 65

Bulletin of the Atomic Scientists (BAS;
  US), 77, 105, 118, 121, 131,
  134, 138, 144, 146, 153, 157,
  158, 162, 163, 167
Bundeswehr, 156
Bünemann, Oskar, 27–8, 60, 78
Burgess, Guy, 95, 108
Burhop, Eric, 121, 131, 134, 144, 159–61
Bush, Vannevar, 51, 56, 106
Byrnes, James, 143

Cadogan, Alexander, 146
Cairo (Egypt), 139
Calder Hall reactors, 8
California, 38, 78
  University of California, Berkeley,
    38, 78, 60
  see also Bünemann, Oskar
Cambridge University, 13, 18, 32–3,
    38, 42–4, 48–9, 80, 144
  Mond Laboratory, 18
  see also Cavendish Laboratory
Campaign for Nuclear Disarmament
    (CND), 11, 171
Camus, Albert, 40
Canada, 9, 14, 30, 25, 26, 27, 28, 29,
    48, 53, 57, 60, 63, 78, 86,
    129, 172
Carnegie Foundation, 21
  see also Fuchs, Klaus
Carnegie Institution, 38
Carson, Cathryn, 156
Cassidy, David, 3
Castle Bravo Test (1954; Pacific), 163
  see also Lucky Dragon incident
  see also nuclear testing
Cavendish Laboratory, 13, 32, 43, 80,
    123
  see also Cambridge University
Central British Fund for German
    Jewry, 20
Central Intelligence Agency (CIA; US),
    175
Cesarani, David, 25
Chackett, K.E., 134
Chadwick, James, 35, 40, 41, 42, 44,
    47, 48, 50–2, 63, 75, 78, 80,
    118, 130, 132, 135, 140,
    147–9

Chalk River (Canada), 60, 78
  see also Freundlich, Herbert
  see also Kemmer, Nikolai
  see also Kowarski, Lew
  see also Manhattan Engineering
    District
  see also von Halban, Hans
Chamberlain, Neville, 26, 41
Champion, F.C., 119, 139, 149, 158,
    160–1
Chatham House, 164
Chernobyl nuclear accident, 6
Cherwell, Lord, 20, 132, 160
Chicago (Illinois; US), 46, 65, 67, 78,
    120, 127, 128, 131, 132
  Atomic Scientists of Chicago, Inc.,
    129, 135
  International Conference on
    Nuclear Physics (1951), 120
  University of Chicago, 46, 65, 127,
    131
    Metallurgical Laboratory
      (Met Lab), 65, 127
    Committee on Social and Political
      Implications, 129
  *see also Bulletin of the Atomic
    Scientists* (BAS; US)
  see also Fermi, Enrico
  see also Franck, James
  see also Szilard, Leo
Christy, Robert, 74, 129
  Christy Gadget, 74
Church of England, *see* Anglican
  Church
Critchfield, Charles, 74
Churchill, Winston, 26, 47, 52, 53,
    128–9, 148
Civil Defence Corps, 24
Clarendon Laboratory, 20, 45, 49,
    132, 151
  see also Oxford
Clark, Ronald, 44
Cockcroft, John, 41, 48, 50, 135, 140,
    147, 149, 164, 165, 166–7,
    169–70
Cold War, 3, 4, 6, 7, 10, 82, 95, 110,
    152, 174, 175, 177
Columbia University (New York; US),
    38, 57, 65, 67, 76

Commoner, Barry, 167
Commonwealth, 9, 15, 17
Communist Party of Germany
    (Kommunistische Partei
    Deutschlands), *see* KPD
Communist Workers' School
    (New York City; US), 115
Communist Youth Association of
    Germany (Kommunistischer
    Jugendverband Deutschlands;
    KJVD), 85
Compton, Arthur, 51, 128
Conant, James B., 53, 56, 65, 106
Condon, Edward U., 67, 167
Connell Szasz, Margaret, 70
Conspirator (1949), 106
Conservative Party, 8, 153, 162
    cross-party consensus on atomic
        bomb, 8
Copenhagen (Denmark), 32, 28
Cornell University (Ithaca, New York;
    US), 63, 64, 69
Corson, Edward, 115
Council for German Jewry, 20
Courant, Richard, 33
Coventry, 130
Curry, John, 139
Czechoslovakia, 4, 12

Daily Express, 39, 97–8, 123
Daily Mail, 113
Daily Mirror, 162
Daily Worker, 99
Dale, Henry, 118, 135, 147
Darwin, Charles Galton, 48, 147
Deacon, Richard, 123
    *The British Connection: Russia's
        Manipulation of British
        Individuals and Institutions*
        (1979), 123
Debye, Peter, 35
Dempsey, John, 40
Department of Scientific and
    Industrial Research (DSIR),
    32, 48, 53, 140
Deprivation of Citizenship
    Committee, 99
    see also Fuchs, Klaus
Deutsch, Martin, 60, 66–7, 71, 114–5

Devons, S., 134, 144
Dietz, David, 137
    Atomic Energy in the Coming Era
        (1945), 137
    Atomic Energy Now and Tomorrow
        (1946), 137
Dirac, Paul, 17, 42, 49, 116
Dispatch, 26
Dobb, Maurice, 100
Dunkirk, 53
DuPont, 77
Dutch International Council of
    Scientific Unions, 135

Economist, 119–20, 159
Edgerton, David, 77
Edinburgh, 15, 17, 21, 24, 25, 26, 29,
    33, 36, 94, 95
    Edinburgh University, 15, 17, 21,
        26, 29, 30, 33, 36, 94
    see also Born, Max
    see also Fuchs, Klaus
*The Effects of the Atomic Bombs at
    Hiroshima and Nagasaki*
    (1946), 136, 138
Eichholz, G., 27
Eidgenössische Technische
    Hoschschule (ETH;
    Switzerland), 32
Einstein, Albert, 18, 19, 33–6, 39, 84,
    129, 131, 145, 153, 157, 163,
    176
    'Einstein Letter' (1939), 36, 39
    Einstein-Russell Manifesto (1955),
        163
    on anti-Semitism in Kiel, 84
    see also Russell, Bertrand
Eisenhower, Dwight D., 77
Elizabeth II, Queen, 8
    see also Calder Hall reactors
Ellis, Charles, 41
Elsaesser, Thomas, 15, 62
Elvey, Maurice, 7
    see also Kellermann, Bernhard
    The Tunnel (1935), 7
Emergency Committee of Atomic
    Scientists (ECAS; US), 131
émigré, concept of, 4
Endeavour, 136–7, 140

Epstein, Samuel, 99
  The Real Book of Spies (1959), 99
  see also Williams, Beryl
Evening Standard, 98
Ewald, Paul, 33

fallout, 163–7
  see also Atomic Scientists'
    Association (ASA)
  see also nuclear testing
  see also strontium-90
Fat Man implosion device, 75
  see also Manhattan Engineering
    District (MED)
  see also Trinity Test (1945; US)
Feather, Norman, 43, 44, 48, 147,
  167
Federal Bureau of Investigation (US;
  FBI), 86, 87, 91, 92, 101–2,
  105–7, 114–15, 117–18, 121
  investigation of Fuchs, Klaus,
    102–8
  investigation of Peierls, Rudolf,
    113–17
  see also Fuchs, Klaus
  see also Military Intelligence,
    Section 5 (MI5)
  see also Peierls, Rudolf
Federation of American Scientists
  (FAS), 121, 131, 132, 135,
  145, 146, 164, 166, 168
Feklisov, Alexander, 92
Fermi, Enrico, 13, 46, 54, 58, 59, 64,
  67, 68, 118
Feynman, Richard, 62, 64, 114, 129,
  152
Finley, Moses I., 100
First World War, 65, 125
Fishenden, R.M., 148
fission, 37–8, 43, 46–7, 60, 63–4, 71,
  78, 72
  see also Arnold, William
  see also Frisch, Otto
  see also Hahn, Otto
  see also Meitner, Lise
  see also Strassmann, Fritz
Food and Agriculture Organization,
  126
Ford, C.E., 165
Foreign Office (FO), 121, 143, 175

Forman, Paul, 155
Frahm, Herbert, 23
France, 12, 13, 19, 29, 44, 85, 109,
  116, 135
  French Atomic Energy Commission,
    146
  Free French movement, 52
Franck, James, 127–9
  Franck Report, 127–8
Frank, Lorenz, 32
Frankel, Stanley, 74
French, Anthony, 79, 130
Freundlich, Herbert, 18, 43, 60
Friedmann, Ronald, 4
Frisch, Otto, 2, 3, 13, 16, 18, 21, 24,
  33, 36–41, 42, 43, 44, 45–7,
  49, 50, 55, 59, 60, 61, 63, 64,
  66, 67–9, 75, 112, 135, 154,
  169, 172–3
  confirmation of fission, 37–8
  see also Arnold, William
  see also Meitner, Lise
  'Frisch-Peierls Memorandum', 2,
    36–41, 42, 45, 46, 50, 55, 69,
    149
  tickling the dragon's tail
    experiment, 63, 72
'Frisch-Peierls Memorandum', *see*
  Frisch, Otto *and* Peierls,
  Rudolf
Fröhlich, Herbert, 12, 18, 19, 21, 27,
  28, 32, 34, 42, 46
Fuchs, Elisabeth, 14–15, 84–5, 107
Fuchs, Emil, 14–15, 83–4, 87, 90
  Christ in Catastrophe (1949), 15
Fuchs, Gerhard, 14–15, 23, 84–5
Fuchs, Klaus
  Atomic Lunch (2005), 175
  award of Order of Patriotic Merit,
    175
  Carnegie Foundation fellowship, 21
  emigration to Britain, 1–3, 12, 13
  as 'enemy alien', 15, 23–30, 34, 42,
    54, 172
  espionage for Soviet Union, 82–109
    codenames 'Charles' and 'Rest', 86
    confession of espionage, 88–90
    impact on Rudolf Peierls and
      other German-speaking
      émigrés, 110–24

impact on MI5 – FBI relations, 101–8
impact on perceptions of homeland security, 93–102
and introduction of 'positive vetting', 96
Rudolf and Genia Peierls's reaction to, 111–12
public reaction to, 97–102
see also Perrin, Michael
see also Skardon, William
detection of espionage, 86–8
see also Arnold, Henry
see also Gardner, Meredith
see also 'Gouzenko Affair'
see also Lamphere, Robert
see also Nunn May, Alan
see also Venona (messages)
value of, 90–93
see also Federal Bureau of Investigation (FBI; US)
see also KGB (Soviet Secret Police)
see also Military Intelligence, Section 5 (MI5)
German background of, 1–4, 14–15, 22–3, 83–6, 172–4
internment of, 25–30
at Los Alamos, 56–81
see also Manhattan Engineering Project (MED)
medical treatment in Bahiva (Soviet Union), 175
as member of the Academy of Science (GDR), 175
as member of Central Committee of the Socialist Unity Party (SED), 175
as member of GDR Research Council, 175
as member of German-Soviet Friendship Society, 175
motivations to engage in work on atom bomb, 35–6
and move to GDR, 3, 11, 86, 175
multiple allegiances of, 100–2
new approach to nuclear science, 68–76
in New York City, 57–8, 91

see also Manhattan Engineering Project (MED)
Nuclear Secrets (2007), 175
as part of political emigration, 12, 22
planned extradition to United States, 107
plutonium implosion bomb, 70–6
political radicalization in Germany, 83–6, 173
professional integration in Britain, 31–55
social integration in Britain, 12–30
trial of, 96–7, 106
see also Goddard, Lord
see also Shawcross, Hartley
and Tube Alloys (TA), 48–54
as victim of National Socialist persecution, 1–2, 12–23, 83–6, 173
see also Fuchs, Elisabeth
see also Fuchs, Emil
see also Fuchs, Gerhard
views on Stalinism, 88–9
visa extension issues, 22–23
see also Atomic Energy Research Establishment (AERE) Harwell
see also Bristol
see also Edinburgh
see also isotope separation
see also Peierls, Rudolf
see also Rossendorf (Germany)
Fukushima (Japan), 6

Gabor, Dennis, 20
Galison, Peter, 70
Gallup poll, 25, 136, 145, 146, 176
Gardner, Meredith, 86
Garrity, Joseph, 112
Gaumont-British Equipments Ltd., 140
Gaumont-British Instructional, 140
Atomic Physics (1947), 140
Geertz, Clifford, 5
Geheime Staatspolizei (Secret State Police; Germany), *see* Gestapo
German Emergency Committee, 20
German physics (*Deutsche Physik*), 1
German-speaking émigré, concept of, 4
see also names of individual scientists

Germany, 1, 2, 4, 8, 12, 13, 14, 15, 16,
    19, 20, 22, 23, 24, 27, 32, 35,
    36, 37, 39, 53, 56, 65, 68, 69,
    77, 82, 83, 85, 87, 88, 89, 92,
    94, 95, 100, 103, 108, 119,
    127, 151, 154, 155, 156, 166,
    168, 174, 175
  Federal Republic of Germany (FRG),
    23, 156, 166
  German Democratic Republic
    (GDR), 3, 4, 11, 87, 92, 103,
    175
  German Empire (*Kaiserreich*), 154,
    155
  monolithic definition of *Kultur*, 8
  Nazi Germany, 35, 39, 41, 53, 68,
    77
  Weimar Republic, 1, 32, 83, 84, 154,
    155, 156
  see also Hitler, Adolf
  see also National Socialism
Gestapo (Secret State Police;
    Germany), 22–3, 85, 93
  Field Office Kiel, 22
Goddard, Lord, 83, 96–7, 106
Gold, Harry (alias 'Raymond'), 86,
    101, 106
Goldsmith, Hyman, 131
Gollancz, Victor, 22
  Shall Our Children Live or Die?
    (1942), 22
Göppert-Mayer, Maria, 58, 59, 60,
    66, 67
Göttingen, 32, 45, 69, 156
  'Göttingen 18', 156
  'Göttingen Manifesto', 156
  Göttingen University, 32, 45, 69
  see also Born, Max
'Gouzenko Affair', 86
Gowing, Margaret, 36, 39, 83
  Britain and Atomic Energy 1939–1945
    (1964), 39
Grant, Matthew, 149
Greenglass, David, 91, 101, 105–6
Gromyko, Andrei, 143–4, 146
Gross, Philipp, 27, 32
Groves, Leslie, 40, 61, 65, 71, 73, 103,
    106, 127
Guernica (Spain), 39

Gusterson, Hugh, 77
gyro compass, 84

H-bomb, 7, 9, 41, 66, 77, 79, 92–3,
    101, 102, 103, 109, 122,
    146–50, 151, 154, 157,
    159–67
  see also Atomic Scientists'
    Association (ASA)
  see also Britain
  see also Castle Bravo Test (1954;
    Pacific)
  see also fallout
  see also *Lucky Dragon* incident
  see also nuclear testing
  see also strontium, 90
  see also Soviet Union
  see also United States
Haber, Fritz, 125
Haddow, A., 165
Hahn, Otto, 37–8, 43
  see also Arnold, William
  see also fission
  see also Frisch, Otto
  see also Hahn, Otto
  see also Meitner, Lise
  see also Strassmann, Fritz
Haldane, John B.S., 99
Hales, Peter Bacon, 78
Hall, Ted, 91
Hamburg (Germany), 13, 28, 45
  Hamburg University, 13
  see also Stern, Otto
Hamish Hamilton (publisher), 98, 123
Hanford (Washington; US), 49, 57,
    60, 78
  see also Manhattan Engineering
    District (MED)
Harrisburg (Pennsylvania; US), 6
Harteck, Paul, 35
Hartree, Douglas R., 28
Harwell, *see* Atomic Energy Research
    Establishment (AERE) Harwell
Harwell (1952), 108
Haworth, William, 41, 44, 49
Hayek, Friedrich August, 119
  The Road to Serfdom (1944), 119
heavy water, 44, 48
Hecht, Gabrielle, 8

Heisenberg, Werner, 17, 34–5, 63,
    155–7, 168
Heitler, Hans, 27
Heitler, Walter, 18, 19, 27–8, 32, 42, 46
Henriksen, Margot, 8
Hennessy, Peter, 83
Hersey, John, 136
    Hiroshima (1946), 136
Herzberg, Gerhard, 36
Hewlett, Richard, 68
Higgins Boat, 125
Higinbotham, William, 146
Hill, D.L., 146
Himsworth, Harold, 165
    see also Medical Research Council
        (MRC)
Hiroshima (Japan), 6, 40, 41, 61, 124,
    128, 129, 136, 137, 138, 141,
    146, 149, 150, 157
Hiss, Alger, 105
Hitler, Adolf, 1, 12, 13, 14, 16, 22, 23,
    26, 34, 37, 65, 77, 127, 176
    see also National Socialism
    see also Nazi Germany
Hobsbawm, Eric, 100
Hoch, Paul K., 70
Hodgson, 163, 164
Hollywood (California, US), 19, 58,
    97, 106
Home Office, 22, 23, 24, 26, 99, 116
    Aliens Department, 22
Hoover, J. Edgar, 92, 102, 105, 106,
    114
Hoselitz, Kurt, 27, 32
The House on 92nd Street (1945), 97
House of Commons, 107, 164
House on Un-American Activities
        Committee (HUAC; US), 21,
        105
    see also McCarthy, Joseph
    see also McCarthyism
Hughes, James, 130
Hungary, 4, 65
Hunner, Jon, 126
Hyde, Douglas, 99
Hyde, Montgomery, 4
Hyde Park (New York; US), 54
    see also Anglo-American relations
    see also Quebec Agreement

Ice Station Zebra (1968), 176
Illustrated London News, 99, 113
Imperial Chemical Industries (ICI),
    20, 21, 45, 48, 49
Infeld, Leopold, 33
international control of atomic
    energy, 6, 11, 119–20,
    129–30, 134, 141–7, 158–60,
    164, 166–8, 174
    see also Atomic Scientists'
        Association (ASA)
    see also Baruch, Bernard
    see also Gromyko, Andrei
International Trade Organization,
    171
internment, *see* Fuchs, Klaus
Isle of Man, 24, 26–7
isotope separation, 24, 43–6, 49, 51,
    54, 57, 63, 76, 78, 91, 151
    gaseous diffusion, 44–6, 51, 63, 91
    thermal diffusion, 43
    see also uranium
Italy, 27, 70

Jewish Refugee Committee, 20
Johns Hopkins University (Baltimore;
        US), 38
Joliot-Curie, Frédéric, 38, 44, 121
    see also Radium Institute (Paris)
Joliot-Curie, Irène, 121
Jones, Gwyn Owain, 134, 135, 147,
    167
Jordan, Pascual, 168
Joyce, William (alias 'Lord Haw Haw'),
    87

Kahle, Hans, 86
Kaiser, David, 105
Kearton, Christopher Frank, 52, 57,
    87
Kellermann, Bernhard, 7
    The Tunnel (1915), 7
    see also Elvey, Maurice
Kellermann, Walter, 26
Kellex Corporation (New York; US),
    57, 76
Kemmer, Nikolai, 18, 43, 46, 60, 134,
    144
KGB (Soviet Secret Police), 86, 87, 92

Kiel (Germany), 14, 22–3, 83–5, 173
    Kiel University, 14, 83–5, 173
    see also Fuchs, Klaus
Kiribati, Republic of, 9
Kistiakowsky, George, 65, 69, 72
Klemperer, Otto, 20
Klopstech, Hannah, 86
Koestler, Arthur, 99
    The God That Failed (1950), 99
Kojevnikov, Alexei, 119
    see also Beyler, Richard
    see also Wang, Jessica
Korean War, 105, 153
Kowarski, Lew, 42, 44, 48, 52, 116,
    123, 146
KPD (Communist Party of Germany),
    15, 23, 85, 86, 88, 95
Kremer, Simon, 86
Krige, John, 121
Kronberger, Hans, 27, 33
Kubrick, Stanley, 175
    Dr Strangelove (1964), 175
Kuczynski, Jürgen, 86
Kuczynski, Ursula, 86
    see also Werner, Ruth
Kuhn, Heinrich, 18, 19, 20, 45, 49
Kungälv (Sweden), 38
Kurti (Kürti), Nicholas, 18, 20, 21, 45,
    49, 57, 112, 134, 135
Kushner, Tony, 25, 26

Labour Party, 8, 99, 107
    cross-party consensus on atomic
        bomb, 8
Lafitte, François, 25
Lamphere, Robert, 86, 87
Landshoff, Rolf, 60, 66, 67
Laurence, William, 40–1
Law for the Restoration oft he
        Career Civil Service (*Gesetz
        zur Wiederherstellung des
        Berufsbeamtentums*), 1
    see also Hitler, Adolf
    see also National Socialism
Lawrence Livermore National
        Laboratory (California; US),
        78
Lee, Sabine, 4
Leiden (the Netherlands), 32

Leipzig University (Germany), 63, 87
    see also Bethe, Hans
    see also Heisenberg, Werner
    see also Peierls, Rudolf
    see also Placzek, George
Leningrad (Soviet Union), 21
Leonard-Jones, John E., 32
Levitt, W.M., 165
Liberty Ships, 125
Life, 98, 103
Lindemann, Frederick, *see* Cherwell,
    Lord
Lindop, Patricia, 165
Lindsay, Franklin A., 144
Liverpool, 16, 27, 28, 30, 34, 38, 40,
    42, 43, 44, 47, 48, 56, 79,
    132, 139, 144
    Central Station, 139
    Liverpool University, 34, 38, 40, 42,
        43, 44, 47, 48, 79, 132, 139,
        144
London, 6, 16, 20, 34, 45, 55, 86, 98,
    118, 123, 134, 141, 161
    University of London, 45, 169
London, Fritz, 18, 19, 20, 35, 45
London, Heinz, 18, 20, 21, 27, 32, 42,
    45, 46, 49
London, Louise, 23
Lonsdale, Kathleen, 146, 147, 157,
    158, 160, 161, 163, 167, 169
Lord Baldwin Fund, 20
Los Alamos (New Mexico; US), 2, 36,
    39, 40, 49, 54, 56–82, 89, 91,
    92, 101, 103, 106, 112, 113,
    115, 122, 123, 126–31, 136,
    141, 142, 150, 152, 173
    Cowpuncher Committee, 65
    Detector Group, 66
    Fermi Division (F-Division), 64, 66
    Gadget Division (G-Division), 63
    German-speaking scientific
        community at, 56–68
    legacy of wartime research, 76–80
    living conditions at, 58–62
    Los Alamos National Laboratory
        (LANL), 78
    Los Alamos Newsletter, 136
    Los Alamos Scientific Laboratory, 78
    Los Alamos Ranch School, 58

Los Alamos Technical Series, 67
and new approach to nuclear
science, 68–70
Theoretical Division (T-Division),
64, 66, 70, 73, 74
work on the uranium fission bomb,
62–71
work on the plutonium implosion
bomb, 70–6
see also Association of Los Alamos
Scientists (ALAS; US)
see also Fuchs, Klaus
see also individual scientists' and
administrators' names
see also Manhattan Engineering
District (MED)
see also Peierls, Rudolf
Losey, Joseph, 100
Loutit, J.F., 169
Lucky Dragon incident (1954), 163
see also Castle Bravo Test (1954;
Pacific)
Lysenko, Trofim, 119
Lysenkoism, 119

Maclean, Donald, 95, 108
Manchester University, 13, 17, 18,
28, 48
Mandl, Franz, 49
Manhattan Engineering District
(MED), 2, 3, 9, 10, 11, 19,
49, 50, 54–81, 60, 71, 75, 93,
103, 105, 106, 108, 128, 129,
131, 142, 152, 172, 173
British Mission, 57, 59, 62, 63–4,
67, 68, 73, 75, 78, 80, 81, 87,
91–2, 126, 129, 130–1
see also Britain
see also Canada
see also Chalk River (Canada)
see also Fuchs, Klaus
see also Hanford (Washington; US)
see also individual scientists' and
administrators' names
see also Los Alamos
(New Mexico; US)
see also Oak Ridge (Tennessee; US)
see also Peierls, Rudolf
see also United States

see also University of California,
Berkeley
Manhattan Project, *see* Manhattan
Engineering District (MED)
Manley, John, 103
Mann, Thomas, 154
Reflections of an Unpolitical Man
(1918), 154
Mao Zedong, 105
Mark, Carson, 89, 146
Marley, William, 130, 134, 135, 139
Marriott, John H., 116
Marshak, Robert, 58
Marshak, Ruth, 58
Marshall, Donald, 130
Marshall Plan, 137
Massey, Harrie, 17, 117, 131, 134,
140, 161, 166, 170
Master, Dexter, 145
One World or None (1946), 145
see also Way, Katherine
Maud Committee, 41–8
naming of, 41
Maude, Angus, 162–3
Max-Planck Institute for Biophysics
(FRG), 166
McCarthy, Joseph, 11, 100, 105, 110
Wheeling speech, 105
see also House on Un-American
Activities Committee
(HUAC)
McCarthyism, 100, 102, 105, 106,
115, 120, 121
McMahon Act (1946; US), 9, 94, 101,
104
Medical Research Council (MRC),
164–6
Meitner, Lise, 35, 37–8, 43, 135
confirmation of fission, 37–8
see also Frisch, Otto
see also Hahn, Otto
see also Strassmann, Fritz
Melchior, Lauritz, 103
Mendelssohn, Kurt, 16, 18, 20, 21, 22,
45, 135
Metropolitan-Vickers Electrical Co.
Ltd., 49, 140
Michiels, J.I., 146, 158
Milan (Italy), 27

Military Intelligence, Section 5 (MI5),
        3, 4, 79, 82–124, 173
    impact of Fuchs case on, 93–102
    investigation of Klaus Fuchs, 86–90
    investigation of Rudolf Peierls,
        113–17
    relations with the FBI, 101–8
    see also Federal Bureau of
        Investigation (FBI; US)
    see also Fuchs, Klaus
    see also Peierls, Rudolf
Ministry of Aircraft Production (MAP),
        41–2, 61
Ministry of Supply, 108, 139, 140, 148
    Department of Atomic Energy, 139
    Directorate of Information, 140
    see also Atomic Energy Research
        Establishment (AERE) Harwell
Ministry of Works, 140
Montreal (Canada), 48, 49, 57, 60, 63,
        78, 89
Moon, Philip, 24, 41, 114, 130, 135,
        139, 142, 144, 147, 167
Moore, Michael, 131, 139
Moorehead, Alan, 4, 83, 97, 98, 100,
        102, 107
    The Traitors (1952), 98
Morrison, Philip, 115
Moscow (Soviet Union), 87, 119, 121,
        171, 175
Moss, Norman, 4
Mott, Nevill, 20, 32, 46, 94, 95, 115,
        134, 146, 166
Munich University (Germany), 32

Nagasaki (Japan), 40, 41, 128, 129,
        136, 138, 141, 149
National Committee on Atomic
        Information (NCAI; US), 131,
        168
National Council for Civil Liberties, 119
National Defense Research Committee
        (NDRC; US), 56
National Health Service (NHS), 8
National Socialism, 1, 2, 12, 13, 18,
        34, 36, 50, 57, 83, 172, 173
    expulsion of scientists from
        Germany and Central
        Europe, 1, 12

    see also Germany
    see also Hitler, Adolf
    see also National Socialist German
        Workers' Party (NSDAP)
National Socialist German Workers'
        Party (Nationalsozialistische
        Deutsche Arbeiter Partei;
        NSDAP), 23, 84, 85
National Women Citizen Association,
        135
Nature, 38, 119, 139, 142, 146, 149, 158
Neddermeyer, Seth, 71, 72
Nehring, Holger, 168
Nelson, Eldred, 74
Nernst, Walther, 20, 34
Netherlands, The, 12, 19, 26, 135
Neufeld, Michael J., 153
New Commonwealth Parliamentary
        Committee, 117, 135
New Commonwealth Society, 138
New Mexico (US), 2, 40, 49, 58, 60,
        61, 75, 78, 130, 142, 173
    see also Los Alamos
    see also Manhattan Engineering
        District (MED)
    see also Trinity Test (1945; US)
New Scientist, 102, 139, 162, 170
New York City (US), 2, 38, 40, 54, 56,
        57, 61, 63, 64, 65, 76, 80, 87,
        91, 115, 129, 131, 158, 173
New York Times, 40, 129
New Zealand, 9, 17
Nordheim, Lothar, 18, 19, 36, 60, 129
North Atlantic Treaty Organization
        (NATO), 137, 171, 174
Norway, 23, 135
Notgemeinschaft Deutscher
        Wissenschaftler im Ausland
        (Emergency Society of
        German Scientists Abroad), 19
Notorious (1946), 97
nuclear testing, 148, 163–7
    see also Atomic Scientists'
        Association (ASA)
    see also Bikini Atoll (South Pacific)
    see also Castle Bravo Test (1954;
        Pacific)
    see also fallout
    see also *Lucky Dragon* incident (1954)

see also strontium-90
see also Trinity Test (1945; US)
Nuclear Weapons Freeze Campaign
    (FREEZE), 174
nuclear weapons development, see
    Britain and Manhattan
    Engineering District (MED)
    and Soviet Union and Tube
    Alloys (TA) and United States
Nunn May, Alan, 86, 89, 90, 91, 98,
    101, 107, 108–9, 118, 131

Oak Ridge (Tennessee; US), 10, 19, 49,
    57, 60, 76, 78, 129, 131
    K-25 plant, 76
    see also uranium separation
Oak Ridge Association of Scientists
    and Engineers, 129
    X-10 project, 60
    see also Manhattan Engineering
        District (MED)
    see also Nordheim, Lothar
Obninsk reactor (Soviet Union), 8
Office of Scientific Research and
    Development (OSRD; US),
    51, 56
Official Secrets Act (1911), 89, 96, 152
Oliphant, Marcus, 10, 17, 41, 48, 69,
    131, 134, 140, 142, 146, 158
One World government, 145
One World movement, 145
Operation Crossroads, see Bikini Atoll
    (South Pacific)
Oppenheimer, J. Robert, 51, 59, 61,
    64, 65, 66, 67, 69, 72, 73,
    75, 77, 93, 114–15, 121, 122,
    125, 127, 152
Oppenheimer Security Hearings
    (1954), 66, 122
Orowan, Egon, 20, 45, 48
Osborn, S.B., 165
Ottawa (Canada), 57, 86
Oxford University, 20, 33, 34, 42, 44,
    45, 49, 51, 132, 134–5, 146
    see also Clarendon Laboratory

Panitz, Eberhard, 4
Paris-Presse, 109
Parsons, William, 65

Pascal, Fania, 116
Pascal, Roy, 116
Pauli, Wolfgang, 13, 168
Pauling, Linus, 121, 167
Pearl Harbor (Hawaii; US), 53, 56, 103
Peenemünde (Germany), 58
Pegram, George, 51, 53
Peierls, Alfred, 13, 24
Peierls, Annie, 13
Peierls, Gaby, 14, 32, 89
Peierls, Genia, 13, 56, 58, 59, 111–12,
    116, 120, 122, 123
    letter to Klaus Fuchs, 111–12
Peierls, Ronnie, 14
Peierls, Rudolf
    Anglo-American nuclear
        cooperation, 49–81
    see also Manhattan Engineering
        District (MED)
    anti-Semitism, 22
    and Atomic Scientists' Association
        (ASA), 3, 6, 11, 117–22,
        125–71, 174
    role in creation of, 125–50
    role in disbandment of, 151–71
    see also Atomic Scientists'
        Association (ASA)
    award of Presidential Medal of
        Merit, 81
    as director of British Pugwash
        Group, 174
    emigration to Britain
    as 'enemy alien', 23–6, 37
    'Frisch-Peierls Memorandum', 2,
        36–41, 42, 45, 46, 50, 55,
        69, 149
    see also Frisch, Otto
    as Fuchs's mentor, 3, 11, 173
    German background of, 1–4, 154–7
    Harwell consultancy, 80, 175
    and ideology of 'objective' science,
        151–9
    knighthood (1968), 175
    at Leipzig University, 63, 87
    at Los Alamos, 56–81
    Maud Committee, 41–8
    member of National Institute of
        Research in Nuclear Science,
        174–5

Peierls, Rudolf – *continued*
  motivations to engage in work on
      atom bomb, 34–6
  naturalization of, 24
  new approach to nuclear science,
      68–76
  in New York City, 57–8
  Nuclear Weapons Freeze Campaign
      (FREEZE), 174
  as part of Jewish emigration, 12
  plutonium implosion bomb, 70–6
  professional integration in Britain,
      31–55
  reaction to Fuchs's confession, 111–12
  responsibility of the atomic
      scientist, 125–71
  Rockefeller fellowship, 1, 13, 32
  scientific internationalism, 117–22
  social integration in Britain, 12–30
  as target of FBI and MI5
      surveillance, 87, 113–22
  Tube Alloys (TA), 48–54
  as victim of National Socialist
      persecution, 13–14
  visa issues, 120–1
  see also Birmingham
  see also Fuchs, Klaus
  see also isotope separation
  see also Los Alamos (New Mexico; US)
Pelican Books, 114
Penguin Books, 114
Penney, William, 79, 81, 119, 130
penicillin, 125
Penrose, L.S., 165
Pentagon (Washington, DC; US), 56
Perrin, Michael, 53–4, 91, 101,109,
      116, 139
  see also Fuchs, Klaus
Perutz, Max, 28
Pictorial, 26
Picture Post, 98, 161
Pilat, Oliver, 4, 98
  The Atom Spies (1952), 98
Pincher, Chapman, 39, 97
Pinnacle Productions Ltd., 99
Placzek, Else, 89
Placzek, George, 36, 59, 63, 64, 66,
      67, 75, 114, 135, 146, 173
  diffusion theory, 64

Planck, Max, 34
Plowden, Edwin, 166
plutonium, 8, 43, 57, 60, 63, 64,
      66, 68, 70–6, 78, 91, 92,
      123, 161
  implosion bomb, 70–6
  see also Los Alamos (New Mexico; US)
Polanyi, Michael, 17, 18
Pontecorvo, Bruno, 95, 98, 102,
      107–8, 112
Poole, Michael, 79
Popper, Karl, 33–4, 119
  The Open Society and Its Enemies
      (1945), 119
Pouchin, E.E., 169
Prague (Czechoslovakia), 15, 16
Price, Matt, 151
Priestley, J.B., 171
Princeton University (New Jersey; US),
      120
  Institute of Advanced Study, 120
proximity fuse, 2, 34, 50, 75, 125
Pryce, Maurice, 134, 147–8, 166, 167
Pugwash Conferences, 171, 174
Pugwash movement, 11, 149, 171, 174

Quebec (Canada), 28, 53–5
Quebec Agreement (1943), 53–4
  see also Anglo-American relations
Quito (Ecuador), 18

Rabinowitch, Eugene (Eugen), 18,
      131, 163
radar, 37
Radiotelevisione Italiana (RAI; Italy),
      122–3
Radium Institute (Paris), 38
  see also Joliot-Curie, Frédéric
Rangoon University (Burma), 19
Reader's Digest, 105
The Red Menace (1949), 106
Refugee Children's Movement, 20
Reichstag fire, 85
Revolutionary Student Group
      (Revolutionary Students'
      Group; RSG; Germany), 85
Rickett, Denis, 143–4
Rieger, Bernhard, 9
Ringer, Fritz, 156

Rio Grande Valley (New Mexico; US), 58

'risk society', *see* Beck, Ulrich

Ritchie, James M., 16

Roberts, A., 146

Robinson, Carson, and His Pleasant Valley Boys, 106

'I'm No Communist' (1952), 106

Rockefeller Institute for Medical Research (US), 166

Rodin, Sidney, 112

Roosevelt, Franklin Delano (FDR), 36, 52, 53, 56, 129

Rosenberg, Ethel, 101

Rosenberg, Julius, 101

Rossendorf (GDR), 175

Rossi, Bruno, 66

Rotblat, Joseph, 42, 43, 78, 90, 112, 127, 131, 132, 134, 135, 139, 147, 160, 163, 164, 165, 167, 169

Royal Armament Research and Development Establishment (RARDE), 78

Royal Institute of International Affairs, 164

Royal Society, 41, 111, 164

Russell, Bertrand, 162–3

Einstein-Russell Manifesto (1955), 163

see also Einstein, Albert

Rutherford, Ernest, 9, 17, 18

Sachs, Alexander, 36

Sack, Robert Arno, 27, 29, 32

Salmon, L.A., 165

Sandia National Laboratories (New Mexico; US), 78

Schirrmacher, Arne, 156

Schleswig-Holstein (Germany), 83–4

see also Kiel

Schlinck, Frederick, 113–14

Schoenberg, David, 45

Schrecker, Ellen, 105

Schrödinger, Erwin, 17, 18, 19, 20, 33, 34, 35

Schücking, Walther, 14

Science Museum (London), 140

Science News, 137

see also Penguin Books

Second World War, 2, 5, 7, 9, 11, 16, 20, 24–5, 34, 37, 45, 52, 54, 69, 70, 72, 78, 85, 95, 96, 99, 114, 121, 123, 125, 126, 128, 149, 150, 161, 173,

Security Service, *see* Military Intelligence, Section 5 (MI5)

Segrè, Emilio, 58, 59, 67, 68, 71

Serber, Robert, 66, 71, 115, 136

Shapin, Steven, 127

see also Thorpe, Charles

Shapiro, M., 146

Shawcross, Hartley, 96–7

Sheppard, P.A., 165

Simon, Franz, 18, 20, 21, 22, 33, 42, 44, 45, 46, 48, 49, 50, 51, 81, 112, 134, 135, 169

Simon, G., 165

Simpson, Esther, 26, 61, 110

see also Academic Assistance Council (AAC)

Skardon, William, 82, 84, 86–8, 91, 100–1, 107, 114

see also Fuchs, Klaus

Skemp, Joseph B., 61

Skinner, Erna, 114, 116

Skinner, Herbert, 134, 144, 146, 158, 167

Skyrme, Tony, 57, 87, 130

Smith, Cyril, 65

Smyth, Henry DeWolf, 61, 91, 138

Atomic Energy for Military Purposes (1945), 9, 61, 138

Social Democratic Party of Germany (Sozialdemokratische Partei Deutschlands; SPD), 23, 84

Socialist Unity Party (Sozialistische Einheitspartei Deutschlands; SED; Germany), 175

Society for the Protection of Science and Learning (SPSL), *see* Academic Assistance Council (AAC)

Society for Cultural Relations with the Soviet Union, 95

Soviet-Finnish War, 95

Soviet Union, 2, 3, 6, 7, 8, 21, 23, 35, 52, 76, 82, 83, 85, 86, 87, 88,

Soviet Union – *continued*
    89, 90, 91, 92, 93, 94, 95, 96,
    99, 100, 101, 102, 103, 105,
    106, 108, 109, 110, 111, 112,
    115, 116, 119, 120, 123, 125,
    134, 143, 144, 146, 149, 164,
    167, 173, 175, 176, 177
  Central Committee of the Soviet
    Communist Party, 134
  first test of own atomic device, 105
  Fuchs's espionage for, 82–109
    see also Fuchs, Klaus
    see also Chernobyl nuclear accident
    see also international control of
      nuclear energy
    see also Gromyko, Andrei
    see also Obninisk reactor
Spanish Civil War, 86, 95
Sponer, Hertha, 36
Stanford University (California; US), 66
Staples, Amy, 126
Statements Relating to the Atomic Bomb
  (1945), 138
Staub, Hans, 4, 60, 62, 65, 66
Stein, Paul, 74
Stern, Otto, 13, 36
  see also Hamburg University
Stimson, Henry, 128, 143, 147
Strassmann, Fritz, 38, 43
  see also fission
  see also Hahn, Otto
  see also Meitner, Lise
Strath Report, 149
Strauss, Lewis, 121
Strauß, Franz Josef, 156
strontium-90, 163–7
Stuewer, Roger H., 70
Stuhlinger, Ernst, 176
Suez crisis, 10
Sunday Express, 26, 38, 98, 99, 112, 159
Switzerland, 4, 15, 135
Szilard, Leo, 18, 36, 37, 43, 45, 54, 65,
  127, 129, 131, 157, 177

Taylor & Francis (publisher), 138
Teller, Edward, 4, 19, 36, 45, 59, 60,
  63, 65–7, 72, 73, 77, 79, 93,
  114, 115, 129, 152, 157, 173,
  177

Teller, Mici, 65
Their Trade Is Treachery (1964), 108
Thomson, George, 27, 41, 42, 47, 140,
  147, 148, 160, 161, 164, 165,
  167, 169, 170
  see also Maud Committee
Thorpe, Charles, 127
  see also Shapin, Steven
Three Mile Island, *see* Harrisburg
The Times, 109, 117, 139, 143, 145,
  147, 149, 153, 160, 166, 167
Tizard, Henry, 41, 50, 135
Tolman, Richard, 71
Toulmin, Stephen, 89
Toronto (Canada), 14
The Treasury, 139
Trinity Test (1945; US), 40, 41, 64, 75,
  91, 126, 127, 141
  see also Los Alamos
  see also Manhattan Engineering
    District (MED)
Truman, Harry S., 40, 102, 103, 104,
  127, 128, 143, 153, 159, 161
Tube Alloys (TA), 2, 3, 10, 27, 47–54,
  61, 80, 86, 108, 109, 110,
  116, 132, 139, 150, 173
  Chemical Panel, 50
  Diffusion Project Committee, 50
  Metal Panel, 50
  Technical Committee, 51
Tuck, James, 73, 75, 130
Tyndall, Arthur M., 27, 29, 42
  see also Bristol University

United Free Church (US), 166
United Kingdom Atomic Energy
  Authority (UKAEA), 166
United Nations (UN), 95, 128, 139,
  140, 142, 143, 144, 146, 148
  Association, 140
  Atomic Energy Commission
    (UNAEC), 142, 143, 144, 146
    see also international control of
      nuclear energy
  Security Council, 144, 145, 148
United States, 2, 4, 5, 6, 7, 9–10, 11,
  13, 17, 18, 19, 21, 36, 37, 39,
  42, 47, 48, 50, 51, 52, 53, 54,
  56, 57, 58, 60, 61, 65, 69, 70,

72, 76, 78, 79, 80, 81, 83, 86, 91, 96, 97, 98, 99, 100, 101, 102, 103, 104, 105, 106, 108, 109, 110, 113, 114, 115, 117, 118, 120, 125, 126, 128, 129, 130, 135, 136, 137, 140, 143, 144, 145, 146, 151, 155, 157, 163, 164, 167, 168, 172, 173, 174, 175, 176, 177

Army, 56, 58, 62, 103
Congress, 144, 145, 148
The Geographical Focal Points of Espionage (1950), 83
McCarran Internal Security Act (1950), 120
see also McMahon Act (1946)
decision to use atomic bomb, 40, 127–8
Destroyers for Bases Deal, 52
H-bomb decision, 102–3, 159, 161
Lend-Lease policy, 52
Navy Department, 140
nuclear cooperation, *see* Britain
as nuclear reference culture, 9–10
State Department, 102, 105, 121
see also Federal Bureau of Investigation (FBI)
see also individual scientists', administrators' and political decision makers' names
see also international control of nuclear energy
see also Manhattan Engineering District (MED)
see also McCarthyism
see also nuclear testing
uranium, 9, 24, 37, 38, 42, 43, 44, 45, 46, 47, 49, 50, 57, 60, 63, 64, 67, 68, 71, 72, 76, 91, 92, 116, 143, 161
cross-sections, 43
uranium 235, 43, 44, 46, 57, 60, 63, 68, 91
uranium 238, 43, 47, 60
see also isotope separation
Urey, Harold, 51, 53

V-1, 58
V-2, 58

Vansittart, Sir Robert, 22
Black Record: Germans Past and Present (1941), 22
'Vansittartism', 22
Venona (messages), 87, 123
Vincent, David, 95, 96, 102
'culture of secrecy', 96
von Ardenne, Manfred, 35
von Braun, Wernher, 58, 153, 176, 177
von Halban, Hans, 42, 44, 48, 51, 52, 54, 60, 123, 135
von Hindenburg, Paul, 84
von Laue, Max, 34
von Neumann, John, 45, 59, 65, 66, 67, 72, 74, 79
von Papen, Franz, 13
von Schleicher, Kurt, 13
von Weizsäcker, Carl Friedrich, 155–7

Walton, G.N., 163
Wang, Jessica, 119
see also Beyler, Richard
see also Kojevnikov, Alexei
War Cabinet, 47, 50, 148
Scientific Advisory Committee, 47
War Office, 88
Warburg Institute, 45
Warburg Library, 45
Ward, J.G., 143, 144
Washington, DC (US), 53, 54, 55, 56, 57, 63, 87, 95, 102, 104, 120
see also United States
Way, Katherine, 145
One World or None (1946), 145
see also Master, Dexter
Wehrmacht, 23
Weimar Republic, *see* Germany
Weinberg, Alvin M., 76
Weisskopf, Victor, 4, 33, 36, 59, 65, 66, 67, 72, 114, 115, 129, 131, 157, 161, 168, 173
Wells, H.G., 7
The World Set Free (1914), 7
Werner, Ruth, 175
Sonjas Rapport (*Sonya's Report*; 1991/2006), 175
see also Banbury
see also Fuchs, Klaus

West, Nigel, 123
West, Rebecca, 4, 97–8, 100, 106
   The Meaning of Treason (1945), 98
   The New Meaning of Treason (1952),
     98
Wheeling (West Virginia; US), 105
The Whip Hand (1951), 106
Whitehall, 10, 24, 25, 29, 52, 53, 83,
       96, 101, 136, 137, 138, 144,
       149, 163, 171, 172
   see also Britain
Wigner, Eugene (Eugen), 36, 45, 65, 67
Williams, Beryl, 99
   The Real Book of Spies (1959), 99
   see also Epstein, Samuel
Williams, Brock, 99

Williams, Robert Chadwell, 4
Willis, Kirk, 5
Wills, Garry, 83
Wills Laboratory, 32, 46
   see also Mott, Nevill
   see also Bristol
Windscale, 6, 8
   nuclear accident, 6
Wirtz, Karl, 35
World Federation of Scientific
     Workers, 133
World Health Organization, 126

Zehden, Walter, 20
Zentrum (Centre Party; Germany), 23
Zürich (Switzerland), 13, 32